Lecture Notes in Computer Science   10149

Commenced Publication in 1973
Founding and Former Series Editors:
Gerhard Goos, Juris Hartmanis, and Jan van Leeuwen

More information about this series at http://www.springer.com/series/7412

Reneta P. Barneva · Valentin E. Brimkov
João Manuel R.S. Tavares (Eds.)

Computational Modeling of Objects Presented in Images

Fundamentals, Methods, and Applications

5th International Symposium, CompIMAGE 2016
Niagara Falls, NY, USA, September 21–23, 2016
Revised Selected Papers

 Springer

Editors
Reneta P. Barneva
State University of New York at Fredonia
Fredonia, NY
USA

Valentin E. Brimkov
SUNY Buffalo State
Buffalo, NY
USA

and

Institute of Mathematics and Informatics
Bulgarian Academy of Sciences
Sofia
Bulgaria

João Manuel R.S. Tavares
Universidade do Porto
Porto
Portugal

ISSN 0302-9743 ISSN 1611-3349 (electronic)
Lecture Notes in Computer Science
ISBN 978-3-319-54608-7 ISBN 978-3-319-54609-4 (eBook)
DOI 10.1007/978-3-319-54609-4

Library of Congress Control Number: 2017933072

LNCS Sublibrary: SL6 – Image Processing, Computer Vision, Pattern Recognition, and Graphics

Printed on acid-free paper

This Springer imprint is published by Springer Nature
The registered company is Springer International Publishing AG
The registered company address is: Gewerbestrasse 11, 6330 Cham, Switzerland

Preface

It is indeed our great pleasure to welcome you to the proceedings of the 5th International Symposium on Computational Modeling of Objects Represented in Images. Fundamentals, Methods and Applications (CompIMAGE 2016), which took place at the Conference and Event Center in Niagara Falls, NY, USA. The previous editions were held in Coimbra (Portugal, 2006), Buffalo, NY (USA, 2010), Rome (Italy, 2012), and Pittsburg, PA (USA, 2014).

The purpose of the symposium is to provide a common forum for researchers, scientists, engineers, and practitioners around the world to present their latest research findings, ideas, developments, and applications in the area of computational modeling of objects presented in images. CompIMAGE 2016 attracted scientists who use various approaches to solve problems that appear in a wide range of areas – as diverse as medicine, robotics, defense, security, material science, and manufacturing. Traditionally, CompIMAGE symposia have been truly international, and this one was not an exception: Symposium participants came from 13 different countries from three continents.

The screening and publication process of CompIMAGE 2016 was in a different format than the previous editions of the symposium. In response to the call for papers, we received not only full-length articles but short communications as well. The page length of the latter was suggested to not exceed four pages. The authors of the accepted short communications had the possibility to present at the symposium and receive feedback for their work. After the symposium, they were given time to extend their papers to full-length articles and submit them for possible inclusion in the symposium proceedings.

All submissions underwent a rigorous double-blind review process by members of the international Program Committee. The most important selection criterion for acceptance or rejection of a paper was the overall score received. Other criteria were: relevance to the symposium topics, correctness, originality, mathematical depth, clarity, and presentation quality.

As a result, 18 contributed papers were selected to be included in this volume, as well as a comprehensive survey on case-based reasoning for signal and image analysis, based on the keynote talk of Petra Perner. We hope that the works are of interest to a broad readership.

We would like to thank all those who contributed to the success of the symposium. First, we would like to express our gratitude to all authors who submitted their works to CompIMAGE 2016. Thanks to their contributions, we succeeded in having a technical program of high scientific quality. Our most sincere thanks go to the Program Committee members whose cooperation in carrying out a rigorous and objective review process was essential in establishing a strong symposium program and high-quality publications. We express our sincere gratitude to the invited speakers Donald P. Greenberg (Cornell University, USA), Jiebo Luo (University of Rochester, USA), Petra Perner (Institute of

Computer Vision and Applied Computer Sciences, Germany), and Kamen Kanev (University of Shizuoka, Japan) for their remarkable presentations and overall contribution to the symposium.

Special thanks go to Mark W. Severson, Dean of the School of Natural and Social Sciences at SUNY Buffalo State, for funding the best paper awards and the gifts for the keynote speakers. We are grateful to our partners SUNY Buffalo State, State University of New York at Fredonia, FEUP – Faculdade de Engenharia da Universidade do Porto, APMTAC – Associação Portuguesa de Mecânica Teórica, Aplicada e Computacional, FCT – Fundação para a Ciência e a Tecnologia, and INEGI – Instituto de Ciência e Inovação em Engenharia Mecânica e Engenharia Industrial. We also thank the personnel of the Conference and Event Center at Niagara Falls, and especially Tom Acara and Alexis Schmitz, for the excellent conditions and service during the workshop. Finally, we wish to thank Springer for the pleasant cooperation in the production of this volume.

December 2016

Reneta P. Barneva
Valentin E. Brimkov
João Manuel R.S. Tavares

Organization

The 5th International Symposium on Computational Modeling of Objects Represented in Images: Fundamentals, Methods, and Applications, CompIMAGE 2016 was held in Niagara Falls, NY, USA, September 21–23, 2016.

General Chairs

Reneta P. Barneva	SUNY Fredonia, USA
Valentin E. Brimkov	SUNY Buffalo State College, USA
João Manuel R.S. Tavares	University of Porto, Portugal

Steering Committee

João Manuel R.S. Tavares	University of Porto, Portugal
Renato M. Natal Jorge	University of Porto, Portugal
Jessica Zhang	Carnegie Mellon University, USA

Tutorial and Industry Chairs

Patrick Hung	University of Ontario Institute of Technology, Canada
Bill Kapralos	University of Ontario Institute of Technology, Canada

Invited Speakers

Donald P. Greenberg	Cornell University, USA
Jiebo Luo	University of Rochester, USA
Petra Perner	Institute of Computer Vision and Applied Computer Sciences, Germany
Kamen Kanev	Shizuoka University, Japan

Program Committee

Lyuba Alboul	Sheffield Hallam University, UK
Constantino Carlos Reyes-Aldasoro	University of Sheffield, Sheffield, UK
Fernando Alonso-Fernandez	Halmstad University, Sweden
Luís Amaral	Polytechnic Institute of Coimbra, Portugal
Jorge Anbrósio	Instituto Superior Técnico, Portugal
Hélder Araújo	University of Coimbra, Portugal
Emmanuel A. Audenaert	Ghent University Hospital, Belgium
Jorge M.G. Barbosa	University of Porto, Portugal

Jorge Manuel Batista	University of Coimbra, Portugal
George Bebis	University of Nevada, USA
Nguyen Dang Binh	Hue University of Sciences, Vietnam
Boris Brimkov	Rice University, USA
Nathan Cahill	Rochester Institute of Technology, USA
Francisco Calheiros	University of Porto, Portugal
Daniela Calvetti	Case Western Reserve University, USA
Begoña Calvo Calzada	University of Zaragoza, Spain
Durval C. Campos	HPP-Medicina Molecular, SA., Portugal
Barbara Caputo	IDIAP Research Institute, Switzerland
Jaime Cardoso	University of Porto, Portugal
M. Emre Celebi	Louisiana State University in Shreveport, USA
Jonathon Chambers	Loughborough University, UK
Ke Chen	University of Liverpool, UK
Laurent Cohen	Université Paris Dauphine, France
Miguel Velhote Correia	University of Porto, Portugal
João Paulo Costeira	Instituto Superior Técnico, Portugal
Alexandre Cunha	California Institute of Technology, USA
Miguel Tavares da Silva	Instituto Superior Técnico, Portugal
Jérôme Darbon	University of California at Los Angeles, USA
Fernao Vistulo de Abreu	University of Aveiro, Portugal
Antoine Deza	McMaster University, Canada
Jorge Manuel Dias	University of Coimbra, Portugal
Manuel Doblaré	University of Zaragoza, Spain
Mahmoud El-Sakka	University of Western Ontario London, Canada
José Augusto Mendes Ferreira	University of Coimbra, Portugal
Isabel N. Figueiredo	University of Coimbra, Portugal
Mário A.T. Figueiredo	Instituto Superior Técnico, Portugal
Paulo Flores	University of Minho, Portugal
Mário M. Freire	University of Beira Interior, Portugal
Diamantino Freitas	University of Porto, Portugal
Irene M. Gamba	University of Texas at Austin, USA
Jose M. García Aznar	University of Zaragoza, Spain
Joaquim Silva Gomes	University of Porto, Portugal
Bernard Gosselin	Faculté Polytechnique de Mons, Belgium
Enrique Alegre Gutiérrez	University of León, Spain
Gerhard A. Holzapfel	Graz University of Technology, Austria
Daniela Iacoviello	Università degli Studi di Roma La Sapienza, Italy
Joaquim A. Jorge	Instituto Superior Técnico, Portugal
Kamen Kanev	Shizuoka University, Japan
Jung Hwan Kim	Baylor College of Medicine, USA
Renato Natal Jorge	University of Porto, Portugal
Constantine Kotropoulos	Aristotle University of Thessaloniki, Greece
Maria Elizete Kunkel	Universität Ulm, Germany
Slimane Larabi	U.S.T.H.B. University, Algeria

Contents

Invited Paper

Model Development and Incremental Learning Based on Case-Based Reasoning for Signal and Image Analysis

Petra Perner[(✉)]

Institute of Computer Vision and Applied Computer Sciences,
IBaI, Leipzig, Germany
pperner@ibai-institut.de
http://www.ibai-institut.de

Abstract. Over the years, image mining and knowledge discovery gained importance to solving problems. They are used in developing systems for automatic signal analysis and interpretation. The issues of model building and adaption, allowing an automatic system to adjust to the changing environments and moving objects, became increasingly important. One method of achieving adaptation in model building and model learning is Case-Based Reasoning (CBR). Case-Based Reasoning can be seen as a reasoning method as well as an incremental learning and knowledge acquisition method. In this paper we provide an overview of the CBR process and its main features: similarity, memory organization, CBR learning, and case-base maintenance. Then we review, based on applications, what has been achieved so far. The applications we are focusing on are meta-learning for parameter selection, image interpretation, incremental prototype-based classification, novelty detection and handling, and 1-D signal interpretation represented by a 0_1 sequence. Finally, we will summarize the overall concept of CBR usage for model development and learning.

Keywords: Model development · Incremental learning · Case-Based Reasoning · Similarity · Signal and image interpretation · Image segmentation · Novelty detection · 1/0 sequence interpretation · Computational intelligence

1 Introduction

Signal interpretation systems are becoming increasingly popular in medical applications as well as in industrial applications. The existing statistical and knowledge-based techniques lack robustness, accuracy and flexibility. New strategies are necessary, which can adapt to the changing environmental conditions, user needs and process requirements. These requirements can be satisfied by the introduction of case-based reasoning (CBR) strategies into image interpretation systems [26]. CBR can help developing a model for signal interpretation and incrementally coming up over time with the final model. CBR gives a flexible and powerful method that allows controlling the signal processing chain in all phases of a signal interpretation system so that the derived resulting information is of the best possible quality. Beyond this, CBR offers different learning capabilities in all phases of a signal interpretation system. These

© Springer International Publishing AG 2017
R.P. Barneva et al. (Eds.): CompIMAGE 2016, LNCS 10149, pp. 3–24, 2017.
DOI: 10.1007/978-3-319-54609-4_1

learning and maintenance capabilities satisfy different needs during the development process of a signal interpretation system. Therefore, they are especially appropriate for incremental model development.

CBR [26] solves problems using the already stored knowledge and acquires new knowledge making it immediately available for solving the next problem. Therefore, CBR can be seen as a method for problem solving and also as a method for capturing new experience and making it readily available for problem solving. It can be seen as an incremental learning and knowledge-discovery approach, since it can capture general knowledge from new experience, such as case classes, prototypes and higher-level concepts.

The CBR paradigm has originally been introduced by the cognitive science community. The CBR community aims at developing computer models that follow this cognitive process. Computer models based on CBR have successfully been developed for many application areas, such as signal/image processing and interpretation tasks, help-desk applications, medical applications and e-commerce-product selling systems.

The paper is organized as follows. In Sect. 2 we will explain the model development problem for signal interpretation tasks. The CBR process scheme is introduced in Sect. 3. We will show what kinds of methods are necessary to provide all the required functions for such a computer model. The CBR process model comprised of the CBR reasoning and CBR maintenance process is given in Sect. 4. Then we will focus on similarity in Sect. 5. Memory organization in a CBR system will be described in Sect. 6. Both similarity and memory organization are related to the learning process in a CBR system. Therefore, in each section an introduction will be given as to what kind of learning can be performed. In Sect. 7 we will describe open topics in CBR research for specific applications. In Sect. 7.1 we will describe meta-learning for parameter selection for data processing systems. CBR based image interpretation will be described in Sect. 7.2 and incremental prototype-based classification in Sect. 7.3. New concepts on novelty detection and handling will be introduced in Sect. 7.4. In Sect. 7.5 we will present 1-D signal interpretation by a 0_1 sequence and similarity-based interpretation. We will conclude with a summary of our concept on CBR in Sect. 8.

2 Problems Related to Model Development of Signal Interpretation Tasks

Several factors influence the quality of the final result of a signal interpretation system. Those are the environmental conditions, the signal device, the noise, and the number of observations from the task domain and the chosen part of the task domain. The variety of influences can often not be taken into account during the system development and many of them will only be discovered during run-time of the system. It even cannot be guaranteed that the task domain is limited. For example, in defect classification for industrial tasks, new defects could occur due to scratches on the surface of the manufactured part caused by the manufacturing tool after prolonged usage. In optical character recognition, imaging defects influencing the recognition results such as heavy/light print or stray marks can occur. A first attempt of a systematic overview of the factors which influence the result of an optical character recognition system, and how different systems deal with it, has been given in [23]. However, it has not been possible to observe all real-world influences as of today, nor could a sufficiently large enough sample set for system development and testing be provided.

A robust signal interpretation system must be able to deal with such influences. It must have intelligent strategies on all levels of a signal interpretation system that can adapt the processing units to these new requirements. A strategy which seems to satisfy these requirements could be the Case-Based Reasoning (CBR), since it does not rely on a well-formulated domain theory, which often is difficult to acquire.

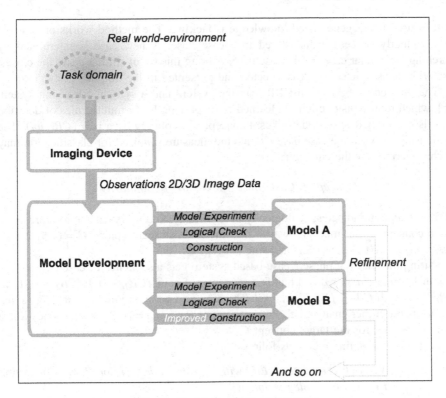

Fig. 1. Model development process.

All that shows that we have to deal with different aspects during system development that are topics in CBR. As we discover new aspects of the environment and the objects during the system usage, an automatic signal interpretation system should be able to incrementally update the model of the system such as the one shown in Fig. 1. Here the maintenance and learning come into play. Beyond that, the issue of the life time of a case plays an important role. Other aspects are concerned with the competence of the system. The range of target problems that a given system or algorithm can solve are often not quite clear to the developer of the interpretation system. Frequently, a researcher would present to the community a new algorithm that could, for example, recognize the shape of an object in a particular image and would claim that a model has been developed. However, if the algorithm is applied to a different image, it would fail. This raises the question: Did the researcher develop a model or did he rather develop a function? Evaluation and testing of the algorithm and systems is an important issue in computer vision and the control of the algorithm so that it fits best to the current problem. CBR strategies can help solving this problem in signal interpretation. What makes CBR so attractive will be discussed next.

3   Case-Based Reasoning

CBR is used when generalized knowledge is lacking. The method works on a set of cases formerly processed and stored in a case base. A new case is interpreted by searching for similar cases in the case base. Among this set of similar cases the closest case with its associated result is selected and presented to the output.

The differences between a CBR learning system and a symbolic learning system [56], which represents explicitly a learned concept, e.g., by formulas, rules or decision trees, is that a CBR system describes a concept C implicitly by a pair (CB, sim). The relationship between the case base CB and the measure sim used for classification may be characterized by the equation:

$$Concept = Case\_Base + Measure\_o\_Similarity \qquad (1)$$

This equation indicates, analogically to arithmetic, that it is possible to represent a given concept C in multiple ways, i.e., there exist many pairs $C = (CB_1, sim_1)$, $(CB_2, sim_2), \ldots, (CB_i, sim_i)$ for the same concept C.

During the learning phase a case-based system gets a sequence of cases X_1, X_2, ..., X_i with $X_i = (x_i,\ class\ (x_i))$ and builds a sequence of pairs $(CB_1,\ sim_1)$, $(CB_2,\ sim_2)$, ..., $(CB_i,\ sim_i)$ with $CB_i \subseteq \{X_1,\ X_2,\ ...,\ X_i\}$. The aim is to get at the end a pair $(CB_n,\ sim_n)$ that needs no further change, i.e., $\exists n\ \forall m \geq n\colon (CB_n,\ sim_n) = (CB_m,\ sim_m)$, since it is a correct classifier for the target concept C.

Formally, we define a case as follows:

Definition 1. A case F is a triple (P,E,L) with a problem description P, an explanation of the solution E, and a problem solution L.

The problem description summarizes the information about a case in the form of attributes or features. Other case representations such as graphs, images or sequences may also be used. The case description is either given a-priori or needs to be elicited during a knowledge acquisition process. Only the most predictive attributes will guarantee finding exactly the most similar cases.

Equation (1) and Definition 1 give a hint as to how a case-based learning system can improve its classification ability. The learning performance of a CBR system is of incremental manner and it can also be considered as on-line learning. In general, there are several possibilities to improve the performance of a case-based system. The system can change the vocabulary V (attributes, features), store new cases in the case base CB, change the measure of similarity sim, or change V, CB and sim in a combinatorial manner.

That brings us to the notion of knowledge containers introduced by Richter [43]. According to Richter, the four knowledge containers are the underlying vocabulary (or features), the similarity measure, the solution transformation, and the cases. The first three represent compiled knowledge, since this knowledge is more stable. The cases are interpreted knowledge. As a consequence, newly added cases can be used directly. This enables a CBR system to deal with dynamic knowledge. In addition, knowledge can be shifted from one container to another container. For instance, in the beginning a simple vocabulary, a rough similarity measure, and no knowledge on solution transformation are used. However, a large number of cases are collected. Over time, the vocabulary can be refined and the similarity measure can be defined in higher degree of accordance with the underlying domain. In addition, it may be possible to reduce the number of cases, because the improved knowledge within the other containers would now enable the CBR system to better differentiate between the available cases.

The generalization of cases into a broader case (concepts, prototypes and case classes) or the learning of the higher-order relation between different cases may reduce the size of the case base and speed up the retrieval phase of the system [49]. It can make the system more robust against noise. More abstract cases which are set in relation to each other will give the domain expert a better understanding about his domain. Therefore, beside the incremental improvement of the system performance through learning, CBR can also be seen as a knowledge-acquisition method that can help getting a better understanding about the domain [3, 10].

The main issues related to the development of a CBR system are answering the questions:

- What makes up a case?
- What is an appropriate similarity measure for the problem?
- How to organize a large number of cases for an efficient retrieval?
- How to acquire and refine a new case for entry in the case base?
- How to generalize a number of specific cases to a case that is applicable to a wide range of situations?

4 Case-Based Reasoning Process Model

The CBR reasoning process is comprised of seven phases (see Fig. 2): Current problem description, problem indexing, retrieval of similar cases, evaluation of candidate cases, modification of selected cases, application to a current problem, and assessment of the system.

The current problem is described by a number of keywords, attributes, features or any abstraction that allows describing the basic properties of a case. Based on this description, indexing of the case base is done. Among a set of similar cases retrieved from the case base the closest case is evaluated as a candidate case. If necessary, this case is modified so that it fits the current problem. The problem solution associated to the current case is applied to the current problem and the result is observed by the user. If the user is not satisfied with the result or if no similar case could be found in the case

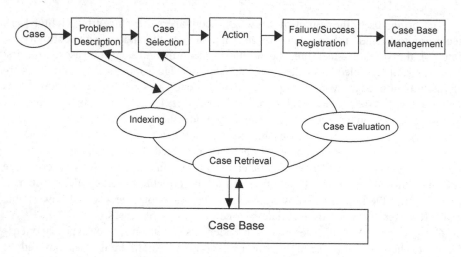

Fig. 2. Case-based reasoning process.

base, the user or the process itself gives feedback to the system. This assessment is used to incrementally improve the system performance by the case-base management process.

The CBR management (see Fig. 3) will operate on new cases as well as on cases already stored in the case base. If a new case has to be stored into the case base, this means there is no similar case in case base. The system has recognized a gap in the case base. A new case has to be inputted into the case base in order to close this gap. From the new case a predetermined case description has to be extracted which should be formatted into the predefined case format. After that, the case is stored into the case base. Selective case registration means that no redundant cases will be stored into the case base and similar cases will be grouped together or generalized by a case that applies to a wider range of problems. Generalization and selective case registration ensure that the case base will not grow too large and that the system can find similar cases fast.

It might also happen that too many non-relevant cases will be retrieved during the CBR reasoning process. Therefore, it might be wise to rethink the case description or to adapt the similarity measure. For the case description, more distinctive attributes should be found that allow putting aside the cases not applicable to the current problem. The weights in the similarity measure might be updated in order to retrieve only a small set of relevant cases.

CBR maintenance is a complex process and works over all knowledge containers (vocabulary, similarity, retrieval, case base) of a CBR system. Consequently, architectures and systems have been developed which support this process [16, 31] and also look into the life-time aspect concerned with case-based maintenance.

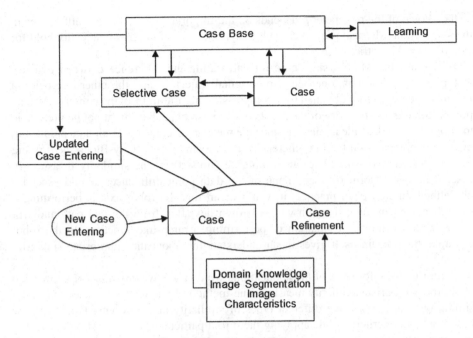

Fig. 3. Case-based maintenance.

5 Similarity

Although similarity is a concept humans like using when reasoning over problems, they usually do not have a good understanding of how similarity is formally expressed. Similarity seems to be a very incoherent concept.

From the cognitive point of view, similarity can be viewed from different perspectives [50]. A red bicycle and a blue bicycle might be similar in terms of the concept "bicycle", but both bicycles are dissimilar when looking at the color. It is important to know what kind of similarity is to be considered when reasoning over two objects. Overall similarity, identity, dissimilarity, and partial similarity need to be modelled by the right flexible control strategy in an intelligent reasoning system. This is especially important in image data bases where the image content can be viewed from different perspectives. Image data bases need to have flexibility and computerized conversational strategies to figure out from what perspective the problem is looked at and what kind of similarity has to be applied to achieve the desired goal. From the mathematical point of view, Minkowski metric is the most used similarity measure for technical problems:

$$d_{ii'}^{(p)} = \left[\sum_{j=1}^{J} |x_{ij} - x_{i'j}|^p \right]^{1/p} \tag{2}$$

as the choice of the parameter p depends on the importance given to the differences in the summation. Metrical properties such as symmetry, identity and unequality hold for the Minkowski metric.

If we use the Minkowski metric for calculating the difference between two trajectories of a robot axis, one being the original trajectory and the other one being a reconstructed trajectory obtained by a compression algorithm from the compressed data points stored in the memory of the robot control system [27], it might not be preferable to choose $p = 2$ (Euclidean metric), since the measure averages over all data points, but puts more emphasis on the big differences. If choosing $p = 1$ (City-Block metric), big and small differences have the same influence (impact) on the similarity measure. In case of the Max-Norm ($p = \infty$) none of the data point differences should exceed a predefined difference. In practice, it would mean that the robot axis is performing a smooth movement over the path with a known deviation from the real path and will never come to the worst situation of performing a ramp-like function. In the robot example, the domain itself gives us an understanding about the appropriate similarity metric.

Unfortunately, for most of the applications we do not have any *a priori* knowledge about the appropriate similarity measure. The method of choice for the selection of the similarity measure is trying different types of similarity and observing their behavior based on quality criteria while applying them to a particular problem. The error rate is the quality criterion that allows selecting the right similarity measure for classification problems. Otherwise, it is possible to measure how well similar objects are grouped together, based on the chosen similarity measure, and at the same time, how well different groups can be distinguished from each other. This approach changes the problem into a categorization one, for which proper category measures are known from clustering [18] and machine learning [13].

In general, distance measures can be classified based on the data-type dimension. There are measures for numerical data, symbolical data, structural data and mixed-data types. Most of the overviews given for similarity measures in various works are based on this view [25, 28, 58]. A more general view to similarity is given in Richter [44].

Other classifications on similarity measures focus on the application. There are measures for time-series [45], similarity measures for shapes [48], graphs [13], music classification [55], and others.

Translation, size, scale, and rotation invariance are another important aspect of similarity as concerns technical systems.

Most real-world applications nowadays are more complex than the robot example given above. They are usually comprised of many attributes that are different in nature. Numerical attributes given by different sensors or technical measurements and categorical attributes that describe meta-knowledge of the application usually make up a case. These n different attribute groups can form partial similarities $Sim_1, Sim_2, \ldots, Sim_n$ that can be calculated based on different similarity measures and may have a contextual meaning for itself. The final similarity might be comprised of all the contextual similarities. The simplest way to calculate the overall similarity is to sum up all contextual similarities: $Sim = \alpha_1 Sim_1 + \alpha_2 Sim_2 \ldots + \alpha_n Sim_n$ and model the influence of the similarities by different importance coefficients α_i. Other schemas for combining similarities are possible as well. The usefulness of such a strategy has been shown for meta-learning of segmentation parameters [29] and for medical diagnosis [61].

The introduction of weights into the similarity measure in Eq. (1) puts a different importance on particular attributes and views similarity not only as global similarity, but also as local similarity. Learning the attribute weights allows building particular similarity metrics for the specific applications. A variety of methods based on linear or stochastic optimization methods [62], heuristics search [57], genetic programming [12], relevance feedback learning [7] and case-ordering [52] or query ordering in NN-classification, have been proposed for attribute-weight learning.

Learning distance function in response to users' feedback is known as relevance feedback [2, 52] and it is very popular in data base and image retrieval. The optimization criterion is the accuracy or performance of the system rather than the individual problem-case pairs. This approach is biased by the learning approach as well as by the case description.

New directions in CBR research build a bridge between the case and the solution [4]. Cases can be ordered based on their solutions by their preference relations [60] or similarity relation [30] given by the users or a priori known from the application. The derived values can be used to learn the similarity metric and the respective features. This means that cases having similar solutions should have similar case descriptions. The set of features as well as the feature weights are optimized until they meet this assumption.

6 Organization of the Case Base

The case base plays a central role in a CBR system. All observed relevant cases are stored in the case base. Ideally, CBR systems start reasoning from an empty memory, and their reasoning capabilities stem from their progressive learning from the cases they process [8].

Consequently, the memory organization and structure are in the focus of a CBR system. Since a CBR system should improve its performance over time, this imposes on the memory of a CBR system to change constantly.

In contrast to research in data base retrieval and nearest-neighbor classification, CBR focuses on conceptual memory structures. While k-d trees [6] are space-partitioning data structures for organizing points in a k-dimensional space, conceptual memory structures [13, 31] are represented by a directed graph in which the root node represents the set of all input instances and the terminal nodes represent individual instances. Internal nodes stand for sets of instances attached to that node and represent a super-concept. The super-concept can be expressed as a generalized representation of the associated set of instances, such as the prototype, the medoid or a user-selected instance. Therefore a concept C, called a class, in the concept hierarchy is represented by an abstract concept description (e.g., the feature names and its values) and a list of pointers to each child concept $M(C) = \{C_1, C_2, ..., C_i, ..., C_n\}$, where C_i is the child concept, called subclass of concept C.

The explicit representation of the concept in each node of the hierarchy is preferred by humans, since it allows understanding the underlying application domain.

While for the construction of a k-d tree only a splitting and deleting operation is needed, conceptual learning methods use more sophisticated operations for the construction of the hierarchy [17]. The most common operations are splitting, merging, adding and deleting. What kind of operation is carried out during the concept hierarchy construction depends on a concept-evaluation function: there are statistical functions known, as well as similarity-based functions.

Because of the variety of construction operators, conceptual hierarchies are not sensitive to the order of the samples. They allow the incremental adding of new examples to the hierarchy by reorganizing the already existing hierarchy. This flexibility is not known for k-d trees, although recent work has led to adaptive k-d trees that allow incorporating new examples.

The concept of generalization and abstraction should make the case base more robust against noise and applicable to a wider range of problems. The concept description, the construction operators as well as the concept evaluation function are in the focus of the research in conceptual memory structure.

The conceptual incremental learning methods for case base organization puts the case base into the dynamic memory view of Schank [46] who required a coherent theory of adaptable memory structures and the need to understand how new information changes the memory.

Memory structures in CBR research are not only pure conceptual structures; hybrid structures incorporating k-d tree methods are studied as well. An overview of recent research in memory organization in CBR is given in [8].

Other work goes into the direction of bridging between implicit and explicit representations of cases [32]. The implicit representations can be based on statistical models and the explicit representation is the case base that keeps the single case as it is. As far as evidence is given, the data are summarized into statistical models based on statistical learning methods such as Minimum Description Length (MDL) or Minimum Message Length (MML) learning. If insufficient data for a class or a concept have been detected by the system, the data are kept in the case base. The case base controls the learning of the statistical models by hierarchically organizing the samples into groups. It allows dynamically learning and changing the statistical models based on the experience (data) seen so far and prevents the model from overfitting and bad influences by singularities.

This concept follows the idea that humans have built up very effective models for standard repetitive tasks and that these models can easily be used without a complex reasoning process. For rare events, the CBR unit takes over the reasoning task and collects experience into its memory.

7 Applications

CBR has been successfully applied to a wide range of problems. Among them are signal [33] and image interpretation tasks [40], medical applications [15], and emerging applications such as geographic information systems, applications in biotechnology and topics in climate research (CBR commentaries) [11, 21]. We are focusing here on hot real-world topics such as meta-learning for parameter selection, image & signal interpretation, prototype-based classification and novelty detection & handling.

7.1 Meta-Learning for Parameter Selection of Data/Signal Processing Algorithms

Meta learning is a subfield of machine learning where automatic learning algorithms are applied on meta-data about machine-learning experiments. The main goal is using such meta-data to understand how automatic learning can become flexible in regards to solving different kinds of learning problems, hence to improve the performance of existing learning algorithms. Another important meta-learning task, but not so widely studied yet, is parameter selection for data or signal processing algorithms. Soares and Brazdil [51] have used this approach for selecting the kernel width of a support-vector machine, while Perner [29] and Frucci et. al [14] have studied this approach for image segmentation.

The meta-learning problem for parameter selection can be formalized as follows: For a given signal that is characterized by specific signal properties A and domain properties B find the parameters of the processing algorithm that ensure the best quality of the resulting output signal:

$$f : A \cup B \rightarrow P_i \tag{3}$$

where P_i is the i-th class of parameters for the given domain.

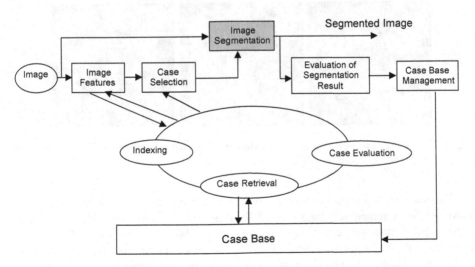

Fig. 4. Meta-learning for parameter selection.

Meta-data for signals are comprised of signal-related meta-data and non-signal related meta-data [29]. Likewise, in the standard system theory [59], the signal-related meta-data should characterize the specific signal properties that influence the result of the signal processing algorithm. In general, the processing of signal-related meta-data from signals should not require too much processing since it is an auxiliary process to achieve the final result.

The architecture of Case-based Reasoning for Image Segmentation is shown in Fig. 4. This architecture has been applied to threshold-based image segmentation [29] and the Watershed Transformation [14]. The resulting better segmentation quality compared to the standard Watershed Segmentation result is shown in Fig. 4 for a biological image.

The signal-related meta-data are for the CBR-based Watershed Transformation statistical grey-level and texture parameters such as mean, standard deviation, entropy, and Haralick's texture-descriptor. The non-signal related meta-data are the category of the images such as biological images, face images, and landscape images. The image segmentation algorithm is the Watershed transformation, where the oversegmentation of the result is controlled by weighted merging rules. The weights and the application of the merging rules are controlled by the CBR unit. This unit selects the weights and rules based on the signal characteristics and the category that should be applied for merging basins obtained by the standard Watershed Transformation in the particular image (see Fig. 5). The output of the segmentation unit is automatically assessed by a specific evaluation measure comparing the input image with the output image and starting the case-base maintenance process if the result is not as good as it should be. Then, the new case is stored with its meta-data and its segmentation parameters in the case base. Case generalization groups similar cases into a more generalized case so that it is applicable to more signals.

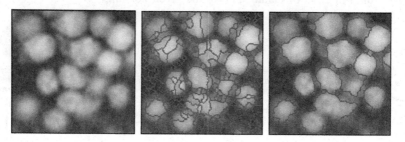

Fig. 5. Image segmentation results. Right: original image. Middle: Watershed Transform. Left: Watershed Transform based on CBR.

The mapping function f can be implemented through any classification algorithm, however the incremental behavior of CBR is most suitable for many data/signal processing problems, where the signals are not available ad-hoc, but rather appear incrementally. The right similarity metric that allows mapping data to similar groups and in the last consequence obtaining good output results should be more extensively studied. Performance measures that allow evaluation of the obtained output and automatic assessment of the system performance are another important issue. From statistical point of view, we need more signal-characterizing parameters for the meta-data that bridge between the signal characteristics and the behavior of the modern, often heuristic signal processing algorithms. Case-generalization for studying domain theory is also related to these tasks and would allow for better understanding of the behavior of many signal processing algorithms that cannot be described to further extent by the standard system theory [59].

7.2 Case-Based Image Interpretation

Image interpretation is the process of mapping the numerical representation of an image into a logical representation which is suitable for scene description. This is a complex process; the image passes through several general processing steps until the final result is obtained. These steps include image preprocessing, image segmentation, image analysis, and image interpretation. Image pre-processing and image segmentation algorithms usually need a lot of parameters to perform well on a specific image. In an image, the automatically extracted objects of interest are first described by primitive image features. Depending on the particular objects and scope, these features can be lines, edges, ribbons, etc. Typically, these low-level features have to be mapped to high-level/symbolic features. A symbolic feature such as a fuzzy margin would be a function of several low-level features.

The image interpretation component identifies an object by finding the class to which it belongs (among the models of the object class). This is done by matching the symbolic description of the object to the model/concept of the object stored in the knowledge base. Most image-interpretation systems run on the basis of a bottom-up control structure. This control structure allows no feedback to preceding processing components if the result of the outcome of the current component is unsatisfactory. A mixture of bottom-up and top-down control would allow the outcome of a component to be refined by returning to the previous component.

CBR is not only applicable as a whole to image interpretation, it is applicable to all the different levels of an image-interpretation system [28, 34] (see Fig. 6) and many of the ideas mentioned in the sections before apply here. CBR-based meta-learning algorithms for parameter selection are preferable for the image pre-processing and segmentation unit [14, 29, 39]. The mapping of the low-level features to the high-level features is a classification task, for which a CBR-based algorithm can be applied. The memory organization [31] of the interpretation unit goes along with the problems discussed for the case base organization in Sect. 5. Different organization structures for image interpretation systems are discussed in [28].

Ideally the system should start working with only a few samples and during usage of the system new cases should be learned and the memory should be updated based on these samples. This view at the usage of a system brings in another topic that is called life-time cycle of a CBR system. Work on this topic takes into account that a system is used for a long time, while experience changes over time. The case structure might change by adding new relevant attributes or deleting attributes that have shown not to be important or have been replaced by other ones. Set of cases might not appear anymore, since these kinds of solutions are not relevant anymore. A methodology and software architecture for handling the life-time cycle problem is needed so that this process can easily be carried out without rebuilding the whole system. It seems to be

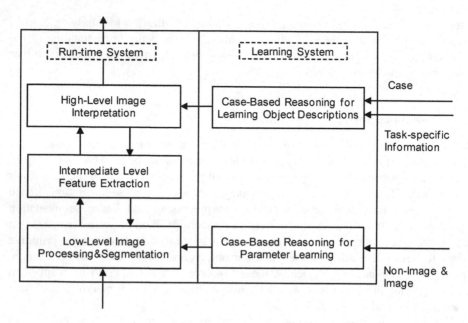

Fig. 6. Architecture of a CBR-based image interpretation system.

more a software engineering task, but has also something to do with evaluation measures for the task of forgetting and relevance of cases that can come from statistics.

7.3 Incremental Prototype-Based Classification

The usage of prototypical cases is very popular in many applications, among them are medical applications [47], Belazzi et al. [5] and by Nilsson and Funk [24], knowledge management systems [9] and image classification tasks [9]. The simple nearest-neighbor-approach [1] as well as hierarchical indexing and retrieval methods [47] have been applied to the problem. It has been shown that an initial reasoning system could be built up based on these cases. The systems are useful in practice and can acquire new cases for further reasoning during utilization of the system.

There are several problems concerned with prototypical CBR. If a large enough set of cases is available, the prototypical case can automatically be calculated as the generalization from a set of similar cases. In medical applications as well as in applications where image catalogues are the development basis for the system, the prototypical cases have been selected or described by humans. This means when building the system, we are starting from the most abstract level (the prototype) and have to collect more specific information about the classes and objects during the usage of the system.

Since the prototypical case has been selected by a human, his decision on the importance of the case might be biased; also picking only one case might be difficult for a human. As for image catalogue-based applications, he can have stored more than one image as a prototypical image. Therefore we need to check the redundancy of the many prototypes for one class before taking them all into the case base.

According to this consideration, the minimal functions that a prototype-based classification system [35] (see Fig. 6) should be able to accomplish are: classifications based on a proper similarity-measure, prototype selection by a redundancy-reduction algorithm, feature weighting to determine the importance of the features for the prototypes and learn the similarity metric, and feature-subset selection to select the relevant features from the whole set of features for the respective domain. Cross validation over the loop of all processing steps can estimate the error rate of such a system (see Fig. 7). However, when the data set is imbalanced, that is a class is underrepresented by samples – something that could always happen in real domain and incrementally collected samples – class-specific error rates have to be calculated to judge the true performance of the system. Otherwise, the overall error rate will turn out to be good, but the underrepresented classes will be classified incorrectly [20]. That means that the learning schema for prototype-selection, feature subset selection and feature weighting cannot rely only on the overall error. More sophisticated learning strategies are necessary, which work incrementally and take into account the class-specific error rate or follow the idea of bridging between the cases and the solutions based on preference or similarity relations mentioned in Sect. 4.

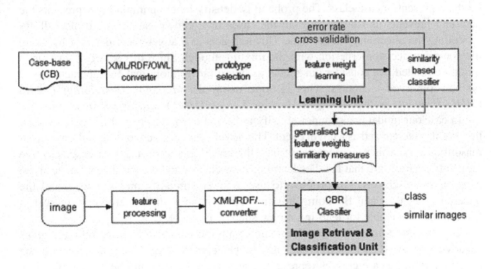

Fig. 7. Architecture of our prototype-based classifier.

Statistical learning methods so far focus on adaptive k-NN approach that adjusts the distance metric by feature weighting or kernel methods or the number k of neighbors. Incremental strategies are used for the nearest-neighbor search, but not for updating the weights, distance metric, and prototype selection. A system for handwriting recognition is described in [53] that can incrementally add data and adapt the solutions to different user writing styles. It should be possible to consider in a k-NN implementation based on functions that can handle data streams by adding data through reorganizing a multi-resolution array data structure and concept drift by realizing a case forgetting strategy [19].

7.4 Novelty Detection by Case-Based Reasoning

Novelty detection [22], i.e., recognizing that an input differs in some respect from previous inputs, can be a useful ability for learning systems.

Novelty detection is particularly useful when an important class is under-represented in the data, so that a classifier cannot be trained to reliably recognize that class. This characteristic is common to numerous problems such as information management, medical diagnosis, fault monitoring and detection, and visual perception.

We propose novelty detection to be regarded as a CBR problem under which we can run different theoretical methods for detecting the novel events and handling the novel events [32]. The detection of novel events is a common subject in the literature. The handling of the novel events for further reasoning is not treated so much in the literature, although this is a hot topic in open-world applications.

The first model we propose is comprised of statistical models and similarity-based models (see Fig. 8) [41, 42]. For now, we assume an attribute-value based representation. Nonetheless, the general framework we propose for novelty detection can be based on any representation. The heart of our novelty detector is a set of statistical models that have been learnt in an off-line phase from a set of observations. Each model represents a case-class. The probability density function implicitly represents the data and prevents us from storing all the cases of a known case-class. It also allows modelling the uncertainty in the data. This unit acts as a novel-event detector by using the Bayesian decision-criterion with the mixture model. Since this set of observations might be limited, we consider our model as being far from optimal and update it based on new observed examples. This is done using the *Minimum Description Length (MDL) Principle* or the *Minimum Message Length (MML) Learning Principle* [54].

In case our model bank cannot classify an actual event into one of the case-classes, this event is recognized as a novel event. The novel event is given to the similarity-based reasoning unit, which incorporates it into the case base according to a case-selective registration procedure that handles learning case-classes and the similarity between the cases and case-classes. We propose to use a fuzzy similarity measure to model the uncertainty in the data. By doing this the unit organizes the novel events in such a fashion that is suitable for learning a new statistical model.

The case-base-maintenance unit interacts with the statistical learning unit and gives an advice as to when a new model has to be learned. The advice is based on the observation that a case-class is represented by a large enough number of samples that are most dissimilar to other classes in the case-base.

The statistical learning unit takes this case class and proves based on the MML-criterion, whether it is suitable to learn the new model or not. In the case that the statistical component recommends to not learn the new model, the case-class is still hosted by the case base maintenance unit and further updated based on newly observed events that might change the inner-class structure as long as there is new evidence to learn a statistical model.

The use of a combination of statistical reasoning and similarity-based reasoning allows implicit and explicit storage of the samples. It allows handling well-represented events as well as rare events.

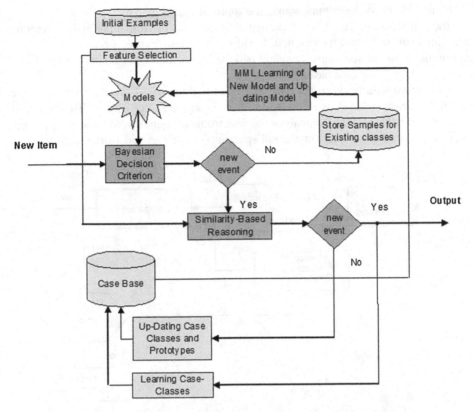

Fig. 8. Architecture of a statistical and similarity-based novelty detector and handling system.

7.5   Representation of 1-D Signals by a 0_1 Sequence and Similarity-Based Interpretation and Learning

Spectrometer signal analysis is an application where 1-D signals have to be analyzed and interpreted. Other 1-D signals are time-series in machine maintenance, speech analysis, and medical signals.

Different spectrometer methods exist that have been developed over time to systems applicable in practice. Researchers in different fields try to apply these methods to various applications especially in the area of chemistry and biology. One of these methods is RAMAN spectroscopy for protein crystallization or *Mid-Infrared Spectroscopy* for biomass identification. Robust and machine learnable automatic signal interpretation methods are required for the applications. These methods should take into account that not so much spectrometer data about the application are available from scratch and that these data need to be learned while using the spectrometer system (see Fig. 9) [36]. That brings us to a CBR approach. We propose to represent the spectrometer signal by a sequence of 0/1 characters obtained from a specific *Delta Modulator* [37].

Figure 10 shows the original signal, the approximated signal by the Delta Modulator and the coded signal. This novel representation prevents us from a particular symbolic description of peaks and background. Besides, the original curve gets smoothed. The interpretation of the spectrometer signal is done by searching for a similar signal in a constantly increasing data base.

The comparison between the two sequences is done based on a flexible syntactic similarity measure that does not need to have background knowledge. The proposed method is flexible enough to analyze the spectrometer signal from different perspectives. We can take into account the full spectrum as well as part of the spectrum.

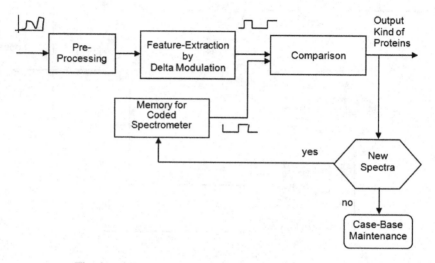

Fig. 9. Architecture of the spectrum interpretation system.

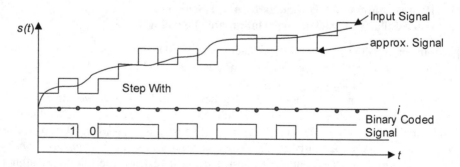

Fig. 10. Diagram with input signal, approximated signal, and binary coded signal.

8   Conclusions

In this paper we have presented our thoughts and work on model development and learning based on CBR [38]. Model building and adaption became an important topic so that an automatic system can adapt to changing environments and image objects.

CBR can help developing the model for signal interpretation and incrementally learning over time the final model. CBR gives a flexible and powerful method that allows controlling the signal processing chain in all phases of a signal interpretation system, so that the resulting information can be derived in the best possible quality. Beyond this CBR offers different learning capabilities in all phases of a signal interpretation system. These learning and maintenance capabilities satisfy the different needs during the development process of a signal interpretation system. Therefore, they are especially appropriate for incremental model development.

CBR solves problems using already stored knowledge, and captures incrementally new knowledge, making it immediately available for solving the next problem. To implement this cognitive model in a computer-based system we need methods known from statistics, pattern recognition, artificial intelligence, machine learning, data base research and other fields. Only the combination of all these methods will give us a system that can efficiently solve practical problems. CBR systems work efficiently in real-world applications, since the CBR method possesses all features of a well-performing and user-friendly system.

We have pointed out that the central aspect of a well-performing system in the real-world is its ability to incrementally collect new experience and reorganize its knowledge based on these new insights. In our opinion, the new challenging research aspects should have its focus on incremental methods for prototype-based classification, meta-learning for parameter selection, complex signals understanding tasks and novelty detection.

Our interest is to build intelligent flexible and robust data-interpreting systems that are inspired by the human CBR process and by doing so to model the human reasoning process when interpreting real-world situations.

References

1. Aha, D.W., Kibler, D., Albert, M.K.: Instance-based learning algorithm. Mach. Learn. **6**(1), 37–66 (1991)
2. Bagherjeiran, A., Eick, C.F.: Distance function learning for supervised similarity assessment. In: Perner, P. (ed.) Case-Based Reasoning on Images and Signals. Studies in Computational Intelligence, pp. 91–126. Springer, Heidelberg (2008)
3. Bergmann, R., Wilke, W.: On the role of abstraction in case-based reasoning. In: Smith, I., Faltings, B. (eds.) EWCBR 1996. LNCS (LNAI), vol. 1168, pp. 28–43. Springer, Heidelberg (1996). doi:10.1007/BFb0020600
4. Bergmann, R., Richter, M., Schmitt, S., Stahl, A., Vollrath, I.: Utility-oriented matching: a new research direction for case-based reasoning. In: Schnurr, H.-P., et al. (eds.) Professionelles Wissensmanagement, pp. 20–30. Shaker Verlag (2001)
5. Bellazzi, R., Montani, S., Portinale, L.: Retrieval in a prototype-based case library: a case study in diabetes therapy revision. In: Smyth, B., Cunningham, P. (eds.) EWCBR 1998. LNCS (LNAI), vol. 1488, pp. 64–75. Springer, Heidelberg (1998). doi:10.1007/BFb0056322
6. Bentley, J.: Multidimensional binary search trees used for associative searching. Commun. ACM **18**(9), 509–517 (1975)

7. Bhanu, B., Dong, A.: Concepts learning with fuzzy clustering and relevance feedback. In: Perner, P. (ed.) MLDM 2001. LNCS (LNAI), vol. 2123, pp. 102–116. Springer, Heidelberg (2001). doi:10.1007/3-540-44596-X_9

8. Bichindaritz, I.: Memory structures and organization in case-based reasoning. In: Perner, P. (ed.) Case-Based Reasoning on Images and Signals. Studies in Computational Intelligence, pp. 175–194. Springer, Heidelberg (2008)

9. Bichindaritz, I.: Mémoire: a framework for semantic interoperability of case-based reasoning systems in biology and medicine. Artif. Intell. Med. 36(2), 177–192 (2006)

10. Branting, L.K.: Integrating generalizations with exemplar-based reasoning. In: Proceedings of the 11th Annual Conference of Cognitive Science Society. Ann Arbor, MI, Lawrence Erlbaum, pp. 129–146 (1989)

11. CBR Commentaries. Knowl. Eng. Rev. 20(3)

12. Craw, S.: Introspective learning to build Case-Based Reasoning (CBR) knowledge containers. In: Perner, P., Rosenfeld, A. (eds.) MLDM 2003. LNCS, vol. 2734, pp. 1–6. Springer, Heidelberg (2003). doi:10.1007/3-540-45065-3_1

13. Fisher, D.H.: Knowledge acquisition via incremental conceptual clustering. Mach. Learn. 2(2), 139–172 (1987). Kluwer Academic Publishers, Hingham, MA, USA

14. Frucci, M., Perner, P., di Baja, G.S.: Case-based reasoning for image segmentation by watershed transformation. In: Perner, P. (ed.) Case-Based Reasoning on Signals and Images, pp. 319–353. Springer, Heidelberg (2007)

15. Holt, A., Bichindaritz, I., Schmidt, R., Perner, P.: Medical applications in case-based reasoning. Knowl. Eng. Rev. 20(3), 289–292 (2005)

16. Iglezakis, I., Reinartz, T., Roth-Berghofer, T.R.: Maintenance memories: beyond concepts and techniques for case base maintenance. In: Funk, P., González Calero, Pedro, A. (eds.) ECCBR 2004. LNCS (LNAI), vol. 3155, pp. 227–241. Springer, Heidelberg (2004). doi:10.1007/978-3-540-28631-8_18

17. Jaenichen, S., Perner, P.: Conceptual clustering and case generalization of two dimensional forms. Comput. Intell. 22(3/4), 177–193 (2006)

18. Jain, A.K., Dubes, R.C.: Algorithms for Clustering Data. Prentice Hall Inc, Upper Saddle River (1988)

19. Law, Y.-N., Zaniolo, C.: An adaptive nearest neighbor classification algorithm for data streams. In: Jorge, A.M., Torgo, L., Brazdil, P., Camacho, R., Gama, J. (eds.) PKDD 2005. LNCS (LNAI), vol. 3721, pp. 108–120. Springer, Heidelberg (2005). doi:10.1007/11564126_15

20. Little, S., Salvetti, O., Perner, P.: Evaluation of feature subset selection, feature weighting, and prototype selection for biomedical applications. J. Softw. Eng. Appl. 3, 39–49 (2010)

21. De Mantaras, R.L., Cunningham, P., Perner, P.: Emergent case-based reasoning applications. Knowl. Eng. Rev. 20(3), 325–328 (2005)

22. Markou, M., Singh, S.: Novelty detection: a review – part 1. Stat. Approaches Sig. Process. 83(12), 2481–2497 (2003)

23. Nagy, G., Nartker, T.H.: Optical Character Recognition: An Illustrated Guide to the Frontier. Kluwer, London (1999)

24. Nilsson, M., Funk, P.: A case-based classification of respiratory sinus arrhythmia. In: Funk, P., González Calero, Pedro, A. (eds.) ECCBR 2004. LNCS (LNAI), vol. 3155, pp. 673–685. Springer, Heidelberg (2004). doi:10.1007/978-3-540-28631-8_49

25. Pekalska, E., Duin, R.: The Dissimilarity Representation for Pattern Recognition. World Scientific, Singapore (2005)

26. Perner, P.: Introduction to case-based reasoning for signals and images. In: Perner, P. (ed.) Case-Based Reasoning on Signals and Images, pp. 1–4. Springer, Heidelberg (2007)

27. Perner, P.: Data Reduction Methods for Industrial Robots with Direct Teach-in-Programing, Second Unchanged Edition. IBAI Publishing, Fockendorf. ISBN 978-3-940501-16-5

28. Perner, P.: Why case-based reasoning is attractive for image interpretation. In: Aha, D.W., Watson, I. (eds.) ICCBR 2001. LNCS (LNAI), vol. 2080, pp. 27–43. Springer, Heidelberg (2001). doi:10.1007/3-540-44593-5_3

29. Perner, P.: An architecture for a CBR image segmentation system. J. Eng. Appl. Artif. Intell. 12(6), 749–759 (1999)

30. Perner, P., Perner, H., Müller, B.: Similarity guided learning of the case description and improvement of the system performance in an image classification system. In: Craw, S., Preece, A. (eds.) ECCBR 2002. LNCS (LNAI), vol. 2416, pp. 604–612. Springer, Heidelberg (2002). doi:10.1007/3-540-46119-1_44

31. Perner, P.: Case-base maintenance by conceptual clustering of graphs. Eng. Appl. Artif. Intell. 19(4), 295–381 (2006)

32. Perner, P.: Concepts for novelty detection and handling based on a case-based reasoning process scheme. In: Perner, P. (ed.) ICDM 2007. LNCS (LNAI), vol. 4597, pp. 21–33. Springer, Heidelberg (2007). doi:10.1007/978-3-540-73435-2_3

33. Perner, P., Holt, A., Richter, M.: Image processing in case-based reasoning. Knowl. Eng. Rev. 20(3), 311–314 (2005)

34. Perner, P.: Using CBR learning for the low-level and high-level unit of a image interpretation system. In: Singh, S. (ed.) Advances in Pattern Recognition, pp. 45–54. Springer, Heidelberg (1998)

35. Perner, P.: Prototype-based classification. Appl. Intell. 28(3), 238–246 (2008)

36. Perner P.: A novel method for the interpretation of spectrometer signals based on delta-modulation and similarity determination. In: Barolli, L., Li, K.F., Enokido, T., Xhafa, F., Takizawa, M. (eds.) Proceedings IEEE 28th International Conference on Advanced Information Networking and Applications AINA 2014, Victoria, Canada, pp. 1154–1160 (2014). doi:10.1109/AINA.2014.44

37. Perner, P.: Representation of 1-D signals by a 0_1 sequence and similarity-based interpretation: a case-based reasoning approach. In: Perner, P. (ed.) Machine Learning and Data Mining in Pattern Recognition. LNCS (LNAI), vol. 9729, pp. 728–739. Springer, Heidelberg (2016). doi:10.1007/978-3-319-41920-6_55

38. Perner, P.: Case-based reasoning and the statistical challenges II. In: Gruca, A., Czachórski, T., Kozielski, S. (eds.). AISC, vol. 242, pp. 17–38. Springer, Heidelberg (2014). doi:10.1007/978-3-319-02309-0_2

39. Perner, P., Attig, A.: Meta-learning for image processing based on case-based reasoning. In: Bichindaritz, I., Vaidya, S., Jain, A., Jain, L.C. (eds.) Computational Intelligence in Healthcare 4. SIC, vol. 309, pp. 229–264. Springer, Heidelberg (2010)

40. Perner, P.: Case-based reasoning for image analysis and interpretation. In: Chen, C., Wang, P.S.P. (eds.) Handbook on Pattern Recognition and Computer Vision, 3rd Edition, pp. 95–114. World Scientific Publisher (2005)

41. Perner, P.: Novelty detection and in-line learning of novel concepts according to a case-based reasoning process schema for high-content image analysis in system biology and medicine. Comput. Intell. 25(3), 250–263 (2009)

42. Perner, P.: Concepts for novelty detection and handling based on a case-based reasoning process scheme. Eng. Appl. Artif. Intell. 22(1), 86–91 (2009)

43. Richter, Michael, M.: Introduction. In: Lenz, Mario, Burkhard, Hans-Dieter, Bartsch-Spörl, Brigitte, Wess, Stefan (eds.). LNCS (LNAI), vol. 1400, pp. 1–15. Springer, Heidelberg (1998). doi:10.1007/3-540-69351-3_1

44. Richter, M.M.: Similarity. In: Perner, P. (ed.) Case-Based Reasoning on Images and Signals. Studies in Computational Intelligence, pp. 1–21. Springer, Heidelberg (2008)

45. Sankoff, D., Kruskal, J.B. (eds.): Time Warps, String Edits, and Macromolecules: The Theory and Practice of Sequence Comparison. Addison-Wesley, Readings (1983)

46. Schank, R.C.: Dynamic Memory. A theory of reminding and learning in computers and people. Cambridge University Press, Cambridge (1982)
47. Schmidt, R., Gierl, L.: Temporal abstractions and case-based reasoning for medical course data: two prognostic applications. In: Perner, P. (ed.) MLDM 2001. LNCS (LNAI), vol. 2123, pp. 23–34. Springer, Heidelberg (2001). doi:10.1007/3-540-44596-X_3
48. Shapiro, L.G., Atmosukarto, I., Cho, H., Lin, H.J., Ruiz-Correa, S.: Similarity-based retrieval for biomedical applications. In: Perner, P. (ed.) Case-Based Reasoning on Signals and Images. SIC, vol. 73, pp. 355–388. Springer, Heidelberg (2007)
49. Smith, E.E., Douglas, L.M.: Categories and Concepts. Harvard University Press, Cambridge (1981)
50. Smith, L.B.: From global similarities to kinds of similarities: the construction of dimensions in development. In: Smith, L.B. (ed.) Similarity and analogical reasoning, pp. 146–178. Cambridge University Press, New York (1989)
51. Soares, C., Brazdil, P.B.: A meta-learning method to select the kernel width in support vector regression. Mach. Learn. 54, 195–209 (2004)
52. Stahl, A.: Learning feature weights from case order feedback. In: Aha, David, W., Watson, I. (eds.) ICCBR 2001. LNCS (LNAI), vol. 2080, pp. 502–516. Springer, Heidelberg (2001). doi:10.1007/3-540-44593-5_35
53. Vuori, V., Laaksonen, I., Oja, E., Kangas, J.: Experiments with adaptation strategies for a prototype-based recognition system for isolated handwritten characters. Int. J. Doc. Anal. Recogn. 3(3), 150–159 (2001)
54. Wallace, C.S.: Statistical and Inductive Inference by Minimum Message Length. Information Science and Statistics. Springer, Series (2005)
55. Weihs, C., Ligges, U., Mörchen, F., Müllensiefen, M.: Classification in music research. J. Adv. Data Anal. Classif. 3(1), 255–291 (2007). Springer
56. Wess, S., Globig, C.: Case-based and symbolic classification. In: Wess, S., Althoff, K.-D., Richter, M.M. (eds.) EWCBR 1993. LNCS, vol. 837, pp. 77–91. Springer, Heidelberg (1994). doi:10.1007/3-540-58330-0_78
57. Wettschereck, D., Aha, D.W., Mohri, T.: A review and empirical evaluation of feature weighting methods for a class of lazy learning algorithms. Artif. Intell. Rev. 11, 273–314 (1997)
58. Wilson, D.R., Martinez, T.R.: Improved heterogeneous distance functions. J. Artif. Intell. Res. 6, 1–34 (1997)
59. Wunsch, G.: Systemtheorie der Informationstechnik. Akademische Verlagsgesellschaft, Leipzig (1971)
60. Xiong, N., Funk, P.: Building similarity metrics reflecting utility in case-based reasoning. J. Intell. Fuzzy Syst. 17(4), 407–416 (2006). IOS Press
61. Xueyan, S., Petrovic, S., Sundar S.: A case-based reasoning approach to dose planning in radiotherapy. In: Wilson, D.C., Khemani, D. (eds.) The seventh international Proceedings of Conference on Case-Based Reasoning, Belfast, Northern Ireland, pp. 348–357 (2007)
62. Zhang, L., Coenen, F., Leng, P.: Formalising optimal feature weight settings in case-based diagnosis as linear programming problems. Knowl.-Based Syst. 15, 298–391 (2002)

Theoretical Contributions

CVT-Based 3D Image Segmentation for Quality Tetrahedral Meshing

Kangkang Hu[1], Yongjie Jessica Zhang[1(✉)], and Guoliang Xu[2]

[1] Department of Mechanical Engineering, Carnegie Mellon University,
Pittsburgh, USA
{kangkanh,jessicaz}@andrew.cmu.edu
[2] LSEC, Institute of Computational Mathematics,
Academy of Mathematics and Systems Science,
Chinese Academy of Sciences, Beijing, China
xuguo@lsec.cc.ac.cn

Abstract. Given an input 3D image, in this paper we first segment it into several clusters by extending the 2D harmonic edge-weighted centroidal Voronoi tessellation (HEWCVT) method to the 3D image domain. The Dual Contouring method is then applied to construct tetrahedral meshes by analyzing both material change edges and interior edges. An anisotropic Giaquinta-Hildebrandt operator (GHO) based geometric flow method is developed to smooth the surface with both volume and surface features preserved. Optimization based smoothing and topological optimizations are also applied to improve the quality of tetrahedral meshes. We have verified our algorithms by applying them to several datasets.

Keywords: Centroidal voronoi tessellation · Image segmentation · Tetrahedral mesh · Quality improvement · Giaquinta-Hildebrandt operator

1 Introduction

Many methods have been developed for 2D/3D image segmentation in the literature [3,13]. Thresholding [1,18] is a very common approach which partitions the image based on the intensity values and a given threshold. Binarization was used in [19] to segment the raw image for accurate 3D reconstruction of the air exchange regions of the lung. K-means clustering [15] groups pixels in an image into non-overlapping clusters through the minimization of the total inter-cluster variance. Watershed [17] segments images into homogeneous regions by using concepts from edge detection and mathematical morphology. In recent years, centroidal Voronoi tessellation (CVT) has been extensively studied for image segmentation [5,6], where the key idea is to partition the image by updating generators with respect to a specific energy function. The edge-weighted CVT (EWCVT) model [20] was proposed by incorporating spatial information into the energy function in order to eliminate the noises and unnecessary details.

© Springer International Publishing AG 2017
R.P. Barneva et al. (Eds.): CompIMAGE 2016, LNCS 10149, pp. 27–42, 2017.
DOI: 10.1007/978-3-319-54609-4_2

The harmonic EWCVT (HEWCVT) model [9] extends EWCVT by introducing a harmonic form of clustering energy to generate more stable and accurate results. Starting from segmented images, the Dual Contouring method [22] generates dual meshes from an octree structure. Tetrahedral meshes for complicated domains with topology ambiguity can be generated by splitting the ambiguous leaf cells into tetrahedra and analyzing the edges of these tetrahedra [25]. A parallel Image-to-Mesh conversion algorithm [7] was proposed to generate quality tetrahedral meshes via dynamic point insertions and removals. However, it is still challenging to generate quality finite element meshes directly from raw images bridging image segmentation and mesh generation.

It is also crucial to improve the mesh quality in order to avoid the ill-conditioned linear systems during the finite element analysis. Smoothing methods improve mesh quality by relocating vertices without changing the connectivity [8]. However, traditional smoothing techniques are heuristic and sometimes invert or degrade the local elements. To address this problem, optimization-based smoothing methods are proposed, where each node is relocated at the optimum location based on the local gradient of the surrounding element quality [2]. Methods based on local curvature and volume preserving geometric flows are developed to identify and preserve the main surface features [11,23]. Topological optimization techniques, such as face swapping and edge removal [10], are utilized to improve the node valence and mesh quality. Although there already exist a variety of mesh denoising methods, research on feature preserving denoising remains active due to its challenging nature.

In this paper, we first extend the HEWCVT [9] from 2D to 3D image segmentation and generate compact and connected segments. Based on the segmented image, the Dual Contouring method [22,25] is then applied to construct tetrahedral meshes by analyzing both material change edges and interior edges. We also develop an anisotropic Giaquinta-Hildebrandt operator (GHO) diffusion flow for surface smoothing and quality improvement, while optimization based smoothing and topological optimizations are applied together. The key contributions of our proposed algorithms include:

1. The 2D HEWCVT [9] is extended to 3D image segmentation, where 3D spatial information is included in order to eliminate the noise effect. By improving the connectivity of each segment, it generates compact and connected segments without leaving isolated voxels and in keeping the connectivity of the structure; and
2. The anisotropic GHO diffusion flow is developed for surface smoothing which preserves surface features while removing the noise with an anisotropic weighting function. Since GHO is defined based on the second fundamental form of the surface, our proposed algorithm is more sensitive to the curvature-related features.

The remainder of this paper is organized as follows: Sect. 2 describes the HEWCVT-based 3D image segmentation. Section 3 explains tetrahedral mesh generation via Dual Contouring method. Section 4 discusses surface smoothing via the anisotropic GHO diffusion flow and explains how to improve the quality of tetrahedral meshes. Section 5 shows some results, and Sect. 6 presents conclusions and future work.

2 CVT-Based 3D Image Segmentation

CVT-based clustering methods [5,6] partition discrete data points into non-overlapping clusters with an initialization of generators. It first constructs Voronoi regions by assigning each point to its nearest generator with certain distance metric. For each Voronoi region, we can iteratively calculate its centroid by minimizing a pre-defined energy function until it coincides with the corresponding generator. Inspired by the HEWCVT method [9] for 2D image segmentation, here we extend it to 3D image segmentation.

The input image I is given in the form of function values, $I = \{I(x, y, z)\}$, where x, y, z are indices of X, Y, Z coordinates. Let the dataset $F = \{f_{P(i)}\}_{i=1}^{n}$ denote all the intensity values $f_{P(i)}$ of the grayscale image, where n is the total number of voxels and $P(i)$ represents the ith voxel in the physical space. Let $C = \{c_l\}_{l=1}^{L}$ denote a set of Voronoi generators with intensity values, where L is the number of clusters. The Voronoi regions $V = \{V_l\}_{l=1}^{L}$ in F corresponding to the generators can be obtained by assigning each voxel to the cluster whose generator is the nearest to it according to the distance metric:

$$V_k = \left\{ f_{P(i)} \in F : \text{dist}\left(f_{P(i)}, c_k \right) \leq \text{dist}\left(f_{P(i)}, c_l \right), \quad \text{for } l = 1, \ldots, L \right\}, \quad (1)$$

where $\text{dist}\left(f_{P(i)}, c_k \right) = \sqrt{\left| f_{P(i)} - c_k \right|^2 + 2\lambda \hat{n}_k(P(i))}$ measures the edge-weighted distance between $f_{P(i)}$ and c_k in the grayscale space. The edge-weighted term $\hat{n}_k(P(i))$ represents the number of voxels that do not belong to the kth cluster within ω-ring spherical neighbours of $P(i)$, which includes the local 3D spatial information in the physical space. Here we choose a relatively small value of ω ($\omega = 3$) for all the examples in order to reduce the computational cost. Given any set of generators $C = \{c_l\}_{l=1}^{L}$ and any partition $U = \{U_l\}_{l=1}^{L}$ of F, we can define the corresponding HEWCVT energy function of $(C; U)$ as

$$E\left(C; U \right) = \sum_{i=1}^{n} \left(L \middle/ \sum_{l=1}^{L} \text{dist}^{-2} \left(f_{P(i)}, c_l \right) \right). \quad (2)$$

Note that the HEWCVT energy function uses the edge-weighted distances to all generators for each voxel. It means that all the generators partially influence the harmonic average for each voxel. By taking into account the physical information and using the harmonic form of energy function, HEWCVT is robust to the initialization and can eliminate the noise in the 3D image during the segmentation. To calculate the updated centroids $\{c_k^*\}_{k=1}^{L}$, we minimize the HEWCVT energy function with respect to the generators c_k ($k = 1, \ldots, L$). If the generators of the Voronoi regions $\{V_l\}_{l=1}^{L}$ of F equal to their corresponding centroids, i.e., $c_l = c_l^*$ for $l = 1, \ldots, L$, then we call the Voronoi tessellation $\{V_l\}_{l=1}^{L}$ a centroidal Voronoi tessellation of F. The detailed implementation of HEWCVT-3D is explained as follows:

Algorithm of HEWCVT-3D

Given a 3D image $F = \{f_{P(i)}\}_{i=1}^{n}$, positive integer L and error tolerance ε ($\varepsilon = 10^{-4}$ in this paper). E_i denotes the HEWCVT energy in the ith iteration. Then perform the following:

1. For each voxel, find its ω-ring neighbouring voxels in advance. Choose L random voxels in the image and take their intensity values as the initialization of the generators $\{c_l\}_{l=1}^{L}$;
2. Determine the edge-weighted Voronoi clusters $\{V_l\}_{l=1}^{L}$ of F associated with $\{c_l\}_{l=1}^{L}$ by (1). For each cluster V_l ($l = 1, \ldots, L$), update the cluster centroid c_l^* by minimizing the HEWCVT clustering energy function;
3. If $\frac{E_{i+1}-E_i}{E_i} < \varepsilon$ is reached, return $(\{c_l\}_{l=1}^{L}; \{V_l\}_{l=1}^{L})$; otherwise, set $c_l = c_l^*$ for $l = 1, \ldots, L$ and return to Step 2.
4. Merge small isolated segments to its neighbouring cluster with the longest boundary.

We have tested our HEWCVT-based 3D image segmentation algorithm on a 3D MRI Brain-1 image ($181 \times 217 \times 181$) from BrainWeb dataset [4], with 3% noise and 20% intensity non-uniformity (INU), shown in Fig. 1. We segmented the 3D image into four clusters in order to extract the gray matter, white matter, cerebrospinal fluid and background. Figure 1(b) shows the HEWCVT-3D segmentation result after 20 iterations, where neighbouring clusters are rendered with different colors. Figure 1(c) and (d) show the segmented white matter and gray matter, respectively. We also compared our result with EWCVT-3D by extending EWCVT [20] to 3D image domain. From the slice 103 and corresponding segmented images in Fig. 1(e–h), we can observe that HEWCVT-3D yields more accurate results in many regions and both of them can eliminate the noise effect. Figure 1(i) shows the energy convergence curves for both EWCVT-3D and HEWCVT-3D under the same initialization. Compared to EWCVT-3D, the HEWCVT-3D energy converges faster to the minimum. We also segmented the image with 100 different random initializations using both EWCVT-3D and HEWCVT-3D, and the minimized energy outputs are shown in Fig. 1(j). We can observe that HEWCVT-3D is much more stable and less sensitive to initializations than the EWCVT-3D.

Remark 2.1. Compared to the EWCVT-3D, the HEWCVT-3D yields more accurate results by imposing a soft membership function with a harmonic average form of energy function. By taking into account the local 3D spatial information of each voxel, HEWCVT-3D is robust to eliminate the noise effect during the segmentation process. By improving the connectivity of each segment, our HEWCVT-3D can automatically and robustly generate compact segments without leaving isolated voxels. The segmented image can be used to generate tetrahedral meshes directly via the Dual Contouring method [22].

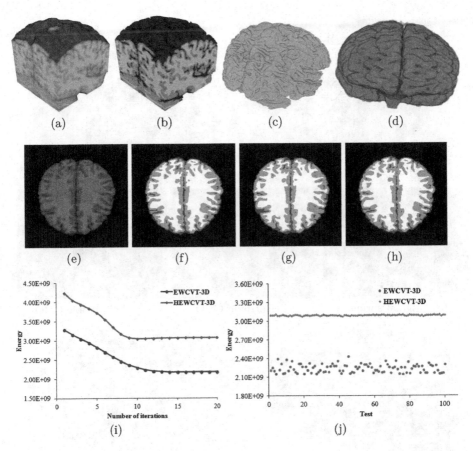

Fig. 1. Segmentation result of the Brain-1 image with 3% noise and 20% INU. (a) Input 3D image; (b) HEWCVT-3D segmentation result; (c) segmented white matter; (d) segmented gray matter; (e–h) the original image, ground truth, EWCVT-3D and HEWCVT-3D results of the slice 103, respectively; (i) energy outputs; and (j) minimized energy outputs of 100 initializations.

3 Tetrahedral Mesh Generation

After the segmentation, we set the image I as a scalar function, $I(x, y, z) \rightarrow J$, where $J = \{0, 1, \ldots, L - 1\}$ is a set of labels where 0 represents the background and $1, \ldots, L - 1$ represent the other materials. Based on the labelled image, we analyze both material changes edges and interior edges to generate tetrahedral meshes by using the Dual Contouring method [22,24,25]. A *material change edge* is defined as an edge whose two end points have different label indices. An *interior edge* is an edge whose two end points have the same label. Each material change edge belongs to a boundary cell, while interior cells only contain interior edges. For each octree cell, a dual vertex is generated and the tetrahedral mesh is constructed by connecting the dual vertices with octree grids. For each

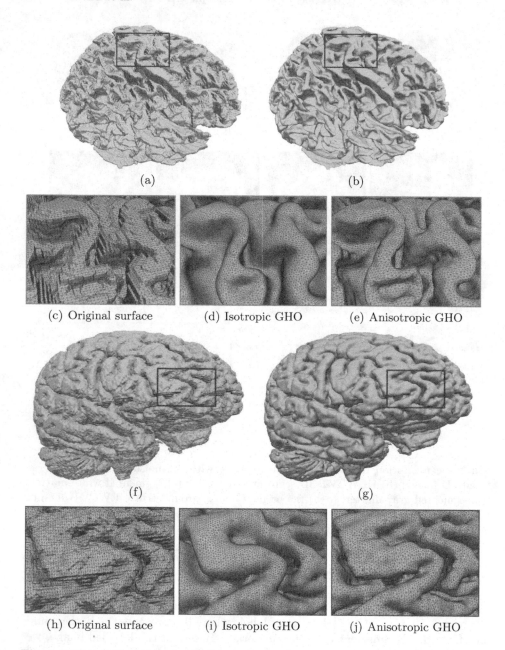

(a)   (b)

(c) Original surface   (d) Isotropic GHO   (e) Anisotropic GHO

(f)   (g)

(h) Original surface   (i) Isotropic GHO   (j) Anisotropic GHO

Fig. 2. Tetrahedral meshes. (a, b) The original and smoothed meshes (using anisotropic GHO) of the white matter, respectively; (c) enlargement of the red window in (a); (d, e) zoom-in pictures showing smoothed results of isotropic and anisotropic GHO flow, respectively; (f, g) the original and smoothed meshes (using anisotropic GHO) of the gray matter, respectively; (h) enlargement of the green window in (f); and (i, j) zoom-in pictures showing smoothed results of isotropic and anisotropic GHO flow, respectively. (Color figure online)

boundary cell, we calculate the mass center as the dual vertex. The mass center is defined as the average of all the middle points of the material change edges in the cell. The cell center is simply selected as the dual vertex for each interior cell. For each material change edge, we first find out all its four surrounding leaf cells and corresponding dual vertices. These four dual vertices and the interior grid point of this edge construct a pyramid. For each interior edge, we also obtain four dual vertices. These four dual vertices and two endpoints of this edge form a diamond. Finally, the pyramids and diamonds can be split into four or two tetrahedra. To handle topology ambiguities, a trilinear function can be introduced to detect the ambiguous cells [25]. The ambiguous cells are split into tetrahedra, and tetrahedral meshes are then generated by analyzing the edges of these tetrahedra. Mesh adaptation can be achieved via an adaptive octree data structure [25]. Since the function values $I(x, y, z)$ of the segmented images are discontinuous, the surfaces of the generated tetrahedral meshes are bumpy. Figure 2(a) and (f) show the initial tetrahedral meshes of the white matter and gray matter of the Brain-1, where red/green windows in Fig. 2(c) and (h) highlight the bumpy surfaces. A surface smoothing technique is needed during the following mesh quality improvement.

4   GHO-Based Geometric Flow and Quality Improvement

In the meshes generated from the above algorithm, some elements around the boundaries may have poor aspect ratio, therefore the mesh quality needs to be improved. There are two kinds of vertices in 3D meshes, boundary vertices and interior vertices. For each boundary vertex, we use geometric flow to denoise the surface and improve the quality. The quality of interior tetrahedra is simultaneously improved by using the optimization-based smoothing and topological optimizations.

Laplacian smoothing is the most commonly used mesh smoothing method which iteratively adjusts the vertex to the geometric center of its neighboring vertices. However, it also produces a shrinking effect and an oversmoothing result. Here we develop a new GHO based geometric flow to smooth the surface, which can preserve the concave/convex features and avoid volume shrinkage. Let $S = \{\mathbf{x}(u, v), (u, v) \in \mathbb{R}^2\}$ be a smooth parametric surface in $\mathbb{R}^3$. Note that (u, v) can also be written as (u^1, u^2) for convenience. The coefficients of the first fundamental form of S are defined as $g_{\alpha\beta} = \langle \mathbf{x}_{u^\alpha}, \mathbf{x}_{u^\beta} \rangle$ $(\alpha, \beta = 1, 2)$, where $\mathbf{x}_{u^\alpha} = \frac{\partial \mathbf{x}}{\partial u^\alpha}$ and $\mathbf{x}_{u^\beta} = \frac{\partial \mathbf{x}}{\partial u^\beta}$. The coefficients of the second fundamental form of S are defined as $b_{\alpha\beta} = \langle \mathbf{n}, \mathbf{x}_{u^\alpha u^\beta} \rangle$, where $\mathbf{x}_{u^\alpha u^\beta} = \frac{\partial^2 \mathbf{x}}{\partial u^\alpha \partial u^\beta}$ and $\mathbf{n} = (\mathbf{x}_u \times \mathbf{x}_v)/\|\mathbf{x}_u \times \mathbf{x}_v\|$. Let $g = \det[g_{\alpha\beta}]$, $[g^{\alpha\beta}] = [g_{\alpha\beta}]^{-1}$, and $[b^{\alpha\beta}] = [b_{\alpha\beta}]^{-1}$. The mean curvature H and Gaussian curvature K can be given by $H = \frac{b_{11}g_{22} - 2b_{12}g_{12} + b_{22}g_{11}}{2g}$ and $K = \frac{b_{11}b_{22} - b_{12}^2}{g}$. Let $f \in C^2(S)$, the Giaquinta-Hildebrandt operator (GHO) acting on f is defined as

$$\Box f = \mathrm{div}(\Diamond f) = \frac{1}{\sqrt{g}} \left[\frac{\partial}{\partial u}, \frac{\partial}{\partial v} \right] \left[\sqrt{g} K \left[b^{\alpha\beta} \right] [f_u, f_v]^T \right], \tag{3}$$

where div($\mathbf{v}$) is the tangential divergence operator acting on a C^1 smooth vector field $\mathbf{v}$ and $\Diamond$ is the second tangential operator (STO) given by

$$\Diamond f = [\mathbf{x}_u, \mathbf{x}_v] \, K \, [b^{\alpha\beta}] \, [f_u, f_v]^T. \tag{4}$$

To preserve the volume, we define a surface diffusion flow using the GHO as

$$\frac{\partial \mathbf{x}}{\partial t} = \mathrm{sign}(K(\mathbf{x})) \Box H(\mathbf{x}) \mathbf{n}(\mathbf{x}). \tag{5}$$

Let $S(t)$ be the smoothed surface at $t \geq 0$. Let $A(t)$ denote the area of $S(t)$, $V(t)$ denote the volume of the region enclosed by $S(t)$. Then we have

$$\frac{dA(t)}{dt} = \int_{S(t)} \Box H H d\sigma, \quad \frac{dV(t)}{dt} = \int_{S(t)} \Box P H d\sigma. \tag{6}$$

Green's Formula [21]: Let $\mathbf{v} = (v_1, v_2, v_3)^T$ be a vector field on S and $f \in C^1(S)$ with compact support. Then

$$\int_S < \mathbf{v}, \nabla f > dA = - \int_S f \mathrm{div}(\mathbf{v}) dA, \tag{7}$$

where $\nabla f = [\mathbf{x}_u, \mathbf{x}_v] \, [g^{\alpha\beta}] \, [f_u, f_v]^T$ is the tangential gradient operator acting on f. According to the Green's formula, we have

$$\frac{dA(t)}{dt} = \int_{S(t)} \Box H H d\sigma = - \int_{S(t)} (\nabla H)^T \Diamond H d\sigma, \tag{8}$$

and

$$\frac{dV(t)}{dt} = \int_{S(t)} \Box H d\sigma = - \int_{S(t)} (\Diamond H)^T \nabla(1) d\sigma = 0. \tag{9}$$

Hence, the proposed geometric flow is volume preserving. Since GHO is defined based on the second fundamental form of the surface, it is more sensitive to the curvature-related features. From the definition of $\Diamond$ and div, we can derive that

$$\Box f = g_u^{\Box} f_u + g_v^{\Box} f_v + g_{uu}^{\Box} f_{uu} + g_{uv}^{\Box} f_{uv} + g_v^{\Box} f_v, \tag{10}$$

where

$$g_u^{\Box} = - \left[b_{11}(g_{22}g_{122} - g_{12}g_{222}) + 2b_{12}(g_{12}g_{212} - g_{22}g_{112}) + b_{22}(g_{22}g_{111} - g_{12}g_{211}) \right] \Big/ g^2,$$

$$g_v^{\Box} = - \left[b_{11}(g_{11}g_{222} - g_{12}g_{122}) + 2b_{12}(g_{12}g_{112} - g_{11}g_{212}) + b_{22}(g_{11}g_{211} - g_{12}g_{111}) \right] \Big/ g^2,$$

$$g_{uu}^{\Box} = b_{22} \Big/ g, \quad g_{uv}^{\Box} = -2b_{12} \Big/ g, \quad g_{vv}^{\Box} = b_{11} \Big/ g,$$

and $g_{\alpha\beta\gamma} = \langle \mathbf{x}_{u^\alpha}, \mathbf{x}_{u^\beta u^\gamma} \rangle$. Since b_{ij} involves the second order derivatives of the surface, a C^2-continuous surface representation is required. In this section, the Loop subdivision basis functions are adopted to evolve the surface.

The above geometric flow smooths the surface by moving the vertex along its normal direction. The isotropic smoothing in Eq. (5) can eliminate noise but also smooth out important features. To preserve surface features while removing the noise, we introduce an anisotropic weight $\chi(\mathbf{x})$ for each vertex by using a function of its two principal curvatures, k_1 and k_2 [12]. In order to improve the aspect ratio of the surface mesh, we also add a tangent movement in Eq. (5),

$$\frac{\partial \mathbf{x}}{\partial t} = \chi(\mathbf{x})\text{sign}(K(\mathbf{x}))\square H(\mathbf{x})\mathbf{n}(\mathbf{x}) + v(\mathbf{x})\mathbf{T}(\mathbf{x}), \tag{11}$$

where

$$\chi(\mathbf{x}) = \begin{cases} 1 & \text{if } |k_1| \leq T \text{ and } |k_2| \leq T & \text{Case 1,} \\ 0 & \text{else if } |k_1| > T \text{ and } |k_2| > T \text{ and } K > 0 & \text{Case 2,} \\ k_1 \big/ (H\,|K|) & \text{else if } |k_1| = \min(|k_1|,|k_2|,|H|) & \text{Case 3,} \\ k_2 \big/ (H\,|K|) & \text{else if } |k_2| = \min(|k_1|,|k_2|,|H|) & \text{Case 4,} \\ 1 \big/ |K| & \text{else if } |H| = \min(|k_1|,|k_2|,|H|) & \text{Case 5,} \end{cases}$$

and T is a user-defined constant (here we select $T = 0.01$). Case 1 is used to detect uniformly noisy regions, which will be smoothed isotropically. Concave/convex features (case 2) will not be smoothed. We also smooth features detected by cases 3–5 with a speed proportional to the minimum curvature and $|K|$ is used to scale the speed of the movement. $v(\mathbf{x})$ is the velocity in the tangent direction $\mathbf{T}(\mathbf{x})$, which controls the strength of the regularization. We first calculate the mass center $m(\mathbf{x})$ for each vertex on the surface, and then project the vector $m(\mathbf{x}) - \mathbf{x}$ onto the tangent plane to obtain $\mathbf{T}(\mathbf{x})$. If the surface has no noise, we can only apply the tangent movement $v(\mathbf{x})\mathbf{T}(\mathbf{x})$ to improve the aspect ratio of the surface while ignoring the vertex normal movement. Equation (11) is solved over triangular surfaces using Loop subdivision based isogeometric analysis [14].

The surface smoothing via GHO-based geometric flow improves the quality of the surface, but the quality of interior mesh also needs to be improved. To measure tetrahedral mesh quality, we choose three metrics [10]: $Q_1 = \theta_{\min}$, the minimal dihedral angle of each element; $Q_2 = \theta_{\max}$, the maximal dihedral angle of each element; and $Q_3 = 8 \cdot 3^{\frac{5}{2}} V \left(\sum_{j=1}^{6} e_j^2 \right)^{-\frac{3}{2}}$, the Joe-Liu parameter, where $\{e_j\}_{j=1}^6$ are six edge lengths, and V is the volume of each tetrahedron. Three techniques are applied to improve the mesh quality: optimization-based mesh smoothing, face swapping and edge removal [10]. The optimization-based smoothing improves all tetrahedra in the mesh by minimizing the objective function $\epsilon = \sum_{\eta \in \tau} \max(\frac{1}{Q_\eta} - q, 0)^p$, where τ is the set of tetrahedra in the mesh, Q_η represents Joe-Liu value of a tetrahedron $\eta \in \tau$, and q and p are parameters. This approach can improve the overall mesh quality efficiently, but some elements still have poor quality because of the bad valence. Face swapping removes edges with valence 3 or 4 by reconnecting vertices of some elements. Edge removal removes poor quality elements by replacing one ring neighboring tetrahedra of the edge with new tetrahedra with higher quality. Figure 2(b) and (g) show the improved tetrahedral meshes of the white matter and gray matter of the Brain-1 model. Both isotropic and anisotropic GHO diffusion flow are applied to denoise the bumpy surface with the same temporal step size ($t = 0.02$) and iteration number (100 iterations). As shown in Fig. 2(d) and (i), the isotropic GHO diffusion flow smooths out the noise but also blurs the surface features. Compared to the isotropic GHO diffusion flow, it is obvious that our anisotropic GHO diffusion flow better preserves surface features while removing the noise, see Fig. 2(e, j).

Remark 4.1. Since GHO is defined based on the second fundamental form of the surface, it is more sensitive to curvature related surface features, such as concave creases and convex ridges. However, isotropic geometric flow smooths out important features while reducing the noise. By introducing an anisotropic weighting function which penalizes surface vertices with a large ratio between their two principal curvatures, the anisotropic GHO diffusion flow preserves concave and convex features such as brain wrinkles while removing the noise.

5   Results and Discussion

In this section, we apply our presented algorithms to eight 3D medical images that are either noise free or corrupted by different types of noises. All the results were computed on a PC equipped with a 2.93 GHz Intel X3470 CPU and 8 GB of Memory. Statistics of all tested models are given in Table 1. For HEWCVT-3D based image segmentation, we need to define two parameters: L, the number of clusters; and λ, the weighting parameter that balances the clustering energy and the edge-weighted energy.

We tested our HEWCVT-based 3D image segmentation algorithm on eight 3D MRI brain images ($181 \times 217 \times 181$) from BrainWeb [4], with four levels of noise ($3\%, 5\%, 7\%, 9\%$) and two levels of INU ($20\%, 40\%$). We segmented each 3D image into four clusters in order to extract the gray matter, white matter, cerebrospinal fluid and background. Figure 3(a) shows the initial Brain-6 3D image with 7% noise and 40% INU, where the slice 106 in both 3D and 2D domains are highlighted in Fig. 3(b). Figure 3(c) shows the HEWCVT-3D based image segmentation, where the green part represent the white matter. We can observe that the noise effect can be well removed during segmentation. We also compared all results with two other methods: k-means [15] and EWCVT-3D [20]; see Fig. 3(d–k). HEWCVT-3D generates better segmentation results without leaving isolated voxels and in keeping the connectivity of the structure,

Table 1. Image segmentation statistics of all tested models.

Image		Brain-1	Brain-2	Brain-3	Brain-4	Brain-5	Brain-6	Brain-7	Brain-8
Noise level		3%	3%	5%	5%	7%	7%	9%	9%
INU level		20%	40%	20%	40%	20%	40%	20%	40%
Number of clusters		4	4	4	4	4	4	4	4
λ		0.05	0.05	0.10	0.10	0.15	0.20	0.30	0.30
Average	k-means	74.38%	73.26%	69.78%	69.24%	68.84%	67.97%	67.46%	66.37%
SA	EWCVT-3D	89.45%	88.68%	84.49%	86.65%	85.13%	83.28%	81.69%	78.43%
	HEWCVT-3D	**93.23%**	**93.12%**	**92.88%**	**92.26%**	**92.24%**	**91.56%**	**90.96%**	**89.26%**
Average	k-means	79.88%	76.96%	74.18%	73.24%	72.89%	72.47%	71.86%	71.64%
BR	EWCVT-3D	91.25%	90.98%	89.69%	89.25%	88.62%	86.29%	83.57%	81.45%
	HEWCVT-3D	**95.28%**	**95.11%**	**94.89%**	**94.66%**	**93.84%**	**92.99%**	**92.16%**	**91.83%**
SCV	k-means	13.79%	13.94%	14.22%	14.79%	15.03%	15.87%	16.24%	16.63%
	EWCVT-3D	11.96%	12.63%	12.99%	13.78%	13.98%	14.77%	15.21%	15.68%
	HEWCVT-3D	**0.78%**	**0.78%**	**0.79%**	**0.82%**	**0.85%**	**0.88%**	**0.90%**	**0.93%**
Average	k-means	35.2	35.4	35.6	34.8	35.2	35.3	34.2	35.4
time	EWCVT-3D	45.2	46.4	45.7	44.9	45.6	45.4	44.9	45.5
(seconds)	HEWCVT-3D	65.8	65.9	66.7	67.8	66.2	68.3	67.9	68.1

while k-means is not robust to handle the noise effect and EWCVT-3D may generate inaccurate results. We first use the segmentation accuracy (SA) [9] to quantitatively evaluate the segmentation results. Given the segmented image B and the ground truth image G obtained from BrainWeb dataset, the SA can be defined as:

$$SA = \frac{N_{Correct}}{N_{Total}} \times 100\%, \qquad (12)$$

where $N_{Correct}$ represents the number of correctly classified voxels and N_{Total} is the total number of voxels in the image. In order to evaluate the accuracy of feature preservation, we also use the boundary recall (BR) [16] to measure the portion of boundary voxels in the ground truth that are also identified as boundary by the segmentation being evaluated. The BR can be defined as:

$$BR = \frac{TP}{TP + FN} \times 100\%, \qquad (13)$$

where TP is the number of boundary voxels in G with at least one boundary voxel in R in range of two voxels, FN is the number of boundary voxels in G with no boundary voxel in R in range of two voxels. Large SA and BR values are usually considered high accuracy. We also introduce another metric named the segmentation coefficient of variation (SCV) to evaluate the stability of different methods. For each brain image, we test N ($N = 100$ in this paper) random initializations of the generators by using k-means, EWCVT-3D and HEWCVT-3D. We can get one minimized energy value for each test and the SCV can be defined as:

$$SCV = \frac{\sqrt{\frac{1}{N} \sum_{i=1}^{N} \left(MinE_i - \overline{MinE} \right)^2}}{\overline{MinE}} \times 100\%, \qquad (14)$$

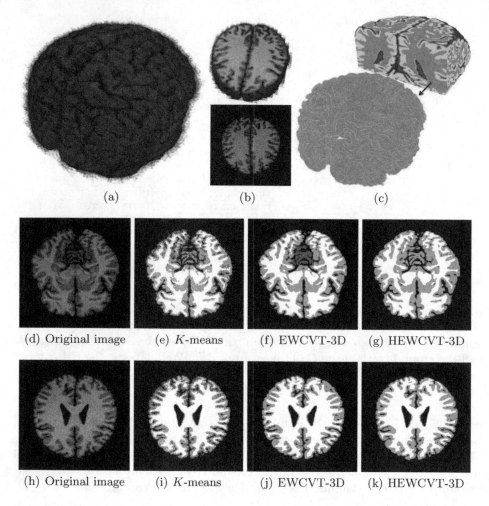

Fig. 3. Brain-6 model. (a) Input image; (b) slice 106; (c) HEWCVT-3D based segmentation; (d–g) and (h–k) from left to right: slices of original data, corresponding slices segmented applying k-means, EWCVT-3D, and HEWCVT-3D.

where $MinE_i$ represents the minimized energy for the ith test, and $\overline{MinE}$ represents the mean of the minimized energy. Large SCV values are usually considered high-variance, otherwise low-variance. The average SA, the average BR and SCV values of the each image with 100 tests are listed in Table 1. We can observe that HEWCVT-3D improves the segmentation accuracy compared to k-means and EWCVT-3D. From the comparison of BR values, it is evident that HEWCVT-3D can also better preserve the connectivity of structure compared to the other two methods. With different initializations, the energy function converges to different values for k-means and EWCVT-3D, while HEWCVT-3D is much more stable and less sensitive to initializations with all $SCVs < 1\%$. In addition, our HEWCVT-3D method is also robust to noise since the SA, BR and SCV values do not change much for different levels of noise and INU. Since HEWCVT-3D updates cluster centroids by calculating distances to all centroids for each voxel, the computational cost is higher than the other two methods.

Tetrahedral meshes consisting of the white matter, gray matter and cerebrospinal fluid are generated and the mesh quality is improved via geometric flow based smoothing and optimization. Table 2 shows the meshing results for each model. Figure 4(b) shows the improved mesh of Brain-6 with the white matter (yellow), gray matter (red) and cerebrospinal fluid (blue). The improved mesh is in good quality with an dihedral angle range of $(15.11°, 167.17°)$. From the zoom-in pictures we can observe that smoothness and regularity of boundary surfaces between different materials are significantly improved. Figure 4(d, e) and (f, g) show the improved meshes of the white matter and gray matter, respectively, with mesh adaptation highlighted in zoom-in pictures. We can observe that surface features are well preserved during the surface denoising via the anisotropic GHO diffusion flow.

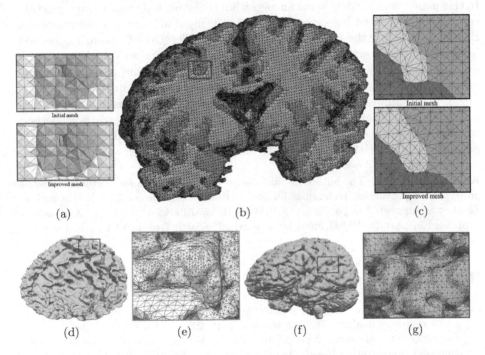

Fig. 4. Tetrahedral mesh of the Brain-6 model. (a, b) Cross section of the final tetrahedral mesh, with zoom-in pictures of the initial and improved meshes of the green box; (c) zoom-in pictures of the initial and improved meshes with three neighboring materials (d) improved tetrahedral mesh of the white matter; (e) enlargement of the red window in (d); (f) improved tetrahedral mesh of the gray matter; and (g) enlargement of the green window in (f). (Color figure online)

Table 2. Mesh statistics of all tested models.

Image	Mesh size (vertices, elements)	Joe-Liu (min, max)	Dihedral angle (min, max)	Time (seconds)
Brain-1	(368,584, 1,796,748)	(0.12, 1.0)	(15.14°, 166.56°)	189.7
Brain-2	(326,576, 1,616,551)	(0.13, 1.0)	(15.10°, 166.94°)	187.8
Brain-3	(332,681, 1,630,137)	(0.12, 1.0)	(15.11°, 167.39°)	184.7
Brain-4	(343,268, 1,682,014)	(0.12, 1.0)	(15.06°, 167.39°)	186.9
Brain-5	(301,298, 1,491,425)	(0.13, 1.0)	(15.08°, 167.76°)	179.9
Brain-6	(282,352, 1,395,796)	(0.11, 1.0)	(15.11°, 167.17°)	179.6
Brain-7	(312,453, 1,534,144)	(0.12, 1.0)	(15.02°, 167.83°)	188.8
Brain-8	(342,683, 1,686,006)	(0.13, 1.0)	(15.01°, 167.89°)	187.7

6   Conclusions and Future Work

In this paper, we have developed an algorithm to segment 3D images and generate tetrahedral meshes with good quality. Given the input 3D image, we first segment the image by using the HEWCVT-3D algorithm. The Dual Contouring method is then used to extract the initial tetrahedral mesh. To smooth out the noise and improve the quality of the tetrahedral mesh, we developed an anisotropic GHO diffusion flow. The quality of the interior tetrahedron is also improved via optimization techniques. We have successfully tested our method using several volumetric imaging datasets. In the future, we will extend our geometric flow method to quadrilateral meshes and investigate more anisotropic schemes. We will also parallelize our algorithms and apply to more real applications.

Acknowledgment. The authors would like to thank Tao Liao for useful discussions on quality improvement techniques for tetrahedral mesh. The work of K. Hu and Y. Zhang was supported in part by NSF CAREER Award OCI-1149591. G. Xu was was supported in part by NSFC Fund for Creative Research Groups of China under the grant 11321061.

References

1. Arifin, A.Z., Asano, A.: Image segmentation by histogram thresholding using hierarchical cluster analysis. Pattern Recogn. Lett. **27**(13), 1515–1521 (2006)
2. Canann, S.A., Tristano, J.R., Staten, M.L.: An approach to combined Laplacian and optimization-based smoothing for triangular, quadrilateral, and quad-dominant meshes. In: 7th International Meshing Roundtable, pp. 479–494 (1998)
3. Chan, T.F., Vese, L.A.: Active contour and segmentation models using geometric PDEs for medical imaging. In: Malladi, R. (ed.) Geometric Methods in Bio-Medical Image Processing, pp. 63–75. Springer, Heidelberg (2002)
4. Cocosco, C.A., Kollokian, V., Kwan, R.K.S., Pike, G.B., Evans, A.C.: BrainWeb: online interface to a 3D MRI simulated brain database. NeuroImage **5**(4), S425 (1997)

5. Du, Q., Faber, V., Gunzburger, M.: Centroidal Voronoi tessellations: applications and algorithms. SIAM Rev. **41**(4), 637–676 (1999)
6. Du, Q., Gunzburger, M., Ju, L., Wang, X.: Centroidal Voronoi tessellation algorithms for image compression, segmentation, and multichannel restoration. J. Math. Imaging Vis. **24**(2), 177–194 (2006)
7. Foteinos, P.A., Chrisochoides, N.P.: High quality real-time image-to-mesh conversion for finite element simulations. J. Parallel Distrib. Comput. **74**(2), 2123–2140 (2014)
8. Freitag, L.A.: On combining Laplacian and optimization-based mesh smoothing techniques. AMD-Vol. 220 Trends in Unstructured Mesh Generation, pp. 37–44 (1997)
9. Hu, K., Zhang, Y.: Image segmentation and adaptive superpixel generation based on harmonic edge-weighted centroidal Voronoi tessellation. Comput. Methods Biomech. Biomed. Eng.: Imaging Vis. **4**(2), 46–60 (2016). The Special Issue of CompIMAGE'14
10. Leng, J., Zhang, Y., Xu, G.: A novel geometric flow approach for quality improvement of multi-component tetrahedral meshes. Comput. Aided Des. **45**(10), 1182–1197 (2013)
11. Liao, T., Li, X., Xu, G., Zhang, Y.: Secondary Laplace operator and generalized Giaquinta-Hildebrandt operator with applications on surface segmentation and smoothing. Comput. Aided Des. **70**, 56–66 (2016). A Special Issue of SIAM Conference on Geometric & Physical Modeling 2015
12. Meyer, M., Desbrun, M., Schröder, P., Barr, A.H.: Discrete differential-geometry operators for triangulated 2-manifolds. In: Hege, H.C., Polthier, K. (eds.) Visualization and Mathematics III, pp. 35–57. Springer, Heidelberg (2003)
13. Pal, N.R., Pal, S.K.: A review on image segmentation techniques. Pattern Recogn. **26**(9), 1277–1294 (1993)
14. Pan, Q., Xu, G., Xu, G., Zhang, Y.: Isogeometric analysis based on extended Loop's subdivision. J. Comput. Phys. **299**, 731–746 (2015)
15. Pappas, T.N.: An adaptive clustering algorithm for image segmentation. IEEE Trans. Sig. Process. **40**(4), 901–914 (1992)
16. Ren, X., Malik, J.: Learning a classification model for segmentation. In: Ninth IEEE International Conference on Computer Vision, pp. 10–17 (2003)
17. Sijbers, J., Scheunders, P., Verhoye, M., der Linden, A.V., Dyck, D.V., Raman, E.: Watershed-based segmentation of 3D MR data for volume quantization. Magn. Reson. Imaging **15**(6), 679–688 (1997)
18. Tobias, O.J., Seara, R.: Image segmentation by histogram thresholding using fuzzy sets. IEEE Trans. Image Process. **11**(12), 1457–1465 (2002)
19. Tsuda, A., Filipovic, N., Haberthür, D., Dickie, R., Matsui, Y., Stampanoni, M., Schittny, J.C.: Finite element 3D reconstruction of the pulmonary acinus imaged by synchrotron X-ray tomography. J. Appl. Physiol. **105**(3), 964–976 (2008)
20. Wang, J., Ju, L., Wang, X.: An edge-weighted centroidal Voronoi tessellation model for image segmentation. IEEE Trans. Image Process. **18**(8), 1844–1858 (2009)
21. Xu, G., Zhang, Q.: A general framework for surface modeling using geometric partial differential equations. Comput. Aided Geom. Des. **25**(3), 181–202 (2008)
22. Zhang, Y., Bajaj, C., Sohn, B.S.: 3D finite element meshing from imaging data. Comput. Methods Appl. Mech. Eng. **194**, 5083–5106 (2005)
23. Zhang, Y., Bajaj, C., Xu, G.: Surface smoothing and quality improvement of quadrilateral/hexahedral meshes with geometric flow. Commun. Numer. Methods Eng. **25**(1), 1–18 (2009)

24. Zhang, Y., Hughes, T., Bajaj, C.: An automatic 3D mesh generation method for domains with multiple materials. Comput. Methods Appl. Mech. Eng. **199**(5–8), 405–415 (2010)
25. Zhang, Y., Qian, J.: Resolving topology ambiguity for multiple-material domains. Comput. Methods Appl. Mech. Eng. **247**, 166–178 (2012)

Structuring Digital Spaces by Path-Partition Induced Closure Operators on Graphs

Josef Šlapal[(⊠)]

IT4Innovations Centre of Excellence,
Brno University of Technology, Brno, Czech Republic
slapal@fme.vutbr.cz

Abstract. We study closure operators on graphs which are induced by path partitions, i.e., certain sets of paths of the same lengths in these graphs. We investigate connectedness with respect to the closure operators studied. In particular, the closure operators are discussed that are induced by path partitions of some natural graphs on the digital spaces $\mathbb{Z}^n$, $n > 0$ a natural number. For the case $n = 2$, i.e., for the digital plane $\mathbb{Z}^2$, the induced closure operators are shown to satisfy an analogue of the Jordan curve theorem, which allows using them as convenient background structures for studying digital images.

Keywords: Graphs · Closure operators · Path-partition · Connectedness · Jordan curve

1 Introduction

The classical approach to digital topology is based on graph theory rather than topology because it uses adjacency graphs for structuring the digital spaces (see, e.g., [6]). In the case of the digital plane $\mathbb{Z}^2$, the most usual of these graphs are the 4-adjacency and 8-adjacency graphs. Unfortunately, neither 4-adjacency nor 8-adjacency itself allows for an analogue of the Jordan curve theorem - cf. [9]. Therefore, one has to use a combination of the two adjacency graphs when studying (the connectivity of) digital images. To eliminate this deficiency, a new, purely topological approach to digital topology was proposed in [5]. This approach utilizes a convenient topology on $\mathbb{Z}^2$, the so-called Khalimsky topology, for structuring the digital plane and has been developed by many authors. For instance, some other convenient topologies on $\mathbb{Z}^2$ were introduced and studied in [14] and, in [13], it was shown that even closure operators, i.e., structures more general than topologies, may be used to advantage for structuring the digital spaces.

Each of the two approaches, classical and topological, has its advantages and it may, therefore, be beneficial to use a combination of them. Being motivated by this idea, we deal with closure operators on graphs. More precisely, we discuss closure operators on (the vertex sets of) graphs that are induced by path partitions of these graphs. The closure operators are studied from the viewpoint

© Springer International Publishing AG 2017
R.P. Barneva et al. (Eds.): CompIMAGE 2016, LNCS 10149, pp. 43–55, 2017.
DOI: 10.1007/978-3-319-54609-4_3

of applications of the results obtained in digital topology for structuring the digital spaces. We will, therefore, focus on the connectedness provided by the closure operators. It will be shown that products of the operators preserve the basic properties of connectedness. Thus, having a path-partition induced closure operator on the digital line $\mathbb{Z}$, the connectedness behavior of the closure operator is also preserved by the corresponding product closure operator on the digital space $\mathbb{Z}^n$, $n > 1$ a natural number. We will discuss special path-partition induced closure operators on $\mathbb{Z}$ which generalize the Khalimsky topology. The products of pairs of copies of these closure operators will be shown to allow for a digital analogue of the Jordan curve theorem, i.e., to provide convenient structures on the digital plane $\mathbb{Z}^2$ for the study of digital images.

Graphs with path partitions were introduced and studied in [15] where it was shown that path partitions provide graphs with special geometric properties that allow for using these graphs in digital topology for the study of digital images. In the present paper, we investigate the topological properties of these graphs via induced closure operators. The results obtained pave the way for further applications of the graphs with path partitions in digital topology.

2 Preliminaries

For the graph-theoretic terminology, we refer to [4]. By a *graph* $G = (V, E)$ we understand an (undirected simple) graph (without loops) with $V \neq \emptyset$ the *vertex* set and $E \subseteq \{\{x, y\};\ x, y \in V,\ x \neq y\}$ the set of *edges*. We will say that G is a graph *on* V. Two vertices $x, y \in V$ are said to be *adjacent* (to each other) if $\{x, y\} \in E$, and an edge $e \in E$ and a vertex $x \in V$ are said to be *adjacent* if $x \in e$. Throughout the paper, natural numbers are considered to be finite ordinals (including 0). Thus, given a natural number n, we write $(x_i|\ i \leq n)$ to denote the (finite) sequence $(x_0, x_1, ..., x_n)$. Similarly, if $n > 0$, we write $(x_i|\ i < n)$ to denote the sequence $(x_0, x_1, ..., x_{n-1})$.

Recall that a *walk* in G is a sequence $(x_i|\ i \leq n)$ of vertices of V such that x_i is adjacent to x_{i+1} whenever $i < n$. The natural number n is called the *length* of the walk $(x_i|\ i \leq n)$. If $C_1 = (x_i|\ i \leq m)$ and $C_2 = (y_i|\ i \leq q)$ are walks in G such that $x_m = y_0$, then we define $C_1 \oplus C_2 = (z_i|\ i \leq m + q)$ where $z_i = x_i$ for all $i \leq m$ and $z_i = y_{i-m}$ for all i with $m < i \leq m + q$. The walk $C_1 \oplus C_2$ (of length $n + m$) is called the *composition* of C_1 and C_2.

A walk $(x_i|\ i \leq n)$ is said to be *closed* if $x_0 = x_n$, and it is said to be a *path* if $x_i \neq x_j$ whenever $i, j \leq n$, $i \neq j$. A closed walk $(x_i|\ i \leq n)$ is called a *circle* if $(x_i|\ i < n)$ is a path. We will often apply set-theoretic operations on walks, in which case the walks will be considered to be just sets (a walk $(x_i|\ i \leq n)$ will be considered to be the set $\{x_i;\ i \leq n\}$). Given graphs $G_1 = (V_1, E_1)$ and $G_2 = (V_2, E_2)$, we say that G_1 is a *subgraph* of G_2 if $V_1 \subseteq V_2$ and $E_1 \subseteq E_2$. If, moreover, $V_1 = V_2$, then G_1 is called a *factor* of G_2.

Given graphs $G_j = (V_j, E_j,)$, $j = 1, 2, ..., m$ ($m > 0$ a natural number), we define their *product* to be the graph $\prod_{j=1}^{m} G_j = (\prod_{j=1}^{m} V_j, E)$ with the set of edges $E = \{\{(x_1, x_2, ..., x_m), (y_1, y_2, ..., y_m)\};$ there exists a nonempty subset $J \subseteq \{1, 2, ..., m\}$ such that $\{x_j, y_j\} \in E_j$ for every $j \in J$ and $x_j = y_j$

for every $j \in \{1, 2, ..., m\} - J\}$. This product differs from the cartesian prod-
uct of G_j, $j = 1, 2, ..., m$, i.e., from the graph $(\prod_{j=1}^{m} V_j, F)$ where $F = \{\{(x_1, x_2, ..., x_3), (y_1, y_2, ..., y_m)\}; \{x_j, y_j\} \in E_j$ for every $j \in \{1, 2, ..., m\}\}$, but
we always have $F \subseteq E$.

By a *closure operator* u on a set X we mean a map $u: \exp X \to \exp X$ (where
$\exp X$ denotes the power set of X) which is

(i) grounded (i.e., $u\emptyset = \emptyset$),
(ii) extensive (i.e., $A \subseteq X \Rightarrow A \subseteq uA$), and
(iii) monotone (i.e., $A \subseteq B \subseteq X \Rightarrow uA \subseteq uB$).

The pair (X, u) is then called a *closure space* and, for every subset $A \subseteq X$, uA
is called the *closure* of A. Closure spaces were studied by Čech in [1] (who called
them topological spaces there).

A closure operator u on X which is

(iv) additive (i.e., $u(A \cup B) = uA \cup uB$ whenever $A, B \subseteq X$) and
(v) idempotent (i.e., $uuA = uA$ whenever $A \subseteq X$)

is called a *Kuratowski closure operator* or a *topology* and the pair (X, u) is called
a *topological space*.

A closure operator u on a set X is said to be *quasi-discrete* (cf. [2]) if the
following condition is satisfied:

$$A \subseteq X \Rightarrow uA = \bigcup_{x \in A} u\{x\}.$$

Thus, any quasi-discrete closure operator is additive and is given by closures
of the singleton subsets. The quasi-discrete closure operators that are idempotent
are called *Alexandroff topologies*.

Many concepts known for topological spaces (see e.g. [3]) may be naturally
extended to closure spaces. Given a closure space (X, u), a subset $A \subseteq X$ is
called *closed* if $uA = A$, and it is called *open* if $X - A$ is closed. A closure
space (X, u) is said to be a *subspace* of a closure space (Y, v) (or, briefly, X is
a subspace of (Y, v)) if $uA = vA \cap X$ for each subset $A \subseteq X$. A closure space
(X, u) is said to be *connected* if $\emptyset$ and X are the only subsets of X which are
both closed and open. A subset $X \subseteq Y$ is connected in a closure space (Y, v) if
the subspace X of (Y, v) is connected. A maximal connected subset of a closure
space is called a *component* of this space. All the basic properties of connected
sets and components in topological spaces are also preserved in closure spaces.

3 Graphs with n-partitions and the Associated Closure Operators

From now on, we suppose that there is given a natural number $n > 0$. For any
graph G, we denote by $\mathcal{P}_n(G)$ the set of all paths of length n in G. Observe that
$\mathcal{P}_n(G)$ is nothing but an n-ary relation on the vertex set of G and so are the
subsets of $\mathcal{P}_n(G)$ (cf. [11]).

Definition 1. Let G be a graph. A subset $\mathcal{B} \subseteq \mathcal{P}_n(G)$ is said to be an *n-path
partition*, briefly an *n-partition*, of G provided that

(i) for every edge e of G, there is a unique path $(x_i| \; i \leq n) \in \mathcal{B}$ with the property that there exists $i \in \{1, 2, ...n\}$ such that e is adjacent to x_{i-1} and x_i,

(ii) every pair of different paths belonging to $\mathcal{B}$ has at most one vertex in common.

Clearly, every graph has a 1-partition. But, for $n > 1$, a graph need not have any n-partition.

Let G be a graph and $\mathcal{B}$ an n-partition of G. Then we define

$\mathcal{B}^{-1} = \{(x_i| \; i \leq n) \in \mathcal{P}_n(G); \; (x_{n-i}| \; i \leq n) \in \mathcal{B}\}$,
$\hat{\mathcal{B}} = \{(x_i| \; i \leq m) \in \mathcal{P}_m(G); \; 0 < m \leq n$ and there exists $(y_i| \; i \leq n) \in \mathcal{B}$ such that $x_i = y_i$ for every $i \leq m\}$ (so that $\mathcal{B} \subseteq \hat{\mathcal{B}}$), and
$\mathcal{B}^* = \hat{\mathcal{B}} \cup \hat{\mathcal{B}}^{-1}$.

Thus, $\mathcal{B}^{-1}$ is obtained by just reversing the sequences belonging to $\mathcal{B}$ and $\hat{\mathcal{B}}$ consists of just the sequences that are initial parts of the sequences belonging to $\mathcal{B}$. The elements of $\mathcal{B}^*$ will be called $\mathcal{B}$-*initial segments* in G - they are just the initial parts of the sequences belonging to $\mathcal{B}$ and the sequences that are their reversals.

Let G be a graph and $\mathcal{B} \subseteq \mathcal{P}_n(G)$. For any subset $X \subseteq V$, we define

$u_{\mathcal{B}}X = X \cup \{x \in V; \text{ there exists } (x_i| \; i \leq m) \in \hat{\mathcal{B}} \text{ with } \{x_i; \; i < m\} \subseteq X \text{ and } x_m = x\}$.

Evidently, $u_{\mathcal{B}}$ is a closure operator on V (which is neither additive nor idempotent in general). It will be said to be *induced* by $\mathcal{B}$. It may easily be shown that $(V, u_{\mathcal{B}})$ is an Alexandroff topological space whenever $u_{\mathcal{B}}$ is idempotent.

Definition 2 ([15]). Let G be a graph with an n-partition $\mathcal{B}$. A finite nonempty sequence $C = (x_i| \; i \leq m)$ ($m > 0$ a natural number) of vertices of G is called a $\mathcal{B}$-*walk* in G if there is a finite increasing sequence $(j_k| \; k \leq p)$ of natural numbers with $j_0 = 0$ and $j_p = m$ such that $j_k - j_{k-1} \leq n$ and $(x_j| \; j_{k-1} \leq j \leq j_k) \in \mathcal{B}^*$ for every k with $k \leq p$. The sequence $(j_k| \; k \leq p)$ is said to be a *binding sequence* of C. If $x_0 = x_m$, then the $\mathcal{B}$-path C is said to be *closed* and, if for any pair i_0, i_1 of natural numbers with $i_0 < i_1 \leq m$ from $x_{i_0} = x_{i_1}$ it follows that $i_0 = 0$ and $i_1 = m$, then C is called a $\mathcal{B}$-*circle*. A $\mathcal{B}$-walk $(x_i| \; i < m)$ in G is said to *connect* points $x, y \in X$ if $\{x, y\} = \{x_0, x_m\}$.

Thus, also one-member sequences are considered to be $\mathcal{B}$-walks (and $\mathcal{B}$-circles) - their binding sequences have just one member, the number 0. It is evident that the composition of $\mathcal{B}$-walks is a $\mathcal{B}$-walk. If $\mathcal{B} \subseteq \mathcal{P}_1(G)$, then every $\mathcal{B}$-path in G is a path in G, and vice versa provided that $\mathcal{B} = \mathcal{P}_1(G)$. By [15], Theorem 1, every $\mathcal{B}$-walk in a graph G has a unique binding sequence.

Theorem 1. *Let $G = (V, E)$ be a graph with an n-partition $\mathcal{B}$. A subset $A \subseteq V$ is connected in the closure space $(V, u_{\mathcal{B}})$ if and only if any two different vertices of G belonging to A can be joined by a $\mathcal{B}$-walk in G contained in A.*

Proof. If A is empty or a singleton, then the statement is trivial. Therefore, suppose that A has at least two elements. Let any two vertices from A can be joined by a $\mathcal{B}$-walk in G contained in A. Then, choosing an arbitrary element $x \in A$, A is the union of the $\mathcal{B}$-walks in G joining x with the elements of $A - \{x\}$. Therefore, A is connected in $(V, u_{\mathcal{B}})$.

Conversely, let A be connected in $(V, u_{\mathcal{B}})$ and suppose that there are vertices $x, y \in A$ which cannot be joined by a $\mathcal{B}$-walk in G contained in A. Let B be the set of all vertices in A which can be joined with x by a $\mathcal{B}$-walk in G contained in A. Let $z \in u_{\mathcal{B}}B \cap A$ be a vertex and assume that $z \notin B$. Then there are a path $(x_i| \ i \leq n) \in \mathcal{B}$ and a natural number i_0, $0 < i_0 \leq n$, such that $z = x_{i_0}$ and $\{x_i; \ i < i_0\} \subseteq B$. Thus, x and x_0 can be joined by a $\mathcal{B}$-walk in G contained in A, and also x_0 and z can be joined by a $\mathcal{B}$-walk in G contained in A - namely by the $\mathcal{B}$-initial segment $(x_i| \ i \leq i_0) \in \mathcal{B}^*$. It follows that x and z can be joined by a $\mathcal{B}$-walk in G contained in A, which is a contradiction. Therefore, $z \in B$, which yields $u_{\mathcal{B}}B \cap A \subseteq B$. As the converse inclusion is evident, we have $u_{\mathcal{B}}B \cap A = B$. Consequently, B is closed in the subspace A of $(V, u_{\mathcal{B}})$. Further, let $z \in u_{\mathcal{B}}(A - B) \cap A$ be a vertex and assume that $z \in B$. Then $z \notin A - B$, thus there are a walk $(x_i| \ i < n) \in \mathcal{B}$ and a natural number i_0, $0 < i_0 \leq n$, such that $z = x_{i_0}$ and $\{x_i; \ i < i_0\} \subseteq A - B$. Since x can be joined with z by a $\mathcal{B}$-walk in G contained in A (because we have assumed that $z \in B$) and z can be joined with x_0 by a $\mathcal{B}$-walk in G contained in A - namely by the initial segment $(x_{i_0-i}| \ i \leq i_0) \in \mathcal{B}^*$, also x and x_0 can be joined by a $\mathcal{B}$-walk in G contained in A. This is a contradiction with $x_0 \notin B$. Thus, $z \notin B$, i.e., $u_{\mathcal{B}}(A-B) \cap A = A - B$. Consequently, $A - B$ is closed in the subspace A of $(V, u_{\mathcal{B}})$. Hence, A is the union of the nonempty disjoint sets B and $A - B$ closed in the subspace A of $(V, u_{\mathcal{B}})$. But this is a contradiction because A is connected in $(V, u_{\mathcal{B}})$. Therefore, any two points of A can be joined by a $\mathcal{B}$-walk in G contained in A.   □

Clearly, every $\mathcal{B}$-segment (and thus also every $\mathcal{B}$-path) in G is a connected subset of $(V, u_{\mathcal{B}})$.

Proposition 1. *Let G_j be a graph, $\mathcal{B}_j$ an n-partition of G_j for every $j = 1, 2, ..., m$ ($m > 0$ a natural number). Then $\prod_{j=1}^{m} \mathcal{B}_j$ is an n-partition of the graph $\prod_{j=1}^{m} G_j$.*

Proof. Clearly, $\prod_{j=1}^{m} \mathcal{B}_j \subseteq \mathcal{P}_n(\prod_{j=1}^{m} G_j)$. Let $e = \{(x_1, x_2, ..., x_m), (y_1, y_2, ..., y_m)\}$ be an edge in the graph $\prod_{j=1}^{m} G_j$. Then there is a nonempty subset $J \subseteq \{1, 2, ..., m\}$ such that $\{x_j, y_j\}$ is an edge in G_j for every $j \in J$ and $x_j = y_j$ for every $j \in \{1, 2, ..., m\} - J$. Thus, for every $j \in J$, there is a unique path $(y_i^j| \ i \leq n)$ with the property that there is $i_j \in \{1, 2, ..., n\}$ such that $\{x_j, y_j\} = \{z_{i_j-1}^j, z_{i_j}^j\}$. For every $i \leq n$, put $t_i^j = z_i^j$ whenever $j \in J$ and $t_i^j = x_j$ whenever $j \in \{1, 2, ..., m\} - J$. Evidently, $((x_i^1, x_i^2, ..., x_i^m)| \ i \leq n) \in \prod_{j=1}^{m} \mathcal{B}_j$ and it is a unique path with the property that there exists $i \in \{1, 2, ..., n\}$ such that $e = \{(x_{i-1}^1, x_{i-1}^2, ..., x_{i-1}^m), (x_i^1, x_i^2, ..., x_i^m)\}$.

Let $C, D \in \prod_{j=1}^{m} \mathcal{B}_j$ be different paths, $C = ((x_i^1, x_i^2, ..., x_i^m)| \ i \leq n)$, $D = ((y_i^1, y_i^2, ..., y_i^m)| \ i \leq n)$. Then there is $J_1 \subseteq \{1, 2, ..., m\}$ such that $(x_i^j| \ i \leq n) \in \mathcal{B}_j$ for every $j \in J_1$ and $(x_i^j| \ i \leq n)$ is a constant sequence for every $j \in \{1, 2, ..., m\} - J_1$. Similarly, there is $J_2 \subseteq \{1, 2, ..., m\}$ such that $(x_i^j| \ i \leq $

$n) \in \mathcal{B}_j$ for every $j \in J_2$ and $(x_i^j | \ i \leq n)$ is a constant sequence for every $j \in \{1, 2, ..., m\} - J_2$. Since C and D are different, there are $j_0 \in \{1, 2, ..., m\}$ and a natural number $i_0 \leq n$ such that $x_{i_0}^{j_0} \neq y_{i_0}^{j_0}$. If $j_0 \in J_1 \cap J_2$, then $(x_i^{j_0} | \ i \leq n) \in \mathcal{B}_{j_0}$ and $(y_i^{j_0} | \ i \leq n) \in \mathcal{B}_{j_0}$ are different paths, so that they may have at most one vertex in common. Therefore, C and D have at most one vertex in common, too. If $j_0 \in J_1 - J_2$, then $(x_i^{j_0} | \ i \leq n) \in \mathcal{B}_{j_0}$ and $(y_i^{j_0} | \ i \leq n)$ is a constant sequence. Hence, C and D have at most one vertex in common. If $j_0 \in J_2 - J_1$, then the situation is analogous. Finally, suppose that $j_0 \notin J_1$ and $j_0 \notin J_2$. Then $(x_i^{j_0} | \ i \leq n)$ and $(y_i^{j_0} | \ i \leq n)$ are different constant sequences, so that C and D have no vertex in common. The proof is complete.   □

Let G_j be a graph and $\mathcal{B}_j \subseteq \mathcal{P}_n(G_j)$ for every $j = 1, 2, ..., m$ ($m > 0$ a natural number). The previous Proposition enables us to define $\prod_{j=1}^{m}(V_j, u_{\mathcal{B}_j}) = (\prod_{j=1}^{m} V_j, u_{\prod_{j=1}^{m} \mathcal{B}_j})$.

Theorem 2. *Let $G_j = (V_j, E_j)$ be a graph, $\mathcal{B}_j \subseteq \mathcal{P}_n(G_j)$, and $Y_j \subseteq V_j$ be a subset for every $j = 1, 2, ..., m$. If Y_j is a connected subset of $(V_j, u_{\mathcal{B}_j})$ for every $i = 1, 2, ..., n$, then $\prod_{j=1}^{m} Y_j$ is a connected subset of $\prod_{j=1}^{m}(V_j, u_{\mathcal{B}_j})$.*

Proof. Let Y_j be a connected subset of $(X_j, u_{\mathcal{B}_j})$ for every $j \in \{1, 2, ..., m\}$ and let $(y_1, y_2, ..., y_m), (z_1, z_2, ..., z_m) \in \prod_{j=1}^{m} Y_j$ be arbitrary points. By Theorem 1, for every $j \in \{1, 2, ..., m\}$, there is a $\mathcal{B}_j$-walk $(x_i^j | \ i \leq p_j)$ in G_j joining the points y_j and z_j which is contained in Y_j. Then the set $\prod_{j=1}^{m} \{x_i^j | \ i \leq p_j\}$ contains the points $(y_1, y_2, ..., y_m)$ and $(z_1, z_2, ..., z_m)$. For each $j = 1, 2, ..., m$, let $(i_k^j | \ k \leq q_j)$ be the binding sequence of $(x_i^j | \ i \leq p_j)$. For every $j = 1, 2, ..., m$, putting $C_k^j = \{x_i^j; \ i_k^j \leq i \leq i_{k+1}^j\}$ for all $k < q_j$, we get $\{x_i^j; \ i \leq p_j\} = \bigcup_{k < q_j} C_k^j$. Therefore, $\prod_{j=1}^{m} \{x_i^j; \ i \leq p_j\} = \bigcup_{k_1 < q_1} \bigcup_{k_2 < q_2} \cdots \bigcup_{k_m < q_m} \prod_{j=1}^{m} C_{k_j}^j$. Put $C_{k_j}^j = \{y_i^j; \ i \leq r_j\}$ for every $j \in \{1, 2, ..., m\}$ and every $k_j < q_j$. For each $j = 1, 2, ..., m$, there is a path $(z_i^j | \ i \leq n) \in \mathcal{B}$ such that $y_i^j = z_i^j$ for all $i \leq r_j$ or $y_i^j = z_{r_j - i}^j$ for all $i \leq r_j$ (because $(y_i^j | \ i \leq r_j)$ is a $\mathcal{B}_j$-initial segment in G_j). Let $y \in \prod_{j=1}^{m} \{y_i^j; \ i \leq r_j\}$ be an arbitrary element. Then, for each $j = 1, 2, ..., m$, there is a natural number s_j, $s_j < r_j$, such that $y = (y_{s_1}^1, y_{s_2}^2, ..., y_{s_m}^m)$. Then either $(y_{s_1 - i}^1 | \ i \leq s_1)$ or $(y_i^1 | \ s_1 \leq i \leq r_1)$ is a $\mathcal{B}_1$-initial segment in G_1 with the first member $y_{s_1}^1$ and the last one x_0^1. Denote this $\mathcal{B}_1$-initial segment by $(t_i^1 | \ i \leq u_1)$ and put $C_1 = ((t_i^1, y_{s_2}^2, y_{s_3}^3, ..., y_{s_m}^m) | \ i \leq u_1)$. Clearly, C_1 is a $\prod_{j=1}^{m} \mathcal{B}_j$-initial segment in $\prod_{j=1}^{m} G_j$ with all members belonging to $\prod_{j=1}^{m} \{y_i^j; \ i \leq r_j\}$, with the first member y, and with $t_{u_1}^1 = x_0^1$. Further, either $(y_{s_2 - i}^2 | \ i \leq s_2)$ or $(y_i^2 | \ s_2 \leq i \leq r_2)$ is a $\mathcal{B}_2$-initial segment in G_2 with the first member $y_{s_2}^2$ and the last one x_0^2. Denote this $\mathcal{B}_2$-initial segment by $(t_i^2 | \ i \leq u_2)$ and put $C_2 = ((z_0^1, t_i^2, y_{s_3}^3, y_{s_4}^4, ..., y_{s_m}^m) | \ i \leq u_2)$. Clearly, C_2 is a $\prod_{j=1}^{m} \mathcal{B}_j$-initial segment in $\prod_{j=1}^{m} G_j$ with all members belonging to $\prod_{j=1}^{m} \{y_i^j; i \leq r_j\}$ such that $t_0^2 = y_{s_2}^2$ and $t_{u_2}^2 = x_0^2$. Thus, $C_1 \oplus C_2$ is a $\prod_{j=1}^{m} \mathcal{B}_j$-walk in $\prod_{j=1}^{m} G_j$ with all members belonging to $\prod_{j=1}^{m} \{y_i^j; \ i \leq r_j\}$, with the first member y, and with the last one

$(z_0^1, z_0^2, y_{s_3}^3, y_{s_4}^4, ..., y_{s_m}^m)$. Repeating this construction m-times, we get $\prod_{j=1}^m \mathcal{B}_j$-initial segments $C_1, C_2,...,C_m$ in $\prod_{j=1}^m G_j$ with the members of each of them belonging to $\prod_{j=1}^m \{y_i^j; \ i \leq r_j\}$ such that $C_1 \oplus C_2 \oplus ... \oplus C_m$ is a $\prod_{j=1}^m \mathcal{B}_j$-walk in $\prod_{j=1}^m G_j$ with the first member y and the last one $(z_0^1, z_0^2, ..., z_0^m)$. Then any point of $\prod_{j=1}^m \{y_i^j; \ i \leq r_j\}$ can be joined with the point $(x_0^1, x_0^2, ..., x_0^m)$ by a $\prod_{j=1}^m \mathcal{B}_j$-walk in $\prod_{j=1}^m G_j$ contained in $\prod_{j=1}^m \{y_i^j; \ i \leq r_j\}$. By Theorem 1, $\prod_{j=1}^m \{y_i^j; \ i \leq r_j\} = \prod_{j=1}^m C_{k_j}^j$ is a connected subset of $\prod_{j=1}^m (V_j, u_{\mathcal{B}_j})$. Thus, for any $k_j < q_j$, $j = 1, 2, ..., m-1$, $(\prod_{j=1}^m C_{k_j}^j | \ k_m < q_m)$ is a finite sequence of connected sets with nonempty intersection of every consecutive pair of them. Hence, the set $\bigcup_{k_m < q_m} \prod_{j=1}^m C_{k_j}^j$ is connected in $\prod_{j=1}^m (V_j, u_{\mathcal{B}_j})$. Consequently, for every k_j with $k_j < q_j, j = 1, 2, ..., m-2$, $(\bigcup_{k_m < q_m} \prod_{j=1}^m C_{k_j}^j | \ k_{m-1} < q_{m-1})$ is a finite sequence of connected sets with nonempty intersection of any consecutive pair of them. Therefore, the set $\bigcup_{k_{m-1} < q_{m-1}} \bigcup_{k_m < q_m} \prod_{j=1}^m C_{k_j}^j$ is connected in $\prod_{j=1}^m (V_j, u_{\mathcal{B}_j})$. After repeating this considerations m-times, we get the conclusion that the set $\bigcup_{k_1 < q_1} \bigcup_{k_2 < q_2} \cdots \bigcup_{k_m < q_m} \prod_{j=1}^m C_{k_j}^j = \prod_{j=1}^m \{x_i^j; \ i \leq p_j\}$ is a connected subset of $\prod_{j=1}^m (V_j, u_{\mathcal{B}_j})$. Thus, by Theorem 1, there is a $\mathcal{B}_j$-walk C in $\prod_{j=1}^m G_j$ joining the points $(y_1, y_2, ..., y_m)$ and $(z_1, z_2, ..., z_m)$ which is contained in $\prod_{j=1}^m \{x_i^j | \ i \leq p_j\}$. Since $\prod_{j=1}^m \{x_i^j | \ i \leq p_j\} \subseteq \prod_{j=1}^m Y_j$, C is contained in $\prod_{j=1}^m Y_j$, too, and so $\prod_{j=1}^m Y_j$ is a connected subset of $\prod_{j=1}^m (V_j, u_{\mathcal{B}_j})$ by Theorem 1. $\qquad \square$

4   Graphs on $\mathbb{Z}^n$ with an n-partition

In this section, we show that, for a number of structures on $\mathbb{Z}^2$ used in digital topology (including the Marcus-Wise and Khalimsky topologies), the connectedness they provide coincides with the connectedness with respect to a closure operator $u_{\mathcal{B}}$ for a suitable n-partition $\mathcal{B}$ of a graph G on $\mathbb{Z}^2$. After discussing some graphs with a 1-partition, we will focus on a certain type of graphs with an n-partition where $n > 1$ may be an arbitrary natural number.

It is well known that closure operators that are more general than the Kuratowski ones have useful applications in computer science. By Theorem 1, connectedness with respect to the closure operators on graphs induced by path partitions is a certain type of path connectedness, which enables us to apply graph-theoretic methods when studying these closure operators. This may especially be useful for applications of the closure operators in digital topology, a branch of digital geometry built for the study of geometric and topological properties of digital images (cf. [7,10]). For such applications, it is desirable that the closure operators satisfy a digital analogue of the Jordan curve theorem. The classical Jordan curve theorem states that every simple closed curve in the real (i.e., Euclidean) plane separates this plane into precisely two connected components.

Recall that the 8-*adjacency graph* and the 4-*adjacency graph* are the graphs $(\mathbb{Z}^2, A_4)$ and $(\mathbb{Z}^2, A_8)$ given by $A_4 = \{\{(x_1, y_1), (x_2, y_2)\}; \ |x_1 - x_2| + |y_1 + y_2| = 1\}$ and $A_8 = \{\{(x_1, y_1), (x_2, y_2)\}; \ (x_1, y_1), (x_2, y_2) \in \mathbb{Z}^2, \max\{|x_1 - x_2|, |y_1 - y_2|\} = 1\}$, respectively. The points adjacent (to each other) in $(\mathbb{Z}^2, A_4)$ are called 4-*adjacent* and those adjacent in $(\mathbb{Z}^2, A_8)$ are called 8-*adjacent*. Given a point $z \in \mathbb{Z}^2$, we denote by $A_4(z)$ the set of points that are 4-adjacent to z and by $A_8(z)$ the set of points that are 8-adjacent to z.

There are two well-known topologies on $\mathbb{Z}^2$ allowing for an analogue of the Jordan curve theorem, which are employed in digital topology. These are the so-called *Marcus-Wyse* and *Khalimsky* topologies, i.e., the Alexandroff topologies s and t, respectively, on $\mathbb{Z}^2$ with the closures of singleton subsets given as follows:

For any $z = (x, y) \in \mathbb{Z}^2$,

$$s\{z\} = \begin{cases} \{z\} \cup A_4(z) & \text{if } x + y \text{ is even,} \\ \{z\} & \text{otherwise} \end{cases}$$

and

$$t\{z\} = \begin{cases} \{z\} \cup A_8(z) & \text{if } x, y \text{ are even,} \\ \{(x+i, y); i \in \{-1, 0, 1\}\} & \text{if } x \text{ is even and } y \text{ is odd,} \\ \{(x, y+j); j \in \{-1, 0, 1\}\} & \text{if } x \text{ is odd and } y \text{ is even,} \\ \{z\} & \text{otherwise.} \end{cases}$$

We will show that both of these topologies may be obtained as closure operators induced by an n-path set in certain graph with the vertex set $\mathbb{Z}^2$.

Let $G = (\mathbb{Z}^2, E)$ where $E = \{\{(x, y), (z, t)\}; (x, y), (z, t) \in \mathbb{Z}^2, |x-z| + |y-t| = 1\}$ and define $\mathcal{B} = \{((x_i, y_i)| \ i \leq 1); (x_i, y_i) \in \mathbb{Z}^2 \text{ for every } i \leq 1, |x_0 - x_1| + |y_0 - y_1| = 1, x_0 + y_0 \text{ even}\}$. Then $\mathcal{B}$ is a 1-partition of G.

It may easily be seen that $(\mathbb{Z}^2, u_{\mathcal{B}})$ is a connected Alexandroff topological space in which the points $(x, y) \in \mathbb{Z}^2$ with $x + y$ even are open while those with $x + y$ odd are closed. The closure operator $u_{\mathcal{B}}$ coincides with the Marcus-Wyse topology.

We denote by $\mathbb{Z}_2$ the 2-*adjacency graph* on $\mathbb{Z}$, i.e., the graph $(\mathbb{Z}, A_2)$ where $A_2 = \{\{p, q\}; p, q \in \mathbb{Z}, |p - q| = 1\}$.

In the rest of this section, for every natural number $n > 0$, $\mathcal{B}_n \subseteq \mathcal{P}_n(\mathbb{Z}_2)$ will denote the n-partition of $\mathbb{Z}_2$ given as follows:

$\mathcal{B}_n = \{(x_i| \ i \leq n) \in \mathcal{P}_n(\mathbb{Z}_2); \text{ there exists an odd number } l \in \mathbb{Z} \text{ such that } x_i = ln + i \text{ for all } i \leq n \text{ or } x_i = ln - i \text{ for all } i \leq n\}$.

Thus, the paths belonging to $\mathcal{B}_n$ are just the arithmetic sequences $(x_i| \ i \leq n)$ of integers with the difference equal to 1 or -1 and with $x_0 = ln$ where $l \in \mathbb{Z}$ is an odd number. Note that each point of $z \in \mathbb{Z}$ belongs to at least one and at most two paths from $\mathcal{B}_n$. It belongs to two (different) paths from $\mathcal{B}_n$ if and only if there is $l \in \mathbb{Z}$ with $z = ln$ (in which case z is the first or last member of each of the two paths if l is odd or even, respectively).

Clearly, $u_{\mathcal{B}_n}$ is additive if and only if $n = 1$. The closure operator $u_{\mathcal{B}_1}$ coincides with the Khalimsky topology on $\mathbb{Z}$ generated by the subbase $\{\{2k - 1, 2k, 2k + 1\}; k \in \mathbb{Z}\}$ - cf. [8].

Theorem 3. $(\mathbb{Z}, u_{\mathcal{B}_n})$ *is a connected closure space.*

Proof. Put $D_l = \{ln + i;\ i \leq n\}$ for each $l \in \mathbb{Z}$. Of course, D_l is connected in $(\mathbb{Z}, u_{\mathcal{B}_n})$ for every $l \in \mathbb{Z}$ (because $(ln + i;\ i \leq n)$ is a $\mathcal{B}_n$-initial segment in $\mathbb{Z}_2$). It is also evident that $D_l \cup D_{l+1}$ is closed in $(\mathbb{Z}, u_{\mathcal{B}_n})$ whenever $l \in \mathbb{Z}$ is even.

Let ω denote the least infinite ordinal and let $(B_i \mid i < \omega)$ be the sequence given by $B_i = D_{\frac{i}{2}}$ whenever i is even and $B_i = D_{-\frac{i+1}{2}}$ whenever i is odd, i.e., $(B_i \mid i < \omega) = (D_0, D_{-1}, D_1, D_{-2}, D_2, ...)$. For each $l \in \mathbb{Z}$ there holds $D_l \cap D_{l+1} = \{(l+1)n)\} \neq \emptyset$. Thus, we have $B_0 \cap B_1 \neq \emptyset$. Let i_0 be a natural number with $i_0 > 1$. Then $B_{i_0} \cap B_{i_0-2} \neq \emptyset$ because $B_{i_0} = D_{\frac{i_0}{2}}$ and $B_{i_0-2} = D_{\frac{i_0}{2}-1}$ whenever i_0 is even, while $B_{i_0} = D_{-\frac{i_0+1}{2}}$ and $B_{i_0-2} = D_{-\frac{i_0+1}{2}+1}$ whenever i_0 is odd. Hence, $(\bigcup_{i<i_0} B_i) \cap B_{i_0} \neq \emptyset$ for each i_0, $0 < i_0 < \omega$. Therefore, $\bigcup_{i<\omega} B_i$ is connected. But $\bigcup_{i<\omega} B_i = \bigcup_{l \in \mathbb{Z}} D_l = \mathbb{Z}$, which proves the statement. $\square$

From now on, m will denote (similarly to n) a natural number with $m > 0$. Using results of the previous section, we may propose new structures on the digital spaces convenient for the study of digital images. Such a structure on $\mathbb{Z}^m$ is obtained as the product of m copies of the 2-adjacency graph $\mathbb{Z}_2$ with the n-partition given by the product of m-copies of the n-partition $\mathcal{B}_n$. More formally, we may consider the graph $G^m = \prod_{j=1}^{m} G_j$ on $\mathbb{Z}^m$, where $G_j = \mathbb{Z}_2$ for every $j \in \{1, 2, ..., m\}$, with the n-partition $\mathcal{B}_n^m \subseteq \mathcal{P}_n(G^m)$ given by $\mathcal{B}_n^m = \prod_{j=1}^{m} \mathcal{B}_j$ where $\mathcal{B}_j = \mathcal{B}_n$ for every $j \in \{1, 2, ..., m\}$. Of course, G^1 is the 2-adjacency graph on $\mathbb{Z}$ and G^2 and G^3 coincide with the 8-adjacency graph on $\mathbb{Z}^2$ and the 26-adjacency graph on $\mathbb{Z}^3$, i.e., the graph $(\mathbb{Z}^3, A_{26})$ where $A_{26} = \{\{(x_1, y_1, z_1), (x_2, y_2, z_2)\};\ (x_1, y_1, z_1), (x_2, y_2, z_2) \in \mathbb{Z}^3,\ \max\{|x_1 - x_2|, |y_1 - y_2|, |z_1 - z_2|\} = 1\}$.

Having defined the graphs G^m with n-partitions $\mathcal{B}_n^m$, we may study their topological properties with respect to the closure operators $u_{\mathcal{B}_n^m}$, i.e., we may investigate behavior of the closure spaces $\prod_{j=1}^{m}(\mathbb{Z}, u_{\mathcal{B}_n}) = (\mathbb{Z}^m, u_{\mathcal{B}_n^m})$. Such an investigation may be based on using results of the previous section concerning the products of closure operators on graphs induced by path partitions. In particular, as an immediate consequence of Theorems 2 and 3, we get:

Theorem 4. $(\mathbb{Z}^m, u_{\mathcal{B}_n^m})$ *is a connected closure space.*

Since the closure operator $u_{\mathcal{B}_n^m}$ coincides with the Khalimsky topology on $\mathbb{Z}^m$ for $n = 1$, we will suppose that $n > 1$ in the sequel. And we will restrict our considerations to $m = 2$ because this case is the most important one with respect to possible applications in digital topology. Thus, we will focuss on the closure spaces $(\mathbb{Z}^2, u_{\mathcal{B}_n^2})$.

We denote by $G(\mathcal{B}_n^2)$ the factor of the 8-adjacency graph on $\mathbb{Z}^2$ whose edges are those $\{(x_1, y_1), (x_2, y_2)\} \in A_8$ that satisfy one of the following four conditions for some $k \in \mathbb{Z}$:

$$x_1 - y_1 = x_2 - y_2 = 2kn,$$
$$x_1 - y_1 = x_2 - y_2 = 2kn,$$
$$x_1 = x_2 = 2kn,$$
$$y_1 = y_2 = 2kn.$$

A section of the graph $G(\mathcal{B}_n^2)$ is demonstrated in Fig. 1 where only the vertices $(2kn, 2ln)$, $k, l \in \mathbb{Z}$, are marked out (by bold dots) and thus, on every edge drawn between two such vertices, there are $2n - 1$ more (non-displayed) vertices, so that the edges represent $2n$ edges in the graph $G(\mathcal{B}_n^2)$. Clearly, every circle C in $G(\mathcal{B}_n^2)$ is a connected subset of the closure space $(\mathbb{Z}^2, u_{\mathcal{B}_n^2})$ because it is a $\mathcal{B}_n^2$-circle in G^2. Indeed, C consists (i.e., is the union) of a finite sequence of paths from $\mathcal{B}_n^2$, hence $\mathcal{B}_n^2$-initial segments, such that every two consecutive paths have a point in common.

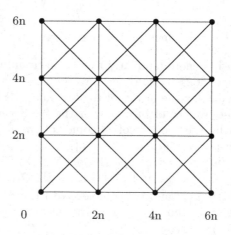

Fig. 1. A portion of the graph $G(\mathcal{B}_n^2)$

Definition 3. A circle J in the graph $G(\mathcal{B}_n^2)$ is said to be

(a) *fundamental* if, whenever $((2k+1)n, (2l+1)n) \in J$ for some $k, l \in \mathbb{Z}$, one of the following two conditions is true:
$$\{((2k+1)n - 1, (2l+1)n - 1), (2k+1)n + 1, (2l+1)n + 1))\} \subseteq J,$$
$$\{((2k+1)n - 1, (2l+1)n + 1), (2k+1)n + 1, (2l+1)n - 1))\} \subseteq J;$$
(b) a *Jordan curve* in $(\mathbb{Z}^2, u_{\mathcal{B}_n^2})$ if the subspace $\mathbb{Z}^2 - J$ of $(\mathbb{Z}^2, u_{\mathcal{B}_n^2})$ consists of two components.

Theorem 5. *Every fundamental circle in the graph $G(\mathcal{B}_n^2)$ is a Jordan curve in the closure space $(\mathbb{Z}^2, u_{\mathcal{B}_n^2})$.*

Proof. For every point $z = ((2k+1)n, (2l+1)n)$, $k, l \in \mathbb{Z}$, each of the following four subsets of $\mathbb{Z}^2$ will be called an n-fundamental triangle (given by z):

$$\{(r, s) \in \mathbb{Z}^2;\ 2kn \leq r \leq (2k+2)n,\ 2ln \leq s \leq (2l+2)n,\ y \leq x + 2ln - 2kn\},$$
$$\{(r, s) \in \mathbb{Z}^2;\ 2kn \leq r \leq (2k+2)n,\ 2ln \leq s \leq (2l+2)n,\ y \geq 4ln + 2kn - x\},$$
$$\{(r, s) \in \mathbb{Z}^2;\ 2kn \leq r \leq (2k+2)n,\ 2ln \leq s \leq (2l+2)n,\ y \geq x + 2ln - 2kn\},$$
$$\{(r, s) \in \mathbb{Z}^2;\ 2kn \leq r \leq (2k+2)n,\ 2ln \leq s \leq (2l+2)n,\ y \leq 4ln + 2kn - x\}.$$

The points of any n-fundamental triangle form a segment of the shape of a (digital) rectangular triangle. Obviously, in each of the four n-fundamental triangles given by z, z is the middle point of the hypotenuse of the triangle. Clearly, the edges of any n-fundamental triangle form a circle in the graph $G(\mathcal{B}_n^2)$, hence a $\mathcal{B}_n^2$-circle in G^2. It may easily be seen that every n-fundamental triangle is connected in $(\mathbb{Z}^2, u_{\mathcal{B}_n^2})$ and so is also every set obtained from an n-fundamental triangle by subtracting some of its edges. We will say that a (finite or infinite) sequence S of n-fundamental triangles is a tiling sequence if the members of S are pairwise different and every member of S, excluding the first one, has an edge in common with at least one of its predecessors. Given a tiling sequence S of n-fundamental triangles, we denote by S' the sequence obtained from S by subtracting, from every member of the sequence, those of its edges that are not shared with any other member of the sequence. It immediately follows that, for every tiling sequence S of n-fundamental triangles, the set $\bigcup\{T;\ T \in S\}$ is connected in $(\mathbb{Z}^2, u_{\mathcal{B}_n^2})$ and the same is true for the set $\bigcup\{T;\ T \in S'\}$.

Let J be a fundamental circle in the graph $G(\mathcal{B}_n^2)$. Then J constitutes the border of a polygon $S_F \subseteq \mathbb{Z}^2$ consisting of n-fundamental triangles. More precisely, S_F is the union of some n-fundamental triangles such that any pair of them is disjoint or meets in just one (common) edge. Let U be a (finite) tiling sequence of the n-fundamental triangles contained in S_F. Then we have $S_F = \bigcup\{T;\ T \in U\}$. Since every n-fundamental triangle $T \in U$ is connected, S_F is connected, too. Similarly, U' is a (finite) sequence with $S_F - J = \bigcup\{T;\ T \in U'\}$ and, since every member of U' is connected, $S_F - J$ is connected, too.

Further, let V be an (infinite) tiling sequence of the n-fundamental triangles which are not contained in S_F. Put $S_I = \bigcup\{T;\ T \in V\}$. Since every n-fundamental triangle $T \in V$ is connected, S_I is connected, too. Similarly, V' is an (infinite) sequence with $S_I - J = \bigcup\{T;\ T \in V'\}$ and, since every member of V' is connected, $S_I - J$ is connected, too.

It may easily be seen that every $\mathcal{B}_n^2$-walk $C = (z_i|\ i \le k)$, $k > 0$ a natural number, in the 8-adjacency graph G^2 on $\mathbb{Z}^2$ connecting a point of $S_F - J$ with a point of $S_I - J$ meets J (i.e., meets an edge of an n-fundamental triangle which is contained in J). Therefore, the set $\mathbb{Z}^2 - J = (S_F - J) \cup (S_I - J)$ is not connected in $(\mathbb{Z}^2, u_{\mathcal{B}_n^2})$. We have shown that $S_F - J$ and $S_I - J$ are components of the subspace $\mathbb{Z}^2 - J$ of $(\mathbb{Z}^2, u_{\mathcal{B}_n^2})$.   □

Remark 1. (a) If C is a fundamental circle in $G(\mathcal{B}_n^2)$, then, by the proof of Theorem 5, one of the components of $\mathbb{Z}^2 - C$ is finite (the inside component $S_F - J$) and the other is infinite (the outside component $S_I - J$).

(b) The famous Jordan curve theorem proved in [5] states that every circle in the graph $G(\mathcal{B}_1^2)$ having at least four points and containing, with each if its points, just two points adjacent to it is a Jordan curve in the Khalimsky topological space $(\mathbb{Z}^2, u_{\mathcal{B}_1^2})$.

(c) A digital analogue of the Jordan curve theorem similar to Theorem 5 may be found in [12] where a different, quite complicated procedure is used to prove it based on employing Jordan curves in the Khalimsky topology on $\mathbb{Z}^2$ and quotient closure spaces.

Example 1. Consider the following (digital picture of a) triangle:

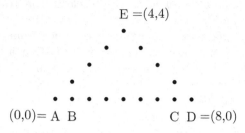

While the triangle ADE is a Jordan curve in $(\mathbb{Z}^2, u_{\mathcal{B}_2^2})$, it is not a Jordan curve in the Khalimsky topological space $(\mathbb{Z}^2, u_{\mathcal{B}_1^2})$. In order that this triangle be a Jordan curve in the Khalimsky topological space, we have to delete the points A,B,C and D. But this will lead to a considerable deformation of the triangle.

5   Conclusion

We introduced and investigated the concept of path-set induced closure operators on graphs. It was shown that connectedness with respect to these closure operators is preserved by a special product of the operators. We applied this result to the products of pairs of copies of the closure operators induced on a graph with the vertex set $\mathbb{Z}$ by certain path partitions. In this way, we obtained closure operators $u_{\mathcal{B}_n^2}$ ($n > 1$ a natural number) on the digital plane $\mathbb{Z}^2$ with the closure spaces $(\mathbb{Z}^2, u_{\mathcal{B}_n^2})$ connected that were then discussed.

One of the basic problems of digital topology is to find structures on the digital plane $\mathbb{Z}^2$ convenient for the study of digital images. A basic criterion of such a convenience is the validity of a digital analogue of the Jordan curve theorem. Namely, in digital images, which are regarded as approximations of real ones, digital versions of simple closed curves form borders of objects. To avoid undesirable paradoxes, it is necessary that the curves (circles) satisfy a digital analogue of the Jordan curve theorem, i.e., separate the digital plane into precisely two components - the inside and the outside of an object. The above Theorem 5 provides such a digital analogue of the Jordan curve theorem, thus making possible to use the closure operators $u_{\mathcal{B}_n^2}$ as convenient background structures on the digital plane $\mathbb{Z}^2$ for the study of digital images. They may also be useful for solving problems of digital image processing closely related to connectedness like pattern recognition, border detection, contour filling, etc.

Acknowledgement. This work was supported by Ministry of Education, Youth and Sports of the Czech Republic from the National Programme of Sustainability (NPU II) project "IT4Innovations Excellence in Science - LQ1602".

References

1. Čech, E.: Topological spaces. In: Topological Papers of Eduard Čech, pp. 436–472. Academia, Prague (1968)

2. Čech, E.: Topological Spaces. Academia, Prague (1966). (revised by Z. Frolík and M. Katětov)
3. Engelking, R.: General Topology. Państwowe Wydawnictwo Naukowe, Warszawa (1977)
4. Harrary, F.: Graph Theory. Addison-Wesley Publ. Comp., Reading (1969)
5. Khalimsky, E.D., Kopperman, R., Meyer, P.R.: Computer graphics and connected topologies on finite ordered sets. Topol. Appl. **36**, 1–17 (1990)
6. Kong, T.Y., Roscoe, W.: A theory of binary digital pictures. Comput. Vis. Graph. Image Process. **32**, 221–243 (1985)
7. Kong, T.Y., Rosenfeld, A.: Digital topology: introduction and survey. Comput. Vis. Graph. Image Process. **48**, 357–393 (1989)
8. Kong, T.Y., Kopperman, R., Meyer, P.R.: A topological approach to digital topology. Am. Math. Mon. **98**, 902–917 (1991)
9. Rosenfeld, A.: Connectivity in digital pictures. J. Assoc. Comput. Mach. **17**, 146–160 (1970)
10. Rosenfeld, A.: Digital topology. Am. Math. Mon. **86**, 621–630 (1979)
11. Šlapal, J.: Direct arithmetics of relational systems. Publ. Math. Debr. **38**, 39–48 (1991)
12. Šlapal, J.: A digital analogue of the Jordan curve theorem. Discret. Appl. Math. **139**, 231–251 (2004)
13. Šlapal, J.: Convenient closure operators on $\mathbb{Z}^2$. In: Wiederhold, P., Barneva, R.P. (eds.) IWCIA 2009. LNCS, vol. 5852, pp. 425–436. Springer, Heidelberg (2009). doi:10.1007/978-3-642-10210-3_33
14. Šlapal, J.: A quotient universal digital topology. Theor. Comput. Sci. **405**, 164–175 (2008)
15. Šlapal, J.: Graphs with a walk partition for structuring digital spaces. Inf. Sci. **233**, 305–312 (2013)

Atypical (Rare) Elements Detection –
A Conditional Nonparametric Approach

Piotr Kulczycki[1,2(✉)], Malgorzata Charytanowicz[1,3],
Piotr A. Kowalski[1,2], and Szymon Lukasik[1,2]

[1] Centre of Information Technology for Data Analysis Methods,
Systems Research Institute, Polish Academy of Sciences, Warsaw, Poland
kulczycki@ibspan.waw.pl
[2] Division for Information Technology and Systems Research,
Faculty of Physics and Applied Computer Science,
AGH University of Science and Technology, Kraków, Poland
kulczycki@agh.edu.pl
[3] Institute of Mathematics and Computer Science, Faculty of Mathematics,
Computer Science and Landscape, Catholic University of Lublin, Lublin, Poland

Abstract. This paper presents a ready-to-use procedure for detecting atypical (rarely occurring) elements, in one- and multidimensional spaces. The issue is considered through a conditional approach. The application of nonparametric concepts frees the investigated procedure from distributions of describing and conditioning variables. Ease of interpretation and completeness of the presented material lend themselves to the use of the worked out method in a wide range of tasks in various applications of data analysis in science and practice, engineering, economy and management, environmental and social issues, biomedicine, and related fields.

Keywords: Rare element · Atypical element · Outlier · Outlier detection · Conditional approach · Distribution free method · Numerical algorithm

1 Introduction

Atypical elements (often casually referred to as outliers) can intuitively be considered as significantly differing from the general population [1, 3, 5, 6]. Their occurrence most commonly results from considerable ("gross") errors arising during the measurement, collection, storage, and processing of data [17]. In practice they hinder the correct utilization of knowledge available and their elimination or correction enables the use of more convenient and more effective methods at later stages of analysis and exploration [2]. What is more, in marketing, atypical elements may represent cases so different from the majority of the population, that any individual decision based on such a group – so different and insignificant – often turns out to be economically unviable. In engineering, the presence of atypical states in dynamic systems may be an evidence of malfunction of a component or the entire device, and proper reaction usually enables any serious consequences to be avoided. The detection of an atypical element may also

© Springer International Publishing AG 2017
R.P. Barneva et al. (Eds.): CompIMAGE 2016, LNCS 10149, pp. 56–64, 2017.
DOI: 10.1007/978-3-319-54609-4_4

signify an attempt to hack into a computer system. On the other hand, in many social and economic problems the appearance of this element could be a positive trait, as it may characterize completely new trends or uncommon phenomena, and their quick discovery allows the appropriate specific action to be taken in anticipation. Therefore, the detection of atypical elements constitutes a natural cognitive challenge of great scientific and practical meaning.

The task of detecting atypical elements is one of very difficult conditioning. Above all most often there is no definition or even criterion indicating which elements should be considered atypical. What is more, we do not have a pattern of atypical elements, and even if we do, it would be – by its nature – small in number, strongly unbalanced with respect to the typical set of elements. For an illustrative example, in the simplest one-dimensional case, where data distribution is unimodal, atypical elements can be considered to be elements distant (according to the basic meaning of the term "outlier") from a median – defining the "center" of a data set – of more than 3/2 of the interquartile range; see [20; Sect. 2.7]. However, a similar approach cannot be taken concerning complex multimodal distributions. In particular, when specific modes are significantly distanced from each other, elements lying in the center between them should be considered as atypical, although they may be located very near to the median, definitely closer than 3/2 of the interquartile range.

This paper assumes as atypical those elements occurring rarely in the population. Thus, having a representative set of data, we highlight regions of lowest distribution density, such that the common probabilities of the elements appearing in those regions are equal to the assumed value, e.g. 0.01, 0.05, 0.1. Such locations can be of any shape, location and number.

In many practical tasks, the data possessed can be significantly refined through the measurement and introduction to a model of the current value of quantity considerably influencing the subject of investigation. In engineering practice, such a factor may often be the current temperature. From a formal point of view, the above aim can be realized by using a conditional probabilistic approach [4]. In this case, the basic attributes, called describing, become dependent on the conditioning attributes, the measured and introduced specific values of which can make substantially more precise the information related to the object under research. This approach is the subject of the present paper.

For defining data characteristics, the nonparametric methodology of kernel estimators is used, which frees the investigated procedures from forms of distributions characterizing both the describing and conditioning quantities. The presented material is complete and ready-to-use without laborious investigations. Particularly valuable is its easy, illustrative interpretation. A broader description of the material investigated here is presented in the paper [15], currently in press.

2 Mathematical Preliminaries: Kernel Estimators

Let the n-dimensional continuous random variable X be given, with a distribution characterized by the density f. Its kernel estimator $\hat{f} : \mathsf{R}^n \to [0, \infty)$, calculated using experimentally obtained values for the m-element random sample

$$x_1, x_2, \ldots, x_m,$$ (1)

in its basic form is defined as

$$\hat{f}(x) = \frac{1}{mh^n} \sum_{i=1}^{m} K\left(\frac{x - x_i}{h}\right),$$ (2)

where $m \in \mathbb{N} \setminus \{0\}$, the coefficient $h > 0$ is called a smoothing parameter, while the measurable function $K : \mathbb{R}^n \to [0, \infty)$ of unit integral $\int_{\mathbb{R}^n} K(x)dx = 1$, symmetrical with respect to zero and having a weak global maximum in this place, takes the name of a kernel. The choice of form of the kernel K and the calculation of the smoothing parameter h is made most often with the criterion of the mean integrated square error [10, 21, 22].

Thus, the choice of the kernel form has – from a statistical point of view – no practical meaning and thanks to this, it becomes possible to take primarily into account properties of the estimator obtained or calculational aspects advantageous from the point of view of the applicational problem under investigation; for broader discussion see the books [10 – Sect. 3.1.3; 22 – Sects. 2.7 and 4.5]. In practice, for the one-dimensional case (i.e., when $n = 1$), the function K is assumed most often to be the density of a common probability distribution. In the multidimensional case, two natural generalizations of the above concept are used: radial and product kernels. However, the former is somewhat more effective, although from an applicational point of view, the difference is insignificant, and the product kernel – significantly more convenient for analysis – is often favored for practical problems. The n-dimensional product kernel K can be expressed as

$$K(x) = K\left(\begin{bmatrix} x_1 \\ x_2 \\ \vdots \\ x_n \end{bmatrix}\right) = K_1(x_1)\, K_2(x_2)\ldots K_n(x_n),$$ (3)

where K_i denotes the previously-mentioned one-dimensional kernels, while the expression h^n appearing in the basic formula (2) should be replaced by the product of the smoothing parameters for particular coordinates $h_1 \cdot h_2 \cdot \ldots \cdot h_n$. For further investigations, the (one-dimensional) Cauchy kernel

$$K(x) = \frac{2}{\pi} \frac{1}{(1+x^2)^2}$$ (4)

is applied.

The fixing of the smoothing parameter h has significant meaning for quality of estimation. Fortunately many suitable procedures for calculating the value of the parameter h on the basis of random sample (1) have been worked out. For broader discussion of this task see the books [10, 21, 22]. In particular, for the one-dimensional case, the effective plug-in method [10 – Sect. 3.1.5; 22 – Sect. 3.6.1] is especially recommended. Of course this method can also be applied in the n-dimensional case when product kernel (3) is used, sequentially n times for each coordinate. One can also apply the simplified method [10 – Sect. 3.1.5; 21 – Sect. 3.4.1; 22 – Sect. 3.2.1], which for Cauchy kernel (4) takes the plain form

$$h = \left(\frac{10}{3\sqrt{\pi m}}\right)^{1/5} \hat{\sigma},$$
(5)

where $\hat{\sigma}$ denotes the estimator of a standard deviation for a given coordinate. The above value may be sufficiently precise for many practical applications, whereas – thanks to its simplicity – this method significantly increases the calculation velocity.

The above concept will now be generalized for the conditional case. Here, besides the basic (sometimes termed the describing) n_Y-dimensional random variable Y, let also be given the n_W-dimensional random variable W, called hereinafter the conditioning random variable. Their composition $X = \begin{bmatrix} Y \\ W \end{bmatrix}$ is a random variable of the dimension $n_Y + n_W$. Assume that distributions of the variables X and, in consequence, W have densities, denoted below as $f_X : \mathsf{R}^{n_Y + n_W} \to [0, \infty)$ and $f_W : \mathsf{R}^{n_W} \to [0, \infty)$, respectively. Let also be given the so-called conditioning value, i.e., the fixed value of conditioning random variable $w^* \in \mathsf{R}^{n_W}$, such that

$$f_W(w^*) > 0.$$
(6)

Then the function $f_{Y|W=w^*} : \mathsf{R}^{n_Y} \to [0, \infty)$ given by

$$f_{Y|W=w^*}(y) = \frac{f_X(y, w^*)}{f_W(w^*)} \quad \text{for every } y \in \mathsf{R}^{n_Y}$$
(7)

constitutes a conditional density of probability distribution of the random variable Y for the conditioning value w^*. The conditional density $f_{Y|W=w^*}$ can, therefore, be treated as a "classic" density, whose form has been made more accurate in practical applications with w^* – a concrete value taken by the conditioning variable W in a given situation.

Let, therefore, the random sample

$$\begin{bmatrix} y_1 \\ w_1 \end{bmatrix}, \begin{bmatrix} y_2 \\ w_2 \end{bmatrix}, \ldots, \begin{bmatrix} y_m \\ w_m \end{bmatrix},$$
(8)

obtained from the variable $X = \begin{bmatrix} Y \\ W \end{bmatrix}$, be given. The particular elements of this sample are interpreted as the values y_i taken in measurements from the random variable Y, when the conditioning variable W assumes the respective values w_i. On the basis of sample (6) one can calculate $\hat{f}_X$, i.e. the kernel estimator of density of the random variable X probability distribution, while the sample

$$w_1, \ w_2, \ \dots, \ w_m \tag{9}$$

gives $\hat{f}_W$ – the kernel density estimator for the conditioning variable W. The kernel estimator of conditional density of the random variable Y distribution for the conditioning value w^*, is defined then – in natural consequence of formula (5) – as the function $\hat{f}_{Y|W=w^*} : \mathbf{R}^{n_Y} \to [0, \infty)$ given by

$$\hat{f}_{Y|W=w^*}(y) = \frac{\hat{f}_X(y, w^*)}{\hat{f}_W(w^*)}. \tag{10}$$

If for the estimator $\hat{f}_W$ one uses a kernel with positive values, then the inequality $\hat{f}_W(w^*) > 0$ implied by condition (4) is fulfilled for any $w^* \in \mathbf{R}^{n_W}$.

In the case when for the estimators $\hat{f}_X$ and $\hat{f}_W$ the product kernel (3) is used, applying in pairs the same kernels to the estimator $\hat{f}_X$ for coordinates which correspond to the vector W and to the estimator $\hat{f}_W$, then the expression for the kernel estimator of conditional density becomes particularly helpful for practical applications. Formula (8) can then be specified to the form

$$\hat{f}_{Y|W=w^*}(y) = \hat{f}_{Y|W=w^*}\left(\begin{bmatrix} y_1 \\ y_2 \\ \vdots \\ y_{n_Y} \end{bmatrix}\right)$$

$$= \frac{\frac{1}{h_1 h_2 \dots h_{n_Y}} \sum_{i=1}^{m} K_1\left(\frac{y_1 - y_{i,1}}{h_1}\right) K_2\left(\frac{y_2 - y_{i,2}}{h_2}\right) \dots K_{n_Y}\left(\frac{y_{n_Y} - y_{i,n_Y}}{h_{n_Y}}\right) K_{n_Y+1}\left(\frac{w_1^* - w_{i,1}}{h_{n_Y+1}}\right) K_{n_Y+2}\left(\frac{w_2^* - w_{i,2}}{h_{n_Y+2}}\right) \dots K_{n_Y+n_W}\left(\frac{w_{n_W}^* - w_{i,n_W}}{h_{n_Y+n_W}}\right)}{\sum_{i=1}^{m} K_{n_Y+1}\left(\frac{w_1^* - w_{i,1}}{h_{n_Y+1}}\right) K_{n_Y+2}\left(\frac{w_2^* - w_{i,2}}{h_{n_Y+2}}\right) \dots K_{n_Y+n_W}\left(\frac{w_{n_W}^* - w_{i,n_W}}{h_{n_Y+n_W}}\right)}, \tag{11}$$

where $h_1, h_2, \dots, h_{n_Y+n_W}$ represent – respectively – smoothing parameters mapped to particular coordinates of the random variable X, while the coordinates of the vectors w^*, y_i and w_i are denoted as

$$w^* = \begin{bmatrix} w_1^* \\ w_2^* \\ \vdots \\ w_{n_W}^* \end{bmatrix} \text{ and } y_i = \begin{bmatrix} y_{i,1} \\ y_{i,2} \\ \vdots \\ y_{i,n_Y} \end{bmatrix}, \quad w_i = \begin{bmatrix} w_{i,1} \\ w_{i,2} \\ \vdots \\ w_{i,n_W} \end{bmatrix} \quad \text{for} \quad i = 1, 2, \dots, m. \tag{12}$$

Define the so-called conditioning parameters d_i for $i = 1, 2, \dots, m$ by the following formula:

$$d_i = K_{n_Y+1}\left(\frac{w_1^* - w_{i,1}}{h_{n_Y+1}}\right) K_{n_Y+2}\left(\frac{w_2^* - w_{i,2}}{h_{n_Y+2}}\right) \dots K_{n_Y+n_W}\left(\frac{w_{n_W}^* - w_{i,n_W}}{h_{n_Y+n_W}}\right). \tag{13}$$

If the values of the kernels K_{n_Y+1}, $K_{n_Y+2}, \ldots, K_{n_Y+n_W}$ are positive, then these parameters are also positive. So the kernel estimator of conditional density (9) can be finally presented in the form

$$
\hat{f}_{Y|W=w^*}(y) = \hat{f}_{Y|W=w^*}\left(\begin{bmatrix} y_1 \\ y_2 \\ \vdots \\ y_{n_Y} \end{bmatrix}\right)
$$
(14)

$$
= \frac{1}{h_1 h_2 \ldots h_{n_Y} \sum\limits_{i=1}^{m} d_i} \sum_{i=1}^{m} d_i K_1\left(\frac{y_1-y_{i,1}}{h_1}\right) K_2\left(\frac{y_2-y_{i,2}}{h_2}\right) \ldots K_{n_Y}\left(\frac{y_{n_Y}-y_{i,n_Y}}{h_{n_Y}}\right).
$$

The value of the parameter d_i – the "distance" of the given conditioning value w^* from w_i – that of the conditioning variable for which the i-th element of the random sample was obtained. Then estimator (12) can be interpreted as the linear combination of kernels mapped to particular elements of a random sample obtained for the variable Y, when the coefficients of this combination characterize how representative these elements are for the given value w^*.

More details concerning kernel estimators can be found in the classic monographs [10, 21, 22]. Sample applications for data analysis tasks are described in the publications [11–13, 16, 18]. See also [19].

3   Conditional Atypical Elements Detection

Consider – in relation to notations introduced in the previous section – the data-set comprised of elements representative for a population $y_1, y_2, \ldots, y_m$, obtained for the conditioning values $w_1, w_2, \ldots, w_m$, respectively. The aim of the developed procedure is the isolation from the set $y_1, y_2, \ldots, y_m$ of elements which are atypical in the sense that they occur most rarely, in conditions when the specific conditioning value w^* appears.

First, fix the number

$$
r \in (0, 1)
$$
(15)

defining a desired proportion of atypical to typical elements, more accurately the share of atypical elements in a population. In practice, the values $r = 0.01, 0.05, 0.1$ are commonly used. In reference to the notations in the previous section, let us treat the set $y_1, y_2, \ldots, y_m$ as the realization of the n_Y-dimensional continuous random variable Y, and the set $w_1, w_2, \ldots, w_m$ as their respective set of realizations of the conditioning random variable W, and then calculate the conditional density $\hat{f}_{Y|W=w^*}$. Next, let us consider the set of its values for the elements of the set $y_1, y_2, \ldots, y_m$, therefore

$$
\hat{f}_{Y|W=w^*}(y_1), \hat{f}_{Y|W=w^*}(y_2), \ldots, \hat{f}_{Y|W=w^*}(y_m) \in \mathsf{R}.
$$
(16)

The value $\hat{f}_{Y|W=w^*}(y_i)$ refers to the probability of occurrence of the element y_i with the assumption that the value of the conditioning variable is w^*. So, the greater the value $\hat{f}_{Y|W=w^*}(y_i)$, the more typical element y_i can be interpreted to be for the given w^*. Let's treat as typical these elements y_i for which $\hat{f}_{Y|W=w^*}(y_i)$ is bigger than a given limit value, while atypical – those y_i for which $\hat{f}_{Y|W=w^*}(y_i)$ is smaller. In accordance with the assumptions made, a quantile of the order r for the condition w^* should be accepted as the above limit value. Thus, the set of elements y_i was hereby divided into $[100 \cdot r]$-percent of elements of lower probability and $[100 \cdot (1-r)]$-percent of those of higher probability.

There remains, however, to calculate the above mentioned value of the quantile. To this end, the kernel estimator scheme presented in the paper [14], fitted to the task investigated here, will be applied. The values (14) will be treated as realizations of the one-dimensional describing random variable Z, obtained, as before, for the realizations $w_1, w_2, \ldots, w_m$ of the n_W-dimensional conditioning random variable W. The kernel estimator of a quantile of the order r for the condition w^* can be effectively calculated on the basis of Newton's algorithm [7] as the limit of the sequence $\{\hat{q}_{r|w^*,j}\}_{j=0}^{\infty}$ defined by

$$\hat{q}_{r|w^*,0} = \frac{\sum_{i=1}^{m} d_i \hat{f}_{Y|W=w^*}(y_i)}{\sum_{i=1}^{m} d_i} \tag{17}$$

$$\hat{q}_{r|w^*,j+1} = \hat{q}_{r|w^*,j} - \frac{L(\hat{q}_{r|w^*,j})}{L'(\hat{q}_{r|w^*,j})} \quad \text{for} \quad j = 0, 1, \ldots, \tag{18}$$

with the functions L and L' being given by dependencies

$$L(\hat{q}_{r|w^*}) = \sum_{i=1}^{m} d_i I \left(\frac{\hat{q}_{r|w^*} - \hat{f}_{Y|W=w^*}(y_i)}{h} \right) - r \sum_{i=1}^{m} d_i \tag{19}$$

$$L'(\hat{q}_{r|w^*}) = \frac{1}{h} \sum_{i=1}^{m} d_i K \left(\frac{\hat{q}_{r|w^*} - \hat{f}_{Y|W=w^*}(y_i)}{h} \right), \tag{20}$$

where $I : \mathbb{R} \to [0, 1]$ means a primitive of the kernel K, i.e., $I(x) = \int_{-\infty}^{x} K(y) \, dy$, whereas a stop criterion takes on the form

$$|\hat{q}_{r|w^*,j} - \hat{q}_{r|w^*,j-1}| \leq 0.01 \, \hat{\sigma}_Z, \tag{21}$$

while $\hat{\sigma}_Z$ denotes the estimator of the standard deviation of the random variable Z, found on the basis of set (16). For Cauchy kernel (4), its primitive is given as

$$I(x) = \frac{1}{\pi}\left[\text{arctg}(x) + \frac{x}{(1+x^2)}\right] + \frac{1}{2}. \tag{22}$$

In formulas (19)–(20), the smoothing parameter h should be calculated for set (16).

Thanks to the use of kernel estimators with strong averaging properties, inference takes place not only for data obtained exactly for w^* (among the values w_i there may be some too small for reliable consideration or even not at all), but also for neighboring values proportional to their "closeness" with respect to w^*.

4 Final Remarks and Conclusion

The procedure presented in this paper has been numerically verified in detail. The obtained results confirmed its correct functioning and full completion of the objectives and goals set out in the Introduction. Particularly, in the case of a positive correlation between the describing and conditioning factors, the greater (or smaller) the value of the conditioning attributes, the greater (or smaller) the values of the describing elements detected to be atypical. For the negative correlation, the above relation is inverse.

The procedure also successfully underwent verification in solving a practical problem in control engineering. Based on the current state of the system, the atypical elements discovered, provided evidence of arising failures of an observed device [8, 9]. The conditioning factors allowed the model used to be significantly refined.

A detailed description of the numerical and empirical verifications can be found in the article [15], currently in press.

Finally, this paper presents the algorithm for atypical (rare) elements also for a multivalued case, with continuous coordinates of describing and conditioning variables. The conditional approach allows in practice for refinement of the model by including the current value of the conditioning factors. Use of the nonparametric concepts frees the worked out procedure from distributions of describing and conditioning attributes. The investigated algorithm is ready for direct use without any additional laborious research or calculations. A full version of the material described here is presented in the paper [15].

Acknowledgments. Our heartfelt thanks go to our colleagues Damian Kruszewski and Cyprian Prochot, with whom we collaborated on the subject presented here.

References

1. Aggarwall, C.C.: Outlier Analysis. Springer, Heidelberg (2013)
2. Aggarwal, C.C.: Data Mining. Springer, Heidelberg (2015)
3. Barnett, V., Lewis, T.: Outliers in Statistical Data. Wiley, Hoboken (1994)
4. Dawid, A.P.: Conditional independence in statistical theory. J. Roy. Stat. Soc. Ser. B **41**, 1–31 (1979)
5. Hawkins, D.M.: Identification of Outliers. Chapman and Hall, London (1980)

6. Hodge, V., Austin, J.: A survey of outlier detection methodologies. Artif. Intell. Rev. **22**, 85–126 (2004)
7. Kincaid, D., Cheney, W.: Numerical Analysis. Brooks/Cole, Pacific Grove (2002)
8. Korbicz, J., Kościelny, J.M., Kowalczuk, Z., Cholewa, W. (eds.): Fault Diagnosis: Models, Artificial Intelligence, Applications. Springer, Heidelberg (2004)
9. Kulczycki, P.: Wykrywanie uszkodzen w systemach zautomatyzowanych metodami statystycznymi. Alfa, Warsaw (1998)
10. Kulczycki, P.: Estymatory jadrowe w analizie systemowej. WNT, Warszawa (2005)
11. Kulczycki, P.: Kernel estimators in industrial applications. In: Prasad, B. (ed.) Soft Computing Applications in Industry, pp. 69–91. Springer, Heidelberg (2008).
12. Kulczycki, P., Charytanowicz, M.: A Complete gradient clustering algorithm formed with kernel estimators. Int. J. Appl. Math. Comput. Sci. **20**, 123–134 (2010)
13. Kulczycki, P., Charytanowicz, M.: Conditional parameter identification with different losses of under- and overestimation. Appl. Math. Model. **37**, 2166–2177 (2013)
14. Kulczycki, P., Charytanowicz, M., Dawidowicz, A.: A Convenient ready-to-use algorithm for a conditional quantile estimator. Appl. Math. Inf. Sci. **9**, 841–850 (2015)
15. Kulczycki P., Charytanowicz M., Kowalski P.A., Lukasik S.: Identification of Atypical (Rare) Elements – A Conditional, Distribution-Free Approach. IMA J. Math. Control I. (2017, in press)
16. Kulczycki, P., Daniel, K.: Metoda wspomagania strategii marketingowej operatora telefonii komorkowej. Przeglad Statystyczny **56**(2), 116–134 (2009). Errata: **56**(3-4), 3
17. Kulczycki, P., Hryniewicz, O., Kacprzyk, J. (eds.): Techniki informacyjne w badaniach systemowych. WNT, Warszawa (2007)
18. Kulczycki P., Kowalski P.A.: Bayes classification for nonstationary patterns. Int. J. Comput. Methods **12** (2015). Article ID: 1550008
19. Kulczycki, P., Lukasik, S.: An algorithm for reducing dimension and size of sample for data exploration procedures. Int. J. Appl. Math. Comput. Sci. **24**, 133–149 (2014)
20. Larose, D.T.: Discovering Knowledge in Data. An Introduction to Data Mining. Wiley, Hoboken (2005)
21. Silverman, B.W.: Density Estimation for Statistics and Data Analysis. Chapman and Hall, London (1986)
22. Wand, M.P., Jones, M.C.: Kernel Smoothing. Chapman and Hall, London (1995)

Finding Shortest Isothetic Path Inside a 3D Digital Object

Debapriya Kundu and Arindam Biswas$^{(\boxtimes)}$

Department of Information Technology, Indian Institute of Engineering Science
and Technology, Shibpur, Howrah 711103, West Bengal, India
debapriyakundu1@gmail.com, barindam@gmail.com

Abstract. The problem of finding shortest isothetic path between two points is well studied in the context of two dimensional objects. But it is relatively less explored in higher dimensions. An algorithm to find a shortest isothetic path between two points of a 3D object is presented in this paper. The object intersects with some axis parallel equi-distant slicing planes giving one or more isothetic polygons. We call these polygons as slices. The slice containing the source and destination points are called source and destination slice respectively. A graph is constructed by checking the overlap among the slices on consecutive planes. We call it slice overlap graph. Our algorithm first finds the source and destination slice. Thereafter, it finds the minimum set of slices Π_{st} from the slice overlap graph, that need to be traversed to find SIP. Finally BFS is applied to find a SIP through these set of slices. The advantage of this procedure is that it does not search the whole object to find a SIP, rather only a part of the object is considered, therefore making the search faster.

Keywords: Unit grid cube (UGC) · Shortest isothetic path (SIP) · Breadth first search (BFS)

1 Introduction

The problem of finding shortest path is a well studied area in computational geometry. Path planning or motion planning in automated robots is a challenging area of robotics. Computing shortest path between different locations is often used in various map services and navigation systems. Various prior works on finding shortest path in 3D have been done. Problems of finding shortest path can be of different types. Some of the examples are finding shortest path on 3D polyhedral surface, in presence of a sequence of obstacles or polyhedral obstacles, finding shortest rectilinear minimum bending path, minimum distance path etc. The various kinds of shortest path problems in three dimension are discussed below.

An efficient parallel solution to the problem of finding shortest Euclidean path between two points in three dimensional space in the presence of polyhedral obstacles is discussed in [1]. It finds the shortest path touching n lines in a

© Springer International Publishing AG 2017
R.P. Barneva et al. (Eds.): CompIMAGE 2016, LNCS 10149, pp. 65–78, 2017.
DOI: 10.1007/978-3-319-54609-4_5

specified order. A more general problem is of finding collision-free optimal path for a given robot system. A solution to the problem of finding shortest path in three dimension with polyhedral obstacles is discussed in [7]. A new approach to solve this problem is discussed in [5] which solves the problem in $O(n^3 v^k)$ time, where n is the number of vertices of the polyhedra, k is the number of obstacles and v is the largest number of vertices on any one obstacle. Recently an algorithm has been proposed on finding shortest path in three-dimensional cluttered environment [11].

A different kind of problem is construction of multilayer obstacle avoiding rectilinear Steiner minimum tree. A Steiner tree problem is a combinatorial optimization problem. For a given set of geometric points, the Steiner tree problem is about finding a minimum length graph interconnecting all these points where the length of the graph is the sum of the lengths of all edges. The points are called Steiner points. When the edges are restricted to be axis parallel the problem reduces to rectilinear minimum Steiner tree problem. An analysis of the existing rectilinear Steiner minimum tree algorithms is given in [10]. This has application in circuit layout, network design, VLSI physical design, wire routing, telecommunications, etc.

A related problem is finding rectilinear minimum link path problem in three dimensions [3,13]. It involves finding a continuous path with axis-parallel line segments, such that none of the line segments intersect the interior of any obstacle. [3] solved the problem in $O(n^2 \log n)$ time with worst case running time of $\Omega(n^3)$. A betterment on the worst case running time has been achieved in [13], where the worst case time complexity is $O(n^{5/2} \log n)$.

Our Contribution: Let A be a 3D digital object and s and t be the source and destination points. Our objective is to find a SIP (π_{st}) between s and t such that the SIP lies entirely inside A and consists of moves along grid edges only. There may exist a number of SIPs between s and t, our algorithm will find one of them. Here by point we mean digital points only i.e., points with integer coordinates. The algorithm proposed here is for genus 0 objects. To the best of our knowledge there exists no suitable algorithmic solution on the geometric problem stated above. For a given grid size the proposed algorithm runs in $O(\sum_{i=1}^{k} y_i)$ time where k is the set of slices in the shortest path from source to destination grid vertex and y_i denotes the number of UGCs in the ith slice of these set of slices. The worst case time complexity of the proposed algorithm is $O(\sum_{i=1}^{k} y_i) = O(N)$, where N is the total number of UGCs of the given digital object. Though the worst case complexity is high still this algorithm is computationally very fast for best and average cases.

2 Definitions and Preliminaries

2.1 Digital Grid

A *digital grid* $\mathbb{G}$ consists of three orthogonal sets of equi-spaced grid lines, $\mathbb{G}_{yz}$, $\mathbb{G}_{zx}$, and $\mathbb{G}_{xy}$, where $\mathbb{G}_{yz} = \{l_x(j \pm ag, k \pm bg) \,|\, a \in \mathbb{Z}, b \in \mathbb{Z}\}$. Similarly, $\mathbb{G}_{zx}$

and $\mathbb{G}_{xy}$ can be represented in terms of l_y and l_z for a grid size $g \in \mathbb{Z}^+$. Here, $l_x(j,k) = \{(x,j,k) : x \in \mathbb{R}\}$, $l_y(i,k) = \{(i,y,k) : y \in \mathbb{R}\}$, and $l_z(i,j) = \{(i,j,z) : z \in \mathbb{R}\}$ denote the *grid lines* (Fig. 1) along x-, y-, and z-axes respectively, where i, j, and k are integer multiples of g. The three orthogonal lines $l_x(j,k)$, $l_y(i,k)$, and $l_z(i,j)$ intersect at the point $(i,j,k) \in \mathbb{Z}^3$, which is called a *grid point*; a shift of $(\pm 0.5g, \pm 0.5g, \pm 0.5g)$ with respect to a grid point designates a *grid vertex*, and a pair of adjacent grid vertices defines a *grid edge* [6]. Therefore, the grid point set is $\mathbb{Z}^3$. A grid square is defined by the four grid edges that form a square. A grid cube is defined by six grid squares that form a cube. It is also called a 3-cell, grid square is a 2-cell, a grid edge is a 1-cell, and a grid vertex is a 0-cell. A *unit grid cube* (UGC) is a (closed) cube of length g whose vertices are *grid vertices*, edges constituted by *grid edges*, and faces constituted by *grid faces*.

A *unit grid cube* (UGC) is grid cube of length g whose vertices are *grid vertices*, edges constituted by *grid edges* and faces constituted by *grid faces*. Each face of a UGC lies on a *face plane* (hence referred as a UGC-face), which is parallel to one of three coordinates planes. A UGC contains $g \times g \times g$ number of voxels. So a UGC is same as a voxel when $g = 1$ [8]. A grid for $g = 2$ with its elements are shown in Fig. 1.

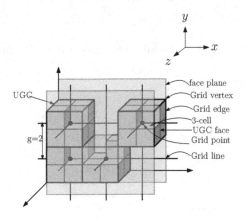

Fig. 1. 3D grid for g = 2.

2.2 Adjacency

Two cells c_1 and c_2 are called α adjacent if $c_1 \neq c_2$ and the intersection contains an α-cell ($\alpha \in 0, 1, 2$). Two 3D grid points $p_1 = (x_1, y_1, z_1)$ and $p_2 = (x_2, y_2, z_2)$ are called 6-adjacent iff $0 < d_e(p_1, p_2) \leqslant 1$, 18 adjacent iff $0 < d_e(p_1, p_2) \leqslant \sqrt{2}$, and 26 adjacent iff $0 < d_e(p_1, p_2) \leqslant \sqrt{3}$.

Let c_1 and c_2 be 3-cells and let p_i be the center of $c_i(i = 1, 2)$. Then c_1 and c_2 are 0-adjacent iff p_1 and p_2 are 26-adjacent iff c_1 and c_2 are not identical but share a grid vertex; c_1, c_2 are 1-adjacent iff p_1 and p_2 are 18 adjacent iff c_1 and c_2 are not identical but share a grid edge; and c_1, c_2 are 2-adjacent iff p_1 and p_2 are 6 adjacent iff c_1 and c_2 are not identical but share a grid square [8] (Fig. 2).

2.3 Digital Object

Let A be a 3D digital object, (referred as an object), which is defined as a finite subset of $\mathbb{Z}^3$, with all its constituent points (i.e., voxels) having integer coordinates and connected in 26-neighborhood [6]. Here, a digital object is a 26-connected component.

2.4 Isothetic Path

An isothetic path from a grid vertex $p \in A$ to a grid vertex $q \in A$ is defined as the sequence of 6-adjacent grid vertices which are all distinct and lie on or inside the digital object A (Fig. 3). An isothetic path of minimum length is called a shortest isothetic path (SIP) [2].

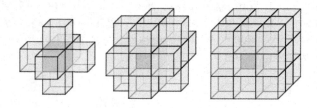

Fig. 2. Left: 6-adjacency; Middle: 18-adjacency; Right: 26-adjacency.

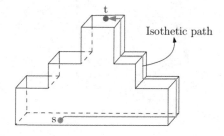

Fig. 3. An isothetic path with source and destination are marked as red and blue respectively. (Color figure online)

3 Slicing

The object A is considered to be a set of voxels. A UGC is said to be object occupied if it contains an object voxel. For slicing the object all the object occupied UGCs need to be found. The object intersects with some grid planes along xy-, yz- and zx- axis. The intersection of the object with a grid plane gives one or more isothetic polygons. We call these polygons as slices. Let us consider $\{\Pi_1, \Pi_2, \Pi_3, \ldots\}$ be the set of slicing planes separated by g parallel to any one of yz-, zx-, or xy-plane. Each slice is uniquely identified using a *faceid*. For slicing an object we used the algorithm proposed in [6]. The entire process of slicing is discussed below in brief.

To start tracing a slice along a slicing plane a start vertex, v_s, is identified. A vertex qualifies as a start vertex if it is unvisited. Each vertex has eight neighboring UGCs. Depending on the object occupancy of these UGCs a starting direction is determined. Tracing a slice starts from v_s and concludes when it comes back to v_s. During traversal all the grid vertices lying in the path of traversal are marked visited. Along a given slicing plane the traversal at some point can be in four possible directions. The direction of traversal depends on the occupancy of the four neighboring UGCs along that plane and the incoming direction of traversal at that point. By applying a set of combinatorial formulas the outgoing direction is found such that the object always lies left. A slice contains the top faces of the adjacent UGCs along that plane. The slicing procedure traces each slice exactly once. An object and its slices are shown in Fig. 4. The set of slices is stored in an adjacency list (L). Each list in L contains the slices along one grid plane. Therefore, the number of lists in L is equal to the number of grid planes the object intersects with the direction of slicing i.e., either xy- or yz- or zx-. Let us consider this count as l_s. The result of slicing bunny is presented in Fig. 5. Here, w.l.o.g., we are slicing the object along zx-plane.

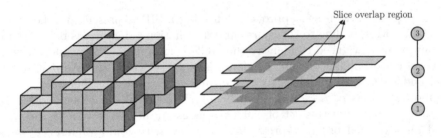

Fig. 4. Left: 3D object A; Middle: Its slices along zx- axis for g = 1. The slices are shown in light grey and the overlapping regions are shown in dark grey; Right: corresponding slice overlap graph.

3.1 Slice Containing a Given Point

To find the slice containing a given point p, the slicing plane containing p is identified first using its coordinates thereafter searching the slices on that plane. Given a point i and a slice S_i, to check whether $i \in S_i$, a LeftOn test [12] is done for each edge of the slice. This technique checks for each edge whether the query point p lies on it or on the left (right) of it, when the slice is traversed in an anti-clockwise (clockwise) manner. The detailed steps of this procedure is discussed in IDENTIFYSLICE (Fig. 6) which identifies the slice S_p containing a given point p. Using this procedure the source and the destination slices corresponding to the source and destination points are obtained respectively. Figure 10 (d, e) shows a given source and destination points with their corresponding UGCs and slices.

Fig. 5. Left: Bunny represented as a set of UGCs for $g = 1$; Middle: Its slices along zx-axis for $g = 1$; Right: Slice overlap graph, (here for clarity of view only a few nodes are shown).

4   Finding the Minimum Set of Slices (Πst) to be Traversed to Move from Source to Destination Slice

4.1   UGC Corresponding to a Point

In this paper, a grid based approach of finding a SIP is presented. Therefore, the UGCs corresponding to the source and destination points needs to be found. For any point i, we need to identify the UGC corresponding to it and for that the voxel corresponding to that point needs to be selected first. We denote the voxel and UGC corresponding to i as v_i and u_i respectively. There are eight neighboring voxels of i. Not necessarily all of them will be object voxel. To choose one voxel from this set of eight voxels as v_i, an object voxel is selected based on a defined order of preference. A voxel with lower preference value will get higher priority. The eight neighboring voxels of a given point with their preference values are shown in Fig. 7(Left). Once v_i is found the UGC containing v_i is selected as u_i. We denote the source and destination points as s and t, hence the source and destination UGCs as u_s and u_t respectively.

4.2   Slice Overlap Graph

Two slices on consecutive planes are said to have overlap if their projection on a plane overlap with each other. A graph is constructed considering each slice of A as a node and an edge between two nodes if there is overlap between the corresponding slices. We call this graph as slice overlap graph. As each slice has an unique *faceid*, so the node corresponding to that slice is uniquely identified using that *faceid*. As the object A is a digital object hence its slices will be orthogonal polygons. To check overlap between two orthogonal polygon (say S_i and S_j) we adopt a simple method, which is discussed below.

S_i and S_j are traced starting from any of their boundary grid vertices. Each grid vertex occurring in the path of traversal of each slice are given label equal to

Procedure IDENTIFYSLICE(L, p)	**Procedure** SLICEOVERLAPGRAPH(L)
01. **for** each slice S_i in L on PlaneNumber(p) 02. **for** all edge $e_{i,j} \in S_i$ 03. **if** (LeftOn($e_{i,j}$, p)) 04. return i 05. **else** 06. break 07. find next slice on planeno(p)	01. **for** each j $\leftarrow$ 2 to L_n 02. i = j-1 03. for each slice polygon in L_i check overlap with all slice polygons in L_j 04. if overlap found then addedge(G, i, j)

Fig. 6. Left: Procedure IDENTIFYSLICE; Right: Procedure SLICEOVERLAPGRAPH.

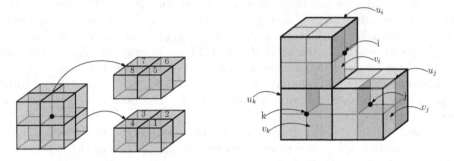

Fig. 7. Left: Order of preference of eight neighboring voxels of a point (the point is shown as a black dot); Right: voxels and UGCs corresponding to some given points. Here, the points are shown in black dots, the voxel corresponding to a point is marked red and the UGC containing that voxel is marked light red. (Color figure online)

the *faceid* of its corresponding slice. Therefore by this process each grid vertex on the boundary of a slice gets a label. Here by slice we mean only object slices, not any hole polygon. To check overlap we start from the slices on plane $\varPi_2$. Let us denote the slices on i^{th} plane as $S_{i,1}$, $S_{i,2}$, $S_{i,3}$, ... Hence, j^{th} slice on i^{th} plane is denoted by $S_{i,j}$. The grid vertices on the boundary of $S_{i,j}$ are denoted by $v_{i,j,1}$, $v_{i,j,2}$, ... Therefore, the kth grid vertex of polygon j on slice i will be denoted by $v_{i,j,k}$. We traverse each grid vertex on the boundary of a slice and for each grid vertex $v_{i,j,k}$ with label l we check label l_1 of $v_{i,j-1,k}$ (w.l.o.g., for slices along zx-plane), if $l_1 \neq 0$, then an edge is added to the slice overlap graph between nodes with *faceid* l and l_1, if this edge does not already exist in the slice overlap graph. This method is able to identify the overlap between projections of two slices if the projections intersect with each other. If the projections of slices do not intersect each other and one is contained within another then the following method is adopted. For two slices (say, $S_{i,j}$ and $S_{i,k}$) between which no edge has been reported by the previous method we find whether $S_{i,j}$ is contained within $S_{i,k}$ or vice-versa. We take any one point from the vertex set of $S_{i,j}$ and

Slice with $\widehat{faceid} = 2$

Slice with $\widehat{faceid} = 1$

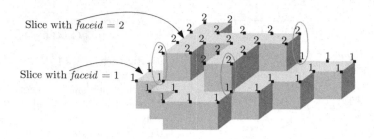

Fig. 8. The labelling of grid vertices on each slice

the method discussed in Sect. 3.1 is applied to find whether it is contained within $S_{i,k}$. If it returns false then the similar method is applied to check whether $S_{i,k}$ is contained within $S_{i,j}$. If there exists a overlap, an edge is added between the nodes corresponding to $S_{i,j}$ and $S_{i,k}$ to the slice overlap graph else no edge is added. Hence by visiting only the boundary grid points of all the slices the complete slice overlap graph is constructed. Figure 8 shows an object with two slices and the labels of the grid vertices on the boundary of each slice.

The procedure SLICEOVERLAPGRAPH is given in Fig. 6. Figures 4, 5 and 10(b, c) shows the slice overlap graph of some objects. The overlapping regions are shown in dark shade.

For genus 0 objects slice overlap graph will be a tree, but for objects of type genus 1 or more it will contain one or more cycles in it. The algorithm proposed in this paper is for genus 0 objects only.

Instead of constructing a slice overlap graph taking all the slices it can be made specific to the source and destination slices. This will require less amount of storage space as well as time to find Π_{st}. This graph construction starts from the source slice and only those slices that overlap with it are further explored and this process continues till the destination slice is visited. This process follows the BFS technique. So this graph will not necessarily contain all the slices of the object, which makes it time and space efficient. But the problem associated with this approach is, as it is specific to a given source and destination slice, the source and destination slices can not be changed dynamically.

4.3 Storage of Slice Overlap Graph

The slice overlap graph is stored in an adjacency list (L') where the number of lists is equal to the number of nodes in the graph. Each list gives information about a particular node i.e., the nodes adjacent to it. Each node of a list contains 2 elements a pointer to the first node of the slice corresponding to it and a pointer to the next node. If a node's *faceid* is 1 it is stored in the 1st list, similarly a node with *faceid* 2 is stored in the 2nd list. Figure 9 shows the memory representation of the slice overlap graph of the object represented in Fig. 4.

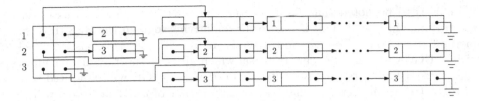

Fig. 9. Representation of slice overlap graph in memory

4.4   Finding the Minimum Set of Slices (Π_{st}) to be Traversed to Find SIP

To find path between two nodes in slice overlap graph BFS is applied. It starts from the source node and continues search until the destination node is visited. The source node is labeled 0 and marked visited. For a node with label i, its adjacent unvisited nodes are labeled i+1 and marked visited. Therefore, the label at a node is its shortest distance from the source node. The shortest path between the source and destination node is found by retracing. This is done in Step 5–6 of the algorithm FINDSIP. The minimum set of nodes or slices found this way, are stored in a linked list (L''). Each node of this list contains the value of the *faceid* of the corresponding slice and a pointer to the next node of Π_{st}.

5   Algorithm FINDSIP

To find a SIP, BFS is applied on the set of UGCs, that are bounded by the slices in the list L''. BFS starts from u_s which is labeled as zero and it continues till u_t is visited. The label of a given UGC gives the count of the minimum number of UGCs that need to be traversed to reach it from the source UGC. Here, BFS is done in 26-neighborhood as the object is connected in 26-neighborhood. Finally SIP is found by retracing the visited UGCs. The complete algorithm to find SIP is given below. A step by step process of finding SIP is shown in Fig. 10. This algorithm follows grid based approach of solving the problem. The BFS done in Step 7 can be considered as an extension of Lee's [9] wavefront technique to three dimension.

Algorithm FINDSIP(A, s, t)

01. L = SLICE(A)
02. S_s = IDENTIFYSLICE(L, s)
03. S_t = IDENTIFYSLICE(L, t)
04. L' = SLICEOVERLAPGRAPH(L)
05. BFS(L')
06. L'' = RETRACE(S_s, S_t)
07. BFS(u_s, u_t, L'')
08. π_{st} = RETRACE(u_t, u_s)

This is to be noted that this algorithm FINDSIP never backtracks during the traversal in Step 7. This is due to the underlying logic in the BFS traversal method. BFS starts from u_s and extends in breadth till u_t is visited. Each visited UGC gets a label during traversal which gives its shortest distance from the source. BFS starts from u_s so its label is 0. As no UGC will be visited after u_t so no UGC will ever get a label greater than that of u_t. Hence there will be no possibility of backtracking during the traversal.

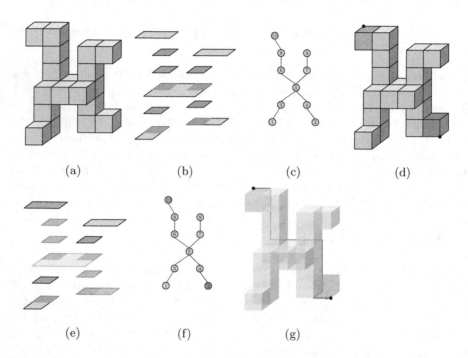

(a)   (b)   (c)   (d)

(e)   (f)   (g)

Fig. 10. Demonstration of various steps in finding SIP. (a) Object represented as a set of UGCs; (b) Object slices along zx-plane. Each of the slices are shown in light grey colour and overlapping regions are shown in dark grey. (c) Corresponding slice overlap graph; (d) Source and destination points with their corresponding UGCs; (e) Source and destination slices are marked red and blue respectively and the set of slices in the path from source to destination are marked light blue; (f) The minimum set of nodes in the path from source to destination node of the slice overlap graph are shown in light blue; (h) Final SIP is shown in red. (Color figure online)

6   Time Complexity

Slicing takes $O(n/g)$ time in best case and $O(ng)$ time in worst case, where n is the total number of voxels constituting the object surface in 26-neighborhood. Therefore, we can consider the complexity of slicing in general as $O(n)$, if $n >> g$.

The procedure IDENTIFYSLICE, in the worst case, traverses edges of all the slices on the plane containing a given point. Therefore, the procedure needs

$O(s_{max}e_{max})$ time, s_{max} is the maximum possible count of slices on a slicing plane and e_{max} is the maximum possible count of edges of a slice.

In procedure SLICEOVERLAPGRAPH the boundary grid vertices of all the slices are visited twice, first time for labeling and second time for checking overlap among the slices. Hence, in this procedure all the grid vertices on the whole object surface are visited twice. The number of grid vertices on an object surface is always in the order of the total number of UGCs constituting that object surface. For an object with n voxels the number of UGCs in it, in best case is $O(n/g^3)$ and in worst case $O(n/g)$. Therefore, the complexity of the procedure SLICEOVERLAPGRAPH becomes $O(n/g^3)$ in best case and $O(n/g)$ in worst case.

The procedure BFS in line 5 of the algorithm FINDSIP starts from the source node of the slice overlap graph and for each node its adjacent nodes are also visited, thereby requiring $O(mk)$ time, where k is total number of slices of the object and m is the maximum possible number of nodes adjacent to a node in the slice overlap graph.

BFS in line 7 visits each UGC bounded by the slices $\in \Pi_{st}$ and for each UGC it visits its 26-adjacent UGCs which are bounded by any of the slices $\in \Pi_{st}$. Hence it requires $O(\sum_{i=1}^{k} y_i)$ time, where y_i denotes the number of UGCs in the ith slice of L''.

Therefore the total time complexity is given by $TC = O(n) + 2 \times O(s_{max}e_{max}) + O(n/g) + O(mk) + O(\sum_{i=1}^{k} y_i)$. As n is the number of voxels on the object surface hence it will have value less than $\sum_{i=1}^{k} y_i$ for most of the objects. Similarly $O(n/g)$ will also have value less than $\sum_{i=1}^{k} y_i$. $s_{max}e_{max}$ can never exceed the value of $\sum_{i=1}^{k} y_i$ and the number of slices of an object will always be much less than the number of UGCs on a set of slices, hence $mk < \sum_{i=1}^{k} y_i$. Therefore, TC $= O(\sum_{i=1}^{k} y_i)$ and in worst case $O(\sum_{i=1}^{k} y_i) = O(N)$. Though the worst case time complexity is high still the algorithm is computationally very fast in best and average cases due to the process of doing BFS only on the UGCs of the selected set of slices. An improvement in terms of complexity can be done using Hadlock's [4] method of traversal (it has to be extended to 3D). This can be considered as a future scope of work.

7   Results and Conclusion

The proposed algorithm has been implemented in C in Linux Fedora Release 7, Kernel 2.6.21.1.3194.fc7, Dual Intel Xeon Processor 2.8 GHz, 800 MHz FSB. It has been tested on several 3D objects for slicing along zx-plane. Some test results are presented in Fig. 11. Some test results found by our algorithm and the results found by doing only BFS to find the SIP between two points has been presented in Fig. 12. For better understanding the UGCs visited by both the methods has been marked blue. The count of voxels visited by each of the methods as well as the percentage of voxels traversed with respect to the total number of object voxel and the respective average CPU times (in milliseconds) for each of the methods has been given under each figure of second and third columns. For the type of cases shown in first four rows of Fig. 12 a significant reduction in the visited voxel count by the proposed algorithm can be observed. However for the type of cases shown in the last row of Fig. 12 worst case occurs, so the visited voxel count is not much reduced by the proposed algorithm.

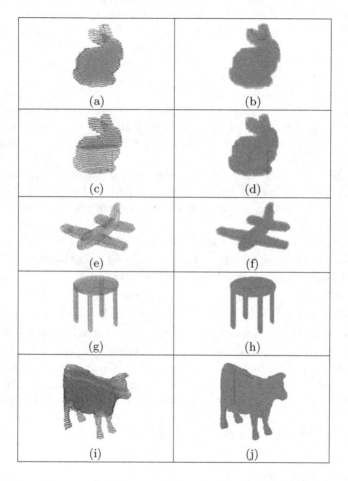

Fig. 11. SIP found by our algorithm on different objects. Left: Object represented as a set of slices. Right: Object represented as a set of UGCs. (a, b, c, d) SIP found in bunny for $g = 2$; (e, f) SIP found in aeroplane for $g = 2$; (g, h) SIP found in table for $g = 2$; (i, j) SIP found in cow for $g = 2$; Here the source and destination points and the corresponding slices are marked in red and pink respectively and the set of slices $\in \Pi_{st}$ are marked blue and the final SIP is shown in red. (Color figure online)

Hence it can be noticed that the algorithm proposed in this paper can give much reduction in the count of visited voxels for best and average cases. This algorithm finds SIP between any pair of points of a 3D digital object of type genus 0. The proposed method can be improved for application on objects of type

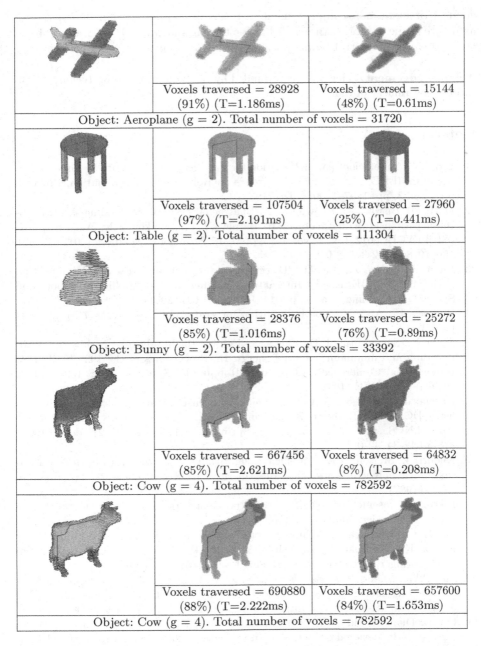

| | Voxels traversed = 28928 (91%) (T=1.186ms) | Voxels traversed = 15144 (48%) (T=0.61ms) |

Object: Aeroplane (g = 2). Total number of voxels = 31720

| | Voxels traversed = 107504 (97%) (T=2.191ms) | Voxels traversed = 27960 (25%) (T=0.441ms) |

Object: Table (g = 2). Total number of voxels = 111304

| | Voxels traversed = 28376 (85%) (T=1.016ms) | Voxels traversed = 25272 (76%) (T=0.89ms) |

Object: Bunny (g = 2). Total number of voxels = 33392

| | Voxels traversed = 667456 (85%) (T=2.621ms) | Voxels traversed = 64832 (8%) (T=0.208ms) |

Object: Cow (g = 4). Total number of voxels = 782592

| | Voxels traversed = 690880 (88%) (T=2.222ms) | Voxels traversed = 657600 (84%) (T=1.653ms) |

Object: Cow (g = 4). Total number of voxels = 782592

Fig. 12. Left: SIP found by algorithm FINDSIP on different objects where the objects are represented as set of slices. The source and destination points and the corresponding slices are marked in red and pink respectively and the set of slices $\in \Pi_{st}$ are marked blue and the final SIP is shown in red. Middle: SIP found by doing BFS starting from the source UGC to the destination UGC. Right: SIP found by algorithm FINDSIP. (Color figure online)

genus 1 or more. Some improvement in terms of complexity can be considered as future scope of work. An algorithm for finding multiple SIPs on a 3D object for a given set of control points remains an open problem.

Acknowledgement. This research is funded by All India Council for Technical Education, Government of India.

References

1. Bajaj, C.: An efficient parallel solution for Euclidean shortest path in three dimensions. In: IEEE International Conference on Robotics and Automation, Proceedings, vol. 3, pp. 1897–1900. IEEE (1986)
2. Dutt, M., Biswas, A., Bhowmick, P., Bhattacharya, B.B.: On finding shortest isothetic path inside a digital object. In: Barneva, R.P., Brimkov, V.E., Aggarwal, J.K. (eds.) IWCIA 2012. LNCS, vol. 7655, pp. 1–15. Springer, Heidelberg (2012). doi:10.1007/978-3-642-34732-0_1
3. Fitch, R., Butler, Z., Rus, D.: 3D rectilinear motion planning with minimum bend paths. In: 2001 IEEE/RSJ International Conference on Intelligent Robots and Systems, Proceedings, vol. 3, pp. 1491–1498. IEEE (2001)
4. Hadlock, F.: A shortest path algorithm for grid graphs. Networks **7**(4), 323–334 (1977)
5. Jiang, K., Seneviratne, L.D., Earles, S.: Finding the 3D shortest path with visibility graph and minimum potential energy. In: Proceedings of the 1993 IEEE/RSJ International Conference on Intelligent Robots and Systems, IROS 1993, vol. 1, pp. 679–684. IEEE (1993)
6. Karmakar, N., Biswas, A., Bhowmick, P.: Fast slicing of orthogonal covers using DCEL. In: Barneva, R.P., Brimkov, V.E., Aggarwal, J.K. (eds.) IWCIA 2012. LNCS, vol. 7655, pp. 16–30. Springer, Heidelberg (2012). doi:10.1007/978-3-642-34732-0_2
7. Khouri, J., Stelson, K.A.: An efficient algorithm for shortest path in three dimensions with polyhedral obstacles. In: American Control Conference, pp. 161–165. IEEE (1987)
8. Klette, R., Rosenfeld, A.: Digital Geometry: Geometric Methods for Digital Picture Analysis. Elsevier, Amsterdam (2004)
9. Lee, D.T.: Rectilinear paths among rectilinear obstacles. In: Ibaraki, T., Inagaki, Y., Iwama, K., Nishizeki, T., Yamashita, M. (eds.) ISAAC 1992. LNCS, vol. 650, pp. 5–20. Springer, Heidelberg (1992). doi:10.1007/3-540-56279-6_53
10. Lin, C.W., Huang, S.L., Hsu, K.C., Lee, M.X., Chang, Y.W.: Multilayer obstacle-avoiding rectilinear steiner tree construction based on spanning graphs. IEEE Trans. Comput.-Aided Des. Integr. Circ. Syst. **27**(11), 2007–2016 (2008)
11. Lu, J., Diaz-Mercado, Y., Egerstedt, M., Zhou, H., Chow, S.N.: Shortest paths through 3-dimensional cluttered environments. In: 2014 IEEE International Conference on Robotics and Automation (ICRA), pp. 6579–6585. IEEE (2014)
12. O'Rourke, J.: Computational Geometry in C. Cambridge University Press, Cambridge (1998)
13. Wagner, D.P., Drysdale, R.S., Stein, C.: An $O(n5/2logn)$ algorithm for the rectilinear minimum link-distance problem in three dimensions. Comput. Geom. **42**(5), 376–387 (2009)

Unified Characterization of *P*-Simple Points in Triangular, Square, and Hexagonal Grids

Péter Kardos and Kálmán Palágyi$^{(\boxtimes)}$

Department of Image Processing and Computer Graphics,
University of Szeged, Szeged, Hungary
{pkardos,palagyi}@inf.u-szeged.hu

Abstract. Topology preservation is a crucial property of topological algorithms working on binary pictures. Bertrand introduced the notion of *P*-simple points on the orthogonal grids, which provides a sufficient condition for topology-preserving reductions. This paper presents both formal and easily visualized characterizations of *P*-simple points in all the three types of regular 2D grids.

Keywords: Digital topology · Regular 2D grids · Topology preservation · *P*-simple points

1 Introduction

The aim of *digital topology* is to study the properties of digital pictures. Special attention is given to *binary pictures*, which are made up of only black and white elements, so-called *points* [7,15,16]. Although such pictures are mostly sampled in the orthogonal grid, triangular and hexagonal grids also attract remarkable scientific interest [1,5,6,15–19,21–23].

A *reduction* transforms a binary picture only by changing some black points to white ones, which is referred to as *deletion* [7]. Reductions have a key role in thinning algorithms [13,15,20]. *Topology preservation* is one of the most crucial issues of reductions. A *simple point* is a point whose deletion preserves the topology. Deleting only simple points ensures topology preservation for *sequential reductions* [15], however, this is not the case for *parallel* operators. First Ronse and Kong proposed some sufficient conditions for topology-preserving 2D parallel reductions in the square grid [13,20], then the authors of this work adapted their results to the hexagonal and triangular cases [9,11,12]. Bertrand proposed an alternative solution to the problem: he introduced the notion of *P-simple points*, whose simultaneous deletion is proved to be topology-preserving, and he also gave a local characterization of *P*-simple points for $(26, 6)$-pictures on the 3D cubic grid [2]. Later, Bertrand and Couprie gave similar characterizations for $(2, 1)$-pictures (see Sect. 2) on the square grid [3]. Note that Bertrand and Couprie linked the concepts critical kernels, minimal nonsimple sets, and *P*-simple points on the 2D, 3D, and 4D orthogonal grids [4]. In this work, we

R.P. Barneva et al. (Eds.): CompIMAGE 2016, LNCS 10149, pp. 79–88, 2017.
DOI: 10.1007/978-3-319-54609-4_6

present a unified characterization of P-simplicity for 2D binary pictures, which equally applies to the triangular, square, and hexagonal grids.

The rest of this paper is organized as follows. In Sect. 2, we review the basic notions and results. In Sect. 3 we present formal and easily visualized characterizations of P-simple points for all the three types of regular 2D grids, and show that deletion of any subset composed solely of P-simple points satisfies some earlier proposed sufficient conditions for topology-preserving reductions.

2   Basic Notions and Results

Here, we give a brief summary of some concepts of digital topology as reviewed by Kong and Rosenfeld in their seminal work [15].

Let us denote by $\mathcal{T}$, $\mathcal{S}$, and $\mathcal{H}$ the triangular, the square, and the hexagonal grids, respectively. Elements of a grid (i.e., regular polygons) are called *points*. Two points p and q are said to be *1-adjacent* if they share an edge, and they are called *2-adjacent* if they share an edge or a vertex. We denote by $\mathcal{N}_j^{\mathcal{V}}(p)$ the set of the points being j-adjacent to point p on the grid $\mathcal{V}$ and let $\mathcal{N}_j^{*\mathcal{V}}(p) = \mathcal{N}_j^{\mathcal{V}}(p) \backslash \{p\}$ ($j = 1, 2$). It is obvious that $\mathcal{N}_1^{\mathcal{V}}(p) \subseteq \mathcal{N}_2^{\mathcal{V}}(p)$. Furthermore, in the hexagonal grid $\mathcal{N}_1^{\mathcal{H}}(p) = \mathcal{N}_2^{\mathcal{H}}(p)$, hence, just one relation is considered (Fig. 1).

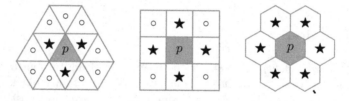

Fig. 1. Adjacency relations on the considered regular grids. Points that are 1-adjacent to the central point p are marked "★", while points being 2-adjacent but not 1-adjacent to p are depicted by "○".

The sequence of distinct points $\langle p_0, p_1, \ldots, p_m \rangle$ is called a j-*path* of length m from p_0 to p_m in a non-empty set of points $\mathcal{X} \subseteq \mathcal{V}$ if each point of the sequence is in $\mathcal{X}$ and p_i is j-adjacent to p_{i-1} for each $i = 1, 2, \ldots, m$ ($j = 1, 2$; $\mathcal{V} = \mathcal{T}, \mathcal{S}, \mathcal{H}$). Note that a single point is a j-path of length 0. A point p is j-*adjacent* to a set of points $\mathcal{X} \subseteq \mathcal{V}$ if there is a $q \in \mathcal{X}$ such that $q \in \mathcal{N}_j^{*\mathcal{V}}(p)$. Two points are said to be j-*connected* in a set $\mathcal{X}$ if there is a j-path in $\mathcal{X}$ between them. A set of points $\mathcal{X}$ is j-*connected* in a set of points $\mathcal{Y} \supseteq \mathcal{X}$ if any two points in $\mathcal{X}$ are j-connected in $\mathcal{Y}$.

A $(k, \bar{k})$ *binary digital picture* is a quadruple $(\mathcal{V}, k, \bar{k}, \mathcal{B})$ [15], where $\mathcal{V}$ is the set of points in the considered grid ($\mathcal{V} = \mathcal{T}, \mathcal{S}, \mathcal{H}$), $(k, \bar{k})$ is an ordered pair of adjacency relations $((k, \bar{k}) = (1, 2), (2, 1))$, and $\mathcal{B} \subseteq \mathcal{V}$ denotes the set of *black points*. Each point in $\mathcal{V} \backslash \mathcal{B}$ is said to be a *white point*. An *object* is a maximal k-connected set of black points, while a *white component* is a maximal $\bar{k}$-connected set of white points.

The maximal set composed by mutually 2-adjacent points in a $\mathcal{P} = (\mathcal{V}, k, \bar{k}, \mathcal{B})$ picture $(\mathcal{V} = \mathcal{T}, \mathcal{S}, \mathcal{H}; \quad (k, \bar{k}) = (1, 2), (2, 1))$ is called a *unit element*. A black component is said to be a *small object* in a $(k, \bar{k})$ picture if it is contained in a unit element, it is not singleton, and it is not formed by two $\bar{k}$-adjacent points. Note that all the possible configurations for unit elements and small objects are depicted in [10]. An *isolated point* is a black point that is not k-adjacent to any other black points, (i.e., it forms a singleton object).

A reduction is *topology-preserving* if each object in the input picture contains exactly one object of the output picture, and each white component in the output picture contains exactly one white component of the input picture [15]. A black point is said to be *simple* if its deletion is a topology-preserving reduction [13, 15]. The authors of this work gave the following characterization of simple points which applies to any of the five considered types of binary pictures.

Theorem 1 [8, 10]. *Let p be a (black or white) point in a picture $(\mathcal{V}, k, \bar{k}, \mathcal{B})$ $(\mathcal{V} = \mathcal{T}, \mathcal{S}, \mathcal{H}; \quad (k, \bar{k}) = (1, 2), (2, 1))$. Then p is a simple point if and only if the following conditions hold:*

1. *p is k-adjacent to exactly one k-component of $\mathcal{N}_2^{*\mathcal{V}}(p) \cap \mathcal{B}$.*
2. *p is $\bar{k}$-adjacent to exactly one $\bar{k}$-component of $\mathcal{N}_2^{\mathcal{V}}(p) \setminus \mathcal{B}$.*

Note that an isolated point is not simple by Condition 1 of Theorem 1.

Reductions generally delete a set of black points and not only a single point. Hence we need to consider what is meant by topology preservation when a number of black points are deleted simultaneously. The authors formulated the following unified sufficient conditions for topology-preserving reductions on the considered five types of pictures:

Theorem 2 [10]. *A reduction R is topology-preserving, if all of the following conditions hold for any picture $(\mathcal{V}, k, \bar{k}, \mathcal{B})$ $(\mathcal{V} = \mathcal{T}, \mathcal{S}, \mathcal{H}; \quad (k, \bar{k}) = (1, 2), (2, 1))$.*

1. *Only simple points in picture $(\mathcal{V}, k, \bar{k}, \mathcal{B})$ are deleted by R.*
2. *For any two $\bar{k}$-adjacent black points, $p, q \in \mathcal{B}$ that are deleted by R, p is simple in $(\mathcal{V}, k, \bar{k}, \mathcal{B} \setminus \{q\})$.*
3. *If $(k, \bar{k}) = (2, 1)$ or $\mathcal{V} = \mathcal{H}$, no small object is deleted completely from picture $(\mathcal{V}, k, \bar{k}, \mathcal{B})$ by R.*

The notion of P-simple points and P-simple sets was first proposed by Bertrand for a type of picture on the orthogonal 3D grid [2]. Here we define those notions for the considered 2D regular grids.

Definition 1. *Let $\mathcal{P} \subset \mathcal{B}$ in a picture $(\mathcal{V}, k, \bar{k}, \mathcal{B})$ $(\mathcal{V} = \mathcal{T}, \mathcal{S}, \mathcal{H}; (k, \bar{k}) = (1, 2), (2, 1))$. A point $p \in \mathcal{P}$ is called as P-simple, if for any $\mathcal{Q} \subseteq \mathcal{P} \setminus \{p\}$, p is simple in picture $(\mathcal{V}, k, \bar{k}, \mathcal{B} \setminus \mathcal{Q})$. Furthermore, the set $\mathcal{P}$ is said to be P-simple, if any point $p \in \mathcal{P}$ is P-simple.*

The above definition evidently implies the following useful properties of P-simple sets for any types of binary pictures:

- All the points in a P-simple set are simple.
- If $\mathcal{P}$ is a P-simple set, then each set $\mathcal{P}' \subset \mathcal{P}$ is also P-simple.
- A reduction that deletes a subset composed solely of P-simple points is topology-preserving.

Note that Bertrand and Couprie gave the following characterization of P-simple points for $(2,1)$-pictures on the square grid.

Theorem 3. *Consider the picture $(\mathcal{S}, 2, 1, \mathcal{B})$ and the set $\mathcal{P} \subset \mathcal{B}$ of simple points in that picture. Set $\mathcal{P}$ is P-simple if and only if the following conditions hold for any point $p \in \mathcal{P}$:*

1. *$\mathcal{N}_2^{*\mathcal{S}}(p) \cap (\mathcal{B} \setminus \mathcal{P})$ contains exactly one 2-component.*
2. *Any point $q \in \mathcal{N}_2^{*\mathcal{S}}(p) \cap \mathcal{P}$ is 2-adjacent to $\mathcal{N}_2^{*\mathcal{S}}(p) \cap (\mathcal{B} \setminus \mathcal{P})$.*

In Sect. 3, we present a unified characterization of P-simple points for all the five kinds of pictures on the regular 2D grids.

3   Characterizations of P-Simple Sets

In this section, we introduce some formal and easily visualized sufficient and necessary conditions of P-simple sets in the five considered types of pictures.

Theorem 4. *The set $\mathcal{P} \subset \mathcal{B}$ in picture $(\mathcal{V}, k, \bar{k}, \mathcal{B})$ $(\mathcal{V} = \mathcal{T}, \mathcal{S}, \mathcal{H}; (k, \bar{k}) = (1,2), (2,1))$ is P-simple if and only if the following conditions hold for any point $p \in \mathcal{P}$:*

1. *Point p is k-adjacent to exactly one k-component of $\mathcal{N}_2^{*\mathcal{V}}(p) \cap (\mathcal{B} \setminus \mathcal{P})$. Let $\mathcal{X}$ denote the set of all (black) points in that component.*
2. *Point p is $\bar{k}$-adjacent to exactly one $\bar{k}$-component of $\mathcal{N}_2^{*\mathcal{V}}(p) \setminus \mathcal{B}$. Let $\mathcal{Y}$ denote the set of all (white) points in that component.*
3. *Any point q in $\mathcal{N}_k^{*\mathcal{V}}(p) \cap (\mathcal{P} \setminus \{p\})$ is k-adjacent to $\mathcal{X}$.*
4. *Any point q in $\mathcal{N}_{\bar{k}}^{*\mathcal{V}}(p) \cap (\mathcal{P} \setminus \{p\})$ is $\bar{k}$-adjacent to $\mathcal{Y}$.*

Proof. For the proof of the sufficiency, let $\mathcal{Q} \subseteq \mathcal{P} \setminus \{p\}$, and consider the simplicity of point p in picture $(\mathcal{V}, k, \bar{k}, \mathcal{B} \setminus \mathcal{Q})$.

Let set $\mathcal{R}_1 = \{r \mid r \in (\mathcal{P} \setminus \mathcal{Q})$ and r is k-adjacent to $p\}$. Then each point in $\mathcal{R}_1$ is k-adjacent to $\mathcal{X}$ by Condition 3 of this theorem. Hence $\mathcal{R}_1 \cup \mathcal{X}$ is a subset of the only (black) k-component in $\mathcal{N}_2^{*\mathcal{V}}(p) \cap (\mathcal{B} \setminus \mathcal{Q})$ that is k-adjacent to p.

Let set $\mathcal{R}_2 = \{r \mid r \in \mathcal{Q}$ and r is $\bar{k}$-adjacent to $p\}$. Then each point in $\mathcal{R}_2$ is $\bar{k}$-adjacent to $\mathcal{Y}$ by Condition 4 of this theorem. Hence $\mathcal{R}_2 \cup \mathcal{Y}$ is a subset of the only (white) $\bar{k}$-component in $\mathcal{N}_2^{*\mathcal{V}}(p) \cap (\mathcal{B} \setminus \mathcal{Q})$ that is $\bar{k}$-adjacent to p.

Since both conditions of Theorem 1 hold, point p is simple in picture $(\mathcal{V}, k, \bar{k}, \mathcal{B} \setminus \mathcal{Q})$.

Now consider the necessity part of the statement. Let us suppose that $\mathcal{P} \subset \mathcal{B}$ is P-simple in picture $(\mathcal{V}, k, \bar{k}, \mathcal{B})$ $(\mathcal{V} = \mathcal{T}, \mathcal{S}, \mathcal{H}; (k, \bar{k}) = (1,2), (2,1))$, which means that for any $\mathcal{Q} \subseteq \mathcal{P} \setminus \{p\}$, point p is simple in picture $(\mathcal{V}, k, \bar{k}, \mathcal{B} \setminus \mathcal{Q})$. Let us examine Conditions 1–4 of this theorem:

(i) Let $\mathcal{Q} = \mathcal{P} \setminus \{p\}$. Since p is simple in picture $(\mathcal{V}, k, \bar{k}, \mathcal{B} \setminus \mathcal{P})$, Condition 1 holds.

(ii) Let $\mathcal{Q} = \emptyset$. Since p is simple in picture $(\mathcal{V}, k, \bar{k}, \mathcal{B})$, Condition 2 is satisfied.

(iii) Let point $q \in \mathcal{P}$ be k-adjacent to p and $\mathcal{Q} = \mathcal{P} \setminus \{p\} \setminus \{q\}$. Condition 1 is satisfied for picture $(\mathcal{V}, k, \bar{k}, \mathcal{B} \setminus (\mathcal{P} \setminus \{p\}))$ (see (i)), and p is simple in picture $(\mathcal{V}, k, \bar{k}, \mathcal{B} \setminus (\mathcal{P} \setminus \{p\}) \setminus \{q\})$, which concludes Condition 3.

(iv) Let $q \in \mathcal{P}$ be $\bar{k}$-adjacent to p and $\mathcal{Q} = \{q\}$. Condition 2 is satisfied for picture $(\mathcal{V}, k, \bar{k}, \mathcal{B})$ (see (ii)), and p is simple in picture $(\mathcal{V}, k, \bar{k}, \mathcal{B} \setminus \{q\})$, thus Condition 4 is satisfied. □

With the help of Theorem 4, we can illustrate P-simple points for the considered five kinds of pictures by some configurations (so-called matching templates). The base matching templates are depicted in Figs. 2, 3, 4, 5 and 6. In addition, all their rotated and reflected versions also match P-simple points. Notations: each black template position matches a black point in $\mathcal{B} \setminus \mathcal{P}$; each white element matches a white point; each position depicted in gray matches any point (i.e., either a white point, a black point in $\mathcal{P}$, or a black point in $\mathcal{B} \setminus \mathcal{P}$). The central position denoted by p represents a black point in $\mathcal{P}$.

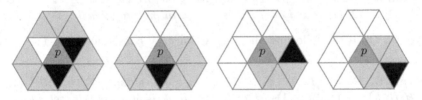

Fig. 2. Base matching templates for characterizing P-simple points on $(2, 1)$ pictures on grid $\mathcal{T}$.

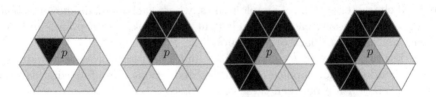

Fig. 3. Base matching templates for characterizing P-simple points on $(1, 2)$ pictures on grid $\mathcal{T}$.

Notice that the matching templates assigned to $(\bar{k}, k)$ pictures can be generated by recoloring of the templates associated with the $(k, \bar{k})$ case (see the pair of figures (Fig. 2, Fig. 3) and (Fig. 4, Fig. 5)). The recoloring means that each black element is changed to white and vice versa. (All elements depicted in gray remain unchanged.) It is due to the following duality theorem:

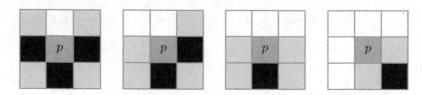

Fig. 4. Base matching templates for characterizing P-simple points on $(2,1)$ pictures on grid $\mathcal{S}$.

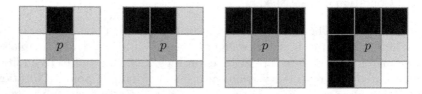

Fig. 5. Base matching templates for characterizing P-simple points on $(1,2)$ pictures on grid $\mathcal{S}$.

Theorem 5 [10]. *A point $p \in \mathcal{B}$ is simple in picture $(\mathcal{V}, k, \bar{k}, \mathcal{B})$ $(\mathcal{V} = \mathcal{T}, \mathcal{S}, \mathcal{H};$ $(k, \bar{k}) = (1,2), (2,1))$ if and only if p is simple in picture $(\mathcal{V}, \bar{k}, k, (\mathcal{V} \setminus \mathcal{B}) \cup \{p\})$.*

We can state the following theorem as an easy consequence of Theorems 4 and 5:

Theorem 6. $\mathcal{P} \subset \mathcal{B}$ *is a P-simple set in picture $(\mathcal{V}, k, \bar{k}, \mathcal{B})$ $(\mathcal{V} = \mathcal{T}, \mathcal{S}, \mathcal{H}; (k, \bar{k}) = (1,2), (2,1))$ if and only if it is a P-simple set in picture $(\mathcal{V}, \bar{k}, k, (\mathcal{V} \setminus \mathcal{B}) \cup \mathcal{P})$.*

Notice that the simplicity of the central point p in Figs. 2, 3, 4, 5 and 6 does not depend on the points depicted in gray. Hence those figures provide easily visualized characterizations of simple points, where each element shown in black matches a black point, each white element matches a white point, and gray elements match either black or white points.

Deletion of a subset of P-simple points and Theorem 2 provide two kinds of sufficient conditions for topology-preserving reductions. The following theorem shows the relationship between these conditions:

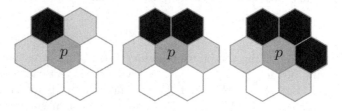

Fig. 6. Base matching templates for characterizing P-simple points on $(2,1) = (1,2)$ pictures on grid $\mathcal{H}$.

Theorem 7. *Let $\mathcal{P} \subset \mathcal{B}$ in picture $(\mathcal{V}, k, \bar{k}, \mathcal{B})$ $(\mathcal{V} = \mathcal{T}, \mathcal{S}, \mathcal{H}; (k, \bar{k}) = (1, 2), (2, 1))$ be P-simple. Then the reduction that deletes all elements of $\mathcal{P}$ satisfies all conditions of Theorem 2 (i.e., our unified sufficient conditions for topology-preserving reductions on the regular 2D grids).*

Proof. Let R be a reduction that removes $\mathcal{P}$. We show that R satisfies the conditions of Theorem 2, which implies the statement to prove:

- By Definition 1, each point in $\mathcal{P}$ is simple, therefore Condition 1 of Theorem 2 holds.
- Let $p, q \in \mathcal{P}$ two $\bar{k}$-adjacent points. As $\{q\} \subset \mathcal{P}$, by Definition 1, p is simple in $(\mathcal{V}, k, \bar{k}, \mathcal{B} \setminus \{q\})$, which means that Condition 2 of Theorem 2 is also fulfilled.
- Let us assume that the set $\mathcal{Q} \cup \{p\} \subseteq \mathcal{P}$ forms a small object that is completely deleted by reduction R. Then p is an isolated (non-simple black) point in picture $(\mathcal{V}, k, \bar{k}, \mathcal{B} \setminus \mathcal{Q})$. Thus we arrived at a contradiction. Hence, Condition 3 of Theorem 2 holds. $\square$

For an easier understanding of Theorem 4, let us see some examples in Figs. 7, 8 and 9. One can easily check that all of the gray points match one of the base configurations or its proper rotation in Figs. 2, 3, 4, 5 and 6, therefore they form P-simple sets. However, note that if we would expand those sets with the points denoted by q, then they would not be P-simple anymore, because q does not match any of the mentioned configurations in Figs. 4, 5 and 6.

For the sake of completeness, note that Kong also introduced the concept of a hereditary simple set [13]: A set of black points $\mathcal{P} \subset \mathcal{B}$ is said to be *hereditary simple* in $\mathcal{B}$ if all subsets of $\mathcal{P}$ (including $\mathcal{P}$ itself) are simple sets in $\mathcal{B}$. Hereditary simple sets are related to P-simple sets: A set of black points $\mathcal{P} \subset \mathcal{B}$ is hereditary simple in $\mathcal{B}$ if and only if every point of $\mathcal{P}$ is P-simple in $\mathcal{B}$ [14]. Hence Theorem 4 is also valid for elements of hereditary simple sets.

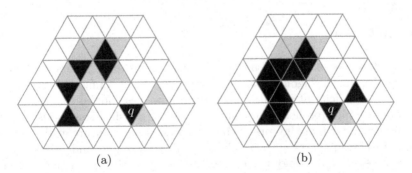

(a)   (b)

Fig. 7. A $(2, 1)$-picture (a) and a $(1, 2)$-picture (b) on grid $\mathcal{T}$. The set of black points $\mathcal{P}$ shown in gray form a P-simple set, but the deletion of set $\mathcal{P} \cup \{q\}$ would completely delete an object, hence $\mathcal{P} \cup \{q\}$ is not P-simple.

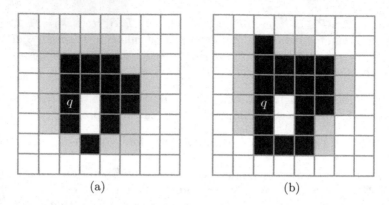

(a)   (b)

Fig. 8. A $(2,1)$-picture (a) and a $(1,2)$-picture (b) on grid $\mathcal{S}$. The set of black points $\mathcal{P}$ shown in gray form a P-simple set, but the deletion of set $\mathcal{P} \cup \{q\}$ would merge two white components, hence $\mathcal{P} \cup \{q\}$ is not P-simple.

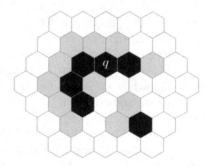

Fig. 9. A $(2,1) = (1,2)$-picture on grid $\mathcal{H}$. The set of black points $\mathcal{P}$ shown in gray form a P-simple set, but the removal of set $\mathcal{P} \cup \{q\}$ would split an object, hence $\mathcal{P} \cup \{q\}$ is not P-simple.

4   Conclusions

In this work we presented the characterization of P-simple points in the triangular, square, and hexagonal grids, and we gave some configurations to easily check the P-simplicity of a point. Furthermore, we also showed that the removal of P-simple points (or hereditary simple points) satisfies some existing sufficient conditions for topology-preserving reductions. As a future goal, we intend to combine P-simple sets with parallel thinning strategies and geometric constraints to generate topology-preserving thinning algorithms on the regular 2D grids.

Acknowledgements. This work was supported by the grant OTKA K112998 of the National Scientific Research Fund.

References

1. Bell, S.B.M., Holroyd, F.C., Mason, D.C.: A digital geometry for hexagonal pixels. Image Vis. Comput. **7**, 194–204 (1989)
2. Bertrand, G.: On P-simple points. Compte Rendu de l'Académie des Sciences de Paris, Série Math. **I**(321), 1077–1084 (1995)
3. Bertrand, G., Couprie, M.: Two-dimensional parallel thinning algorithms based on critical kernels. J. Math. Imaging Vis. **31**, 35–56 (2008)
4. Bertrand, G., Couprie, M.: On parallel thinning algorithms: minimal non-simple Sets, P-simple points and critical kernels. J. Math. Imaging Vis. **35**, 23–35 (2009)
5. Brimkov, V.E., Barneva, R.P.: Analytical honeycomb geometry for raster and volume graphics. Comput. J. **48**, 180–199 (2005)
6. Gaspar, F.J., Gracia, J.L., Lisbona, F.J., Rodrigo, C.: On geometric multigrid methods for triangular grids using three-coarsening strategy. Appl. Numer. Math. **59**, 1693–1708 (2009)
7. Hall, R.W.: Parallel connectivity-preserving thinning algorithms. In: Kong, T.Y., Rosenfeld, A. (eds.) Topological Algorithms for Digital Image Processing, pp. 145–179. Elsevier Science, Amsterdam (1996)
8. Kardos, P., Palágyi, K.: On topology preservation of mixed operators in triangular, square, and hexagonal grids. Discret. Appl. Math., in press. doi:10.1016/j.dam.2015.10.033
9. Kardos, P., Palágyi, K.: Topology preservation on the triangular grid. Ann. Math. Artif. Intell. **75**, 53–68 (2015)
10. Kardos, P., Palágyi, K.: On topology preservation in triangular, square, and hexagonal grids. In: Proceedings of the 8th International Symposium on Image and Signal Processing and Analysis, ISPA 2013, pp. 782–787 (2013)
11. Kardos, P., Palágyi, K.: Topology-preserving hexagonal thinning. Int. J. Comput. Math. **90**, 1607–1617 (2013)
12. Kardos, P., Palágyi, K.: On topology preservation for triangular thinning algorithms. In: Barneva, R.P., Brimkov, V.E., Aggarwal, J.K. (eds.) IWCIA 2012. LNCS, vol. 7655, pp. 128–142. Springer, Heidelberg (2012). doi:10.1007/978-3-642-34732-0_10
13. Kong, T.Y.: On topology preservation in 2-D and 3-D thinning. Int. J. Pattern Recog. Artif. Intell. **9**, 813–844 (1995)
14. Kong, T.Y., Gau, C.-J.: Minimal non-simple sets in 4-dimensional binary images with (8,80)-adjacency. In: Klette, R., Žunić, J. (eds.) IWCIA 2004. LNCS, vol. 3322, pp. 318–333. Springer, Heidelberg (2004). doi:10.1007/978-3-540-30503-3_24
15. Kong, T.Y., Rosenfeld, A.: Digital topology: introduction and survey. Comput. Vis. Graph. Image Process. **48**, 357–393 (1989)
16. Marchand-Maillet, S., Sharaiha, Y.M.: Binary Digital Image Processing - A Discrete Approach. Academic Press, Cambridge (2000)
17. Middleton, L., Sivaswamy, J.: Hexagonal Image Processing: A Practical Approach. Advances Pattern Recognition. Springer, Heidelberg (2005)
18. Nagy, B.: Characterization of digital circles in triangular grid. Pattern Recogn. Lett. **25**, 1231–1242 (2004)
19. Nagy, B., Mir-Mohammad-Sadeghi, H.: Digital disks by weighted distances in the triangular grid. In: Normand, N., Guédon, J., Autrusseau, F. (eds.) DGCI 2016. LNCS, vol. 9647, pp. 385–397. Springer, Heidelberg (2016). doi:10.1007/978-3-319-32360-2_30

20. Ronse, C.: Minimal test patterns for connectivity preservation in parallel thinning algorithms for binary digital images. Discret. Appl. Math. **21**, 67–79 (1988)
21. Sarkar, A., Biswas, A., Mondal, S., Dutt, M.: Finding shortest triangular path in a digital object. In: Normand, N., Guédon, J., Autrusseau, F. (eds.) DGCI 2016. LNCS, vol. 9647, pp. 206–218. Springer, Heidelberg (2016). doi:10.1007/978-3-319-32360-2_16
22. Serra, J.: Image Analysis and Mathematical Morphology. Academic Press, Cambridge (1982)
23. Wuthrich, C., Stucki, P.: An algorithm comparison between square- and hexagonal-based grids. Graph. Models Image Process. **53**, 324–339 (1991)

Concepts of Binary Morphological Operations Dilation and Erosion on the Triangular Grid

Mohsen Abdalla and Benedek Nagy[(✉)]

Faculty of Arts and Sciences, Department of Mathematics,
Eastern Mediterranean University, Mersin-10, Famagusta, North Cyprus, Turkey
moxxom88@gmail.com, nbenedek.inf@gmail.com

Abstract. In this paper, basic concepts of digital binary morphological operations, i.e., dilation and erosion are investigated on a triangular grid. Every triangle pixel is addressed by a unique coordinate triplet with sum zero (even pixels) or one (odd pixels). Even and odd pixels have different orientations. The triangular grid is not a lattice, that is, not every translation with a grid vector maps the grid to itself. Therefore, to extend the morphological operations to the triangular grid is not straightforward. We introduce three types of definition for both of dilation and erosion. Various examples and properties of the considered dilation and erosion are analyzed on the triangular grid.

Keywords: Digital image processing · Mathematical morphology · Binary morphology · Dilation · Erosion · Triangular grid · Non-traditional grids

1 Introduction

The prior work on Mathematical Morphology has been done by Minkowski [12], Dineen [2], Kirsch [8] and others. Mathematical morphology has been formalized since the 1960s. Matheron [11] and Serra [20], at the Centre de Morphologies, found out geometry and edge properties of ores (see also [3, 5, 22]). Notice that already in the 1960s the hexagonal lattice was involved to this field by Golay [4]. There are several places where digital image processing and mathematical morphology are applied, such as various medical fields, remote sensing, robot vision, video processing, color processing, image sharpening, and restoration. Mathematical morphology provides a quantitative description of geometric structure and shape of the objects. In computer imaging and in computer vision, mathematical morphology is applied for extracting image components that are useful in description and representation of the shapes of the objects [21].

In the 2d square grid, a binary image is a finite set of elements of Z^2. The elements of the grid are called pixels and are addressed with two coordinate values (x, y). Each pixel has value zero or one. Pixels with value one form the foreground and they are usually represented by black color; while value zero (background) pixels are represented by white color. The basic binary morphological operations, the dilation and the erosion have been defined on the square grid by using two images. One of them is the active (or current) image; it is the image of the object we are interested in. The other one is called structuring element, and the active image is examined/modified by its

R.P. Barneva et al. (Eds.): CompIMAGE 2016, LNCS 10149, pp. 89–104, 2017.
DOI: 10.1007/978-3-319-54609-4_7

help [21]. The structuring element has its size and shape, and it is used to redefine the shape of the original image.

In digital geometry, pixels (sometimes they are also referred as points) are addressed by integer coordinate values. The square grid is often used in various applications, since the well-known Cartesian coordinate system (that is a part of standard elementary mathematics) describes it. To use another grid in image processing and/or in computer graphics one needs a good, manageable coordinate system. Image processing on the hexagonal lattice is also investigated due to some of its pleasant properties, e.g., there is only one type of usual neighbor relation among pixels (they are also called hexels in this case), and the grid has better symmetric properties than the square grid. The pixels of the hexagonal grid can be addressed with two integers [10], but there is a more elegant solution using three coordinate values and obtaining a symmetric description: addressing hexagons by zero-sum triplets [6]. Similarly, the triangular grid is described by three coordinates [13–16, 18] as it is shown in Fig. 1. There are two orientations of the used triangle pixels, the sum of their coordinate values is zero and one, and they are called even and odd points (pixels), respectively. There are three types of usual neighborhood relations on the triangular grid [1], for formal description see Sect. 3.

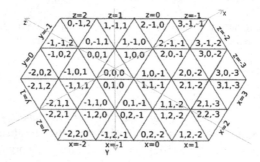

Fig. 1. Symmetric coordinate frame for triangle pixels on a triangular grid.

In digital binary image processing, apart from the traditional square grid, non-traditional grids, such as the triangular grid, are also used [1, 7, 9, 17, 18, 20]. The hexagonal and the triangular grids have more symmetry axes than the square grid has. In addition, rotations with a smaller angle (60°) can transform these non-traditional grids to themselves than the angle (90°) needed for a similar transformation on the square grid. Due to these better symmetric properties, the triangular grid could over-perform the traditional square grid in various applications. Another reason to support the triangular grid is that the various types of neighbor relations give more freedom in applications. As we have already mentioned, morphological operations are studied and used on the square and on the hexagonal grid, however, up to our knowledge, there is a lack of the theory on the triangular grid. Here we will study the concepts of dilation and erosion extrapolating them to the triangular grid. We use the general approach given, e.g., in [21]. Morphological binary image processing deals with two-dimensional pictures (arrays) containing zeros and ones representing pixels with values either white or black. Since the square grid is a lattice, there are various equivalent expressions for dilation and erosion, the roles of pixels and grid vectors can easily be interchanged, one could add the vectors (addresses) of any two points resulting a point in the grid. Opposite to this, in the triangular grid, the situation is different; this grid is not a lattice. In this paper, we will give various types of approaches, different definitions for both dilation and erosion. In next section, we recall the basic definitions used in the square grid. In Sect. 3 we will give a formal description of the triangular grid. In Sect. 4 we will provide the definitions and results on dilation in the triangular grid. In Sect. 5 we give the essential definitions and results on erosion. In Sect. 6 various properties of dilations and erosions are shown. Finally, conclusions close the paper.

2 Preliminaries

In this section, we recall the core definitions of binary morphology on traditional grids (using the terminology of [3, 5, 21]). These definitions are apt for lattices. An *image* was delineated in Euclidean space as a set of a correspondent vectors. Let E^N be N-dimensional Euclidean space (it could be discrete Z^N or continuous R^N) where E^N is the set of all points $p = (x_1, \ldots, x_N)$ in E^N. Since we are interested in digital images, i.e., images in the discrete space, we may consider in most cases that our space is Z^N. A binary image A is a subset of the binary space E^N, where the value of each point p of E^N is either black or white: p is black if and only if $p \in A$, otherwise p is white.

Definition 1. Let $A \subset E^2$ be a binary image. If $A = \emptyset$, then A is empty (sometimes is also called null). Otherwise, a pixel of the image $a \in A$ is addressed by a vector (x, y).

Definition 2. Let $A \subset E^2$ be a binary image, the complement of A is also a binary image, it is defined by $A^c = \{p : p \notin A\}$, i.e., it is obtained by interchanging the roles of black and white pixels.

Definition 3. Let $A \subset E^2$, the reflection of an image A is denoted by $\hat{A}$ and it is defined by $\hat{A} = \{p : p = -a, \forall a \in A\}$.

Note here the important fact that $\hat{A} \subset E^2$ on the square grid.

Definition 4. The union of two binary images $A, B \subset E^2$ is a binary image such that a pixel is black if it is black in A or in B, formally, $A \cup B = \{p : p \in A \text{ or } p \in B\}$.

Definition 5. The intersection of two binary images $A, B \subset E^2$ is a binary image containing those pixels that are black both in A and in B, i.e., $A \cap B = \{p : p \in A, p \in B\}$.

Definition 6. Let $A \subset E^N$, $b \in E^N$, then the translation of A by b, denoted by $(A)_b$ is defined as $(A)_b = \{p \in E^N : p = a + b, \text{for some } a \in A\}$.

Definition 7. Let $A, B \subset E^N$, the binary dilation of A by structuring element B is denoted by $A \oplus B$, $A \oplus B = \{p : p = a + b, \exists a \in A, b \in B\}$. It can also be defined by $A \oplus B = \bigcup_{b \in B} A_b = \bigcup_{a \in A} B_a$.

Note here the similar and interchangeable role of the image A and the structuring element B on the square grid.

Definition 8. Let $A, B \subset E^N$, the binary erosion of A by structuring element B is denoted by $A \ominus B$. Formally, it is $A \ominus B = \{p : p + b \in A, \forall b \in B\}$, it can be written, equivalently, $A \ominus B = \bigcap_{b \in B} A_{-b}$.

3 Definitions and Notation on Triangular Grid

The triangular grid, based on [13, 14, 16, 18], is described by a symmetric coordinate system addressing every pixel by a coordinate triplet. The origin, as a pixel, is addressed by $(0, 0, 0)$. The coordinate axes are lines cutting this pixel to halves (see Fig. 1). They are directed such that their angles are 120°. Every pixel, as a triangle, has three closest neighbor pixels sharing one of the sides of the triangle. Notice that, although each pixel is a triangle, there are two different orientations of them: $\triangle, \nabla$. A pixel and its closest neighbors have opposite orientations. From a pixel having coordinates (x, y, z) with $x + y + z = 0$, its closest neighbor pixels can be reached by a step to the direction of one of the coordinate axes. Consequently, the respective coordinate value is increased by one: the three neighbors are addressed by $(x + 1, y, z)$, $(x, y + 1, z)$, and $(x, y, z + 1)$, respectively. For a pixel (x, y, z) with $x + y + z = 1$, its closest neighbor pixels can be reached by a step to the direction opposite to one of the coordinate axes, and thus, the three neighbors are addressed by $(x - 1, y, z)$, $(x, y - 1, z)$, and $(x, y, z - 1)$. In this way, all pixels of orientation $\triangle$ are addressed by triplets with zero sum (they are called even pixels), while the pixels of orientation ∇ have triplets where the sum of the coordinate values is one (they are the odd pixels). There are three types of neighborhood relations [1]: two triangles are 1-neighbours if they share a side, i.e., an edge of the grid. They are exactly the closest neighbors. Two triangles are strict 2-neighbours if they have a common 1-neighbor triangle. Two pixels are strict 3-neighbours if they share exactly one point on their boundaries (vertex of the grid) but they are not 2-neighbors. See also, Fig. 2, where these three types of neighbors are shown for an even pixel. Each pixel has three 1-neighbours, nine 2-neighbours (including 1-neighbours and six strict 2-neighbours) and twelve 3-neighbours (the nine 2-neighbours and three strict 3-neighbours). Actually, the coordinate triplets of two strict k-neighbor ($k = 1, 2, 3$) pixels mismatch exactly in k places, and the difference in each mismatch position is ± 1. Triplets of two k-neighbor

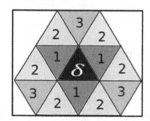

Fig. 2. Various neighborhood relations on the triangular grid of a pixel δ.

pixels could mismatch at most in k places and the difference in each mismatch position is ± 1. As an example, consider the pixel having triplet $(1, 1, -1)$. As one can also observe in Fig. 1, its three 1-neighbors are addressed by the coordinate triplets $(0, 1, -1)$, $(1, 0, -1)$ and $(1, 1, -2)$. The strict 2-neighbors are $(0, 1, 0)$, $(1, 0, 0)$, $(2, 0, -1)$, $(2, 1, -2)$, $(1, 2, -2)$ and $(0, 2, -1)$. Further, $(0, 0, 0)$, $(2, 0, -2)$ and $(0, 2, -2)$ are addressing the strict 3-neighbors of the pixel $(1, 1, -1)$.

We will use the notation G for the triangular grid, we also use the notation G^Δ for the set of even pixels of G ($G^\Delta \subset G$), and G^∇ for the set of odd pixels of G($G^\nabla \subset G$). We should notice that $\emptyset = G^\Delta \cap G^\nabla$ and $G = G^\Delta \cup G^\nabla$. Notice that the triangular grid can also be seen as a special subset of the cubic lattice, i.e., $G \subset Z^3$ [14, 15]. We may also call vectors the elements of Z^3, and specially, even and odd vectors for those that are also elements of G.

Translations on Triangular Grid

First, we recall an important observation from [16] reflecting the fact that the triangular grid is not a lattice.

Proposition 1. Translation with vector $v(x, y, z)$ maps the grid to itself if and only if $x + y + z = 0$, i.e., the coordinate sum of the vector equals to zero.

One can easily establish the following simple, but important facts.

Proposition 2. Addition of two points $p_1(x_1, y_1, z_1)$, $p_2(x_2, y_2, z_2)$ in G results $p_1(x_1, y_1, z_1) + p_2(x_2, y_2, z_2) = p(x_1 + x_2, y_1 + y_2, z_1 + z_2)$. The resulting point is in the triangular grid if

1. $p_1, p_2 \in G^\Delta$, then $p_1 + p_2 \in G^\Delta$.
2. $p_1 \in G^\nabla, p_2 \in G^\Delta$, then $p_1 + p_2 \in G^\nabla$.
3. $p_1 \in G^\Delta, p_2 \in G^\nabla$, then $p_1 + p_2 \in G^\nabla$.

Proposition 3.

1. Let $p_1 \in G^\Delta$, then $-(p_1) \in G^\Delta$.
2. Let $p_2 \in G^\nabla$, then $-(p_2) \notin G$, but if it is allowed to use it, $-(-(p_2)) = p_2$.
3. If $p_1 \in G^\Delta, p_3 \in G^\Delta$, then $-(p_1) + p_3 = p_3 + (-(p_1)) = p \in G^\Delta$.
4. If $p_2 \in G^\nabla, p_1 \in G^\Delta$, then $-(p_2) + p_1 = p_1 + (-(p_2)) = -(p)$, with $p \in G^\nabla$.
5. If $p_1 \in G^\Delta, p_2 \in G^\nabla$, then $-(p_1) + p_2 = p_2 + (-(p_1)) = p$, where $p \in G^\nabla$.
6. If $p_2, p_4 \in G^\nabla$, then $-(p_2) + p_4 = p_4 + (-(p_2)) = p$, where $p \in G^\Delta$.

Remark 1. If $p_1 \in G^\nabla, p_2 \in G^\nabla$, then $p(x, y, z) = p_1(x_1 + y_1 + z_1) + p_2(x_2 + y_2 + z_2)$ $\notin G$ since the sum of coordinate values of p is equal to 2 ($x + y + z = 2$). This operation can be allowed but keeping in mind that the resulted point is not in the grid.

Example 1. Let $p_1 = (-1, 1, 1)$, $p_2 = (0, 1, 0)$, then $p_1 + p_2 = (-1, 2, 1) \notin G$.

The result of a translation of an odd point by another odd point is not in the grid. This type of translations does not map the grid into itself: this grid is not a lattice and therefore, the extensions of the morphological operations to the triangular grid are not straightforward. In next section, we recommend some possible definitions for dilation and erosion solving various ways the above problem.

4   Concepts of Dilation on the Triangular Grid

Since the grid is not a lattice, the types of points of the structuring element play importance. By the first and simplest solution, it is allowed to use only such transformations of the image which gives the resulted points inside the grid: the image points can be translated only by even point(s). See the case of *strict* dilation. The second option is when the translation is allowed by odd points which may produce point(s) outside of the grid (see *weak* and *strong* dilations which allow that option also).

Definition 9. Let $A \subset G, B \subset G^{\Delta}$, then the *strict dilation* of A by set B is defined as $A \oplus_{\Delta} B = \{p \in G : p(x,y,z), x = x_1 + x_2, y = y_1 + y_2, z = z_1 + z_2, \exists\, p_1(x_1,y_1,z_1) \in A, p_2(x_2,y_2,z_2) \in B\}$. The notation refers for the fact that the structuring element must contain only even vectors.

Example   2. Let $A = \{(0,2,-1), (0,2,-2), (0,3,-2), (-1,2,-1), (-1,3,-2), (-1,3,-1)\}$ be a binary image (its points have the value equal to one, their color is black). Now, let $B = \{(0,-1,1), (1,-1,0)\}$, then $A \oplus_{\Delta} B = \{(-1,1,0), (0,2,-1), (1,1,-1), (1,1,-2), (0,2,-2), (0,2,-1), (-1,2,-1), (0,1,0), (-1,2,0), (0,1,-1), (1,2,-2)\}$. See Fig. 3.

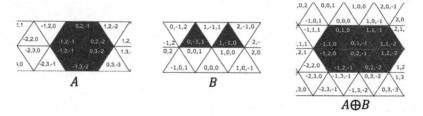

Fig. 3. Strict dilation (Example 2) with the structuring element $B \subset G^{\Delta}$.

The idea of strict dilation is a restriction on the structuring element B: to force the resulted pixels to be in G, the condition $B \subset G^{\Delta}$ is applied.

Definition 10. Let $A, B \subset G$, then the *weak dilation* of A by the set B is defined by $A \oplus_w B = (A \oplus B) \cap G$ where $A \oplus B = \bigcup_{p \in A} B_p = \bigcup_{p \in B} A_p$ as vector addition of 3-dimensional vectors.

In weak dilation we keep only those resulted vectors that belong to the grid G.

Example 3. Let $A = \{(2,-1,0), (1,-1,0), (1,0,0)\}$, $B = \{(-2,1,1), (-1,1,1), (-1,1,0)\}$. Then $A \oplus B = \{(0,1,1), (-1,1,1), (1,0,0), (0,0,1), (1,0,1), (-1,0,1), (0,0,0), (0,1,0)\}$. We have here points, e.g., $(0,1,1)$ such that the sum of its coordinate values equals to 2. These points are removed and do not appear in $A \oplus_w B$, thus the result is $A \oplus_w B = \{(-1, 1, 1), (1, 0, 0), (0, 0, 1), (-1, 0, 1), (0, 0, 0), (0, 1, 0)\}$ (Fig. 4).

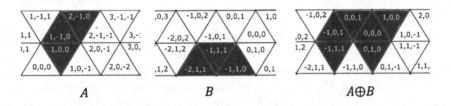

Fig. 4. Example for weak dilation by resulting only points belong to G and deleting the other points that do not belong to G, i.e., $(1, 0, 1)$ and $(0, 1, 1)$.

Notice that in weak dilation some points could be lost, and thus, information can also be lost. Not to lose information because of this phenomenon, we introduce:

Definition 11. Let $A, B \subset Z^3$, then the *strong dilation* of A by the set B is defined by $A \oplus_s B = A \oplus B$, keeping also the resulted points outside of G, but displaying only points which belong to grid G: $p \in (A \oplus B) \cap G$.

Example 4. Let $B = \{(0, 0, 0)(0, 1, 0)\}$, $A = \{(6, 0, -6), (7, 0, -6), (7, 0, -7), (7, 1, -7), (6, 1, -7)\}$. Then, $A \oplus_s B = \{(6, 0, -6), (7, 0, -6), (7, 0, -7), (7, 1, -7), (6, 1, -7), (6, 1, -6), (7, 1, -6), (7, 2, -7), (6, 2, -7)\}$ (Fig. 5).

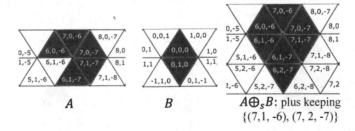

$A \oplus_s B$: plus keeping
$\{(7, 1, -6), (7, 2, -7)\}$

Fig. 5. An example of strong dilation by displaying resulted points $p \in G$ and keep the other points $p \notin G$ in the result of $A \oplus_s B$.

5 Concepts of Erosion on the Triangular Grid

Erosion is an operation of mathematical morphology that combines two sets by using vector subtraction. The original aim of the erosion is to shrink the active image (or simply image) by using a structuring element. Thus, it is a kind of translation that shifts the structuring element totally inside the active image. Originally, there are two types of definitions. The Minkowski decomposition is defined by $A \ominus B = \{p : B_p \subset A\} = \{p \in G : a + b \in A, \forall b \in B\}$; the other definition is $A \ominus B = \bigcap_{b \in B} A_{-b}$, this latter one is more general in the sense that it gives all possible solutions or possible positions that make B inside A. We should notice that working on the triangular grid, some of these positions might not be inside the grid (see Sect. 3). Now we give various definitions of erosion according to possible types of transformations.

Remark 2. For any point (vector), $p(x, y, z) \in G$ we are using the notation $-p(-x, -y, -z)$ for its *inverse*, as a kind of 3-dimensional reflection. However, we should notice that it is not a rotation by 180 degrees, as the analogous transformation was on the square grid. Moreover, it is not even a transformation of the triangular grid, since it maps odd points to triplets with sum -1, and they are clearly not points in G. Even the inverse of an odd element is not in G, we could use it, e.g., in a strong dilation to have some effect. We note here that in the triangular grid rotation having center in the Origin (at the meeting point of the axes) can be used only if the degree is a multiplier of $120°$ (and $180°$ is not like that). There are also other types of mirroring and rotations on the triangular grid, e.g., with the center in the corner of a pixel [16].

Definition 12. Let $A, B \subset G$ where $B \subset G^\triangle$, then the *strict erosion* of the image A by structuring element B is defined by $A \ominus_\triangle B = \{p : p + b \in A, \forall b \in B\}$.

In strict erosion it is guaranteed that each vector p obtained in this way is automatically belonging to the grid G, i.e. $p \in G$. To guarantee this fact a restriction is used for the structuring element, namely its inverse also belonging to the grid.

Remark 3. In this case, we can use the alternative definition, the necessary translations are defined, $A \ominus_\triangle B = \bigcap_{b \in B} A_{-b}$: the resulting image is in G.

Example 5. Let $B = \{(0, 0, 0), (0, 1, -1)\}$ and let $A = \{(-1, 0, 1), (-1, 1, 1), (-2, 1, 1), (-1, 1, 0)\}$. Then $A \ominus_\triangle B = \{(-1, 0, 1)\}$ (Fig. 6).

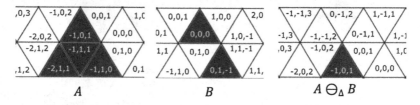

Fig. 6. Example 5: strict erosion with property $A \ominus_\triangle B \subset A$.

Definition 13. Let $A, B \subset G$, then the *weak erosion* of the image A with structuring element B is defined by $A \ominus_w B = (A \ominus B) \cap G$, where $A \ominus B = \{p : p + b \in A, \forall b \in B\}$.

We keep and display only points that belong to G and satisfy $p + b \in A$, for $\forall b \in B$ and delete the other points of $A \ominus B$ that do not belong to G.

Remark 4. There could be vectors p not belonging to G, but satisfying $p + b \in A$. If the result of $A \ominus B$ contains vectors with the sum -1, then by weak erosion, these points are lost and there might be nothing to display (see Example 6).

Example 6. Let $A = \{(-3, 1, 2), (2, 0, -2), (3, 0, -3)\}$, $B = \{(1, 0, 0)\}$. Then, $A \ominus B = \{(-4, 1, 2), (1, 0, -2), (2, 0, -3)\}$ and $A \ominus_w B = \emptyset$.

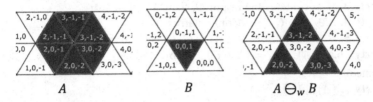

$$A \qquad\qquad B \qquad\qquad A \ominus_w B$$

Fig. 7. Example 7 is shown: a weak erosion.

Example 7. Let $A = \{(3, -1, -1), (2, -1, -1), (2, 0, -1), (2, 0, -2), (3, 0, -2), (3, -1, -2)\}$, $B = \{(0, 0, 1)\}$. Then, $A \ominus B = \{(3, -1, -2), \mathbf{(2, -1, -2)}, (2, 0, -2), \mathbf{(2, 0, -3)}, (3, 0, -3), \mathbf{(3, -1, -3)}\}$ and thus, $A \ominus_w B = \{(3, -1, -2), (2, 0, -2), (3, 0, -3)\}$. (The shaded elements of $A \ominus B$ have coordinate sum -1, they are lost.) (Fig. 7)

Not to lose some of the information one can use the strong erosion:

Definition 14. Let A, $B \subset Z^3$ then the *strong erosion* is defined by $A \ominus_s B = \{p : p + b \in A, \forall b \in B\}$. It can be written as $A \ominus_s B = \{p \in G : p + b \in A, \forall b \in B\} \cup \{p \notin G : p + b \in A, \forall b \in B\}$ by having the pixels that can be displayed (points which belong to G and satisfy $p + b \in A$ for every $b \in B$) and keeping also all the other points which do not belong to G, but satisfy $p + b \in A$ for every $b \in B$.

Notice that in strong erosion, similarly to the strong dilation, it is not required that A and B are in G, instead all integer triplets are allowed, but only those triplets can be displayed that are element of G.

Example 8. In Example 7 we have seen that $A \ominus B = \{(3, -1, -2), \mathbf{(2, -1, -2)}, (2, 0, -2), \mathbf{(2, 0, -3)}, (3, 0, -3), \mathbf{(3, -1, -3)}\}$. Here we keep vectors outside of the grid, e.g., with sum of coordinate values equal to -1: $(2, 0, -3), (3, -1, -3), (2, -1, -2)$, but we can display only those resulting points which are in the grid ($p \in G$).

6 Properties of Dilation and Erosion

In this section we list the basic properties of dilation and erosion. These properties are well known on the Euclidean space [3, 5, 20–22].

Let $A, B, C, D \subset E$ where E is the Euclidean space, assume that all of these four sets can play the role of the active (e.g., input) image and also the role of the structuring element. (This is a usual and valid assumption on lattices.)

Let O be the origin of E and let $p, t \in E$.

Dilation Properties

We should know the following facts: for any arbitrary subset A of E, and for the empty set $\emptyset$, where E is the whole Euclidean space, $A \oplus \emptyset = \emptyset$, $E \oplus A = A \oplus E = E$. Moreover, if $A \oplus B = \emptyset$, then at least one of A or B is the empty set $\emptyset$.

Property 1.a $A \oplus \{O\} = A$. **1.b** $A \oplus \{O\} = \{O\} \oplus A$. (Unit element)

Property 2.a $A \oplus \{p\} = \{p\} \oplus A$. **2.b** $A \oplus \{p\} = A_p$.

Property 3. If D contains the origin O, then $A \subseteq A \oplus D$.

Property 4. $A \oplus C = C \oplus A$. (Commutativity)

Property 5. $B \oplus (A \oplus C) = (B \oplus A) \oplus C$. (Associativity)

Property 6.a $(A)_p \oplus C = (A \oplus C)_p$. **6.b** $A \oplus (C)_p = (A \oplus C)_p$. (Translation invariance)

Property 7. $(A)_p \oplus (C)_{-p} = A \oplus C$.

Property 8. If $A \subseteq B$, then $A \oplus C \subseteq B \oplus C$. (Increasing property, monotonicity)

Property 9. $(A \cap B) \oplus C \subseteq (A \oplus C) \cap (B \oplus C)$.

Property 10. $(A \cup B) \oplus C = (A \oplus C) \cup (B \oplus C)$. (Distributivity over union of images)

Property 11. $A \oplus (C \cup D) = (A \oplus C) \cup (A \oplus D)$.

Property 12. $A \oplus (C \cap D) \subseteq (A \oplus C) \cap (A \oplus D)$.

Erosion Properties

We should know also that for an arbitrary subset A of E, $A \ominus \emptyset = E$, $\emptyset \ominus A = \emptyset$, $E \ominus A = E$. Let us see the other well-known facts involving erosion.

Property 1. If C contains the origin, then $A \ominus C \subseteq A$.

Property 2. $A \ominus C \neq C \ominus A$. (Not commutative)

Property 3. $(A)_t \ominus C = (A \ominus C)_t$. (Translation invariance)

Property 4. If $A \subseteq B$, then $A \ominus C \subseteq B \ominus C$. (Increasing property, monotonicity)

Property 5. $B \ominus (A \oplus C) = (B \ominus A) \ominus C$.

Property 6. $B \oplus (A \ominus C) \subseteq (B \oplus A) \ominus C$.

Property 7. $(A \cap B) \ominus C = (A \ominus C) \cap (B \ominus C)$. (Distributivity over intersection)

Property 8. $(A \cup B) \ominus C \supseteq (A \ominus C) \cup (B \ominus C)$.

Property 9. $A \ominus (C \cup D) = (A \ominus C) \cap (A \ominus D)$.

Property 10. $A \ominus (C \cap D) \supseteq (A \ominus C) \cup (A \ominus D)$.

Property 11. $A \ominus (C)_t = (A \ominus C)_{-t}$.

6.1   Properties of Dilation and Erosion on the Triangular Grid

Tables 1 and 2 summarize which properties of the operations are inherited to the triangular grid using various definitions of the dilations and erosions, respectively. Let A and B be images, C and D be structuring elements based on various definitions of dilation and erosion on the triangular grid. Let p and t are arbitrary three-dimensional vectors. In some cases, to have well defined expressions, conditions are applied. As we can see, some of the properties hold only with some additional conditions.

Table 1. Summary of dilation properties according to their definitions.

Properties type of dilation	1	2	3	4	5	6	7	8	9	10	11	12
Strict	a. ✔ b. Δ	a. Δ b. p^Δ	✔	Δ	Δ	p^Δ	p^Δ	✔	✔	✔	✔	✔
Weak	✔	+	✔	+	+	+	p, +	+	+	+	+	+
Strong	✔	✔	✔	✔	✔	✔	✔	✔	✔	✔	✔	✔

- ✔ the property holds on the triangular grid without any additional conditions.
- Δ the strict dilation is defined only with the condition $A \subset G^\Delta$. With this condition all the expressions are defined and the property holds.
- p^Δ the translation is defined with the condition that p is even pixel. If it is so, then the property holds.
- + the property is fulfilled, but the result could be the empty set.
- p the property holds with the condition $p \in G^\Delta$.

Note that in strict dilation the definition already has a constraint on the structuring element: it can contain only even pixels (zero-sum coordinate triplets). While in case of weak dilation, any subset of G can be used as structural element; and the most general definition, the strong dilation allows any set of three-dimensional vectors (a constraint is applied only on the displayable part of the sets).

Since strict dilation is defined only with structural elements having no odd pixels, some of the expressions in some properties are well formed, i.e., defined only for images consisting only even pixels, e.g., properties 1.b, 2.a. We note here that constraining the work to G^Δ, i.e., to even pixels, one gets a structure that is equivalent to the hexagonal tessellation: the three-dimensional vectors with zero-sum describes exactly the hexagonal grid [6].

At weak dilations we allow to use only the elements of the grid G. Consequently, to allow using both p and $-p$, these translations must be by zero-sum vectors. These vectors transform the grid into itself, and thus this condition is applied at property 7, similarly to the strict case. The other properties of dilation hold without any additional condition for the weak dilations, however in most cases it may happen that an expression results the empty set, even in case when both the image and the structuring element contain some pixels. It could happen if odd pixels appear in the expressions. We show some examples.

In property 2 the empty set is obtained if and only if $A \subset G^\nabla, p \in G^\nabla$. For example, let $A = \{(-1, 0, 2), (0, 0, 1)\}$, $p = \{(1, 0, 0)\}$, then $A \oplus p = \{(0, 0, 2), (1, 0, 1)\} \not\subset G$, moreover, $(A \oplus p) \cap G = \emptyset$.

At property 4, the result is the empty set if and only if $A, C \subset G^\nabla$. Similar conditions can be established for the other properties. We show one additional example.

At property 9, the expression on the left hand side defines the empty set if and only if $A \cap B, C \subset G^\nabla$, i.e., there is no common even pixels of the images A and B and neither the structuring element C contains even pixels. For example, let $A = \{(2, -3, 2), (1, -3, 3)\}$, $B = \{(1, -3, 3)\}$, $C = \{(1, -1, 1)\}$, then $A \cap B = \{(1, -3, 3)\}$, and $(A \cap B) \oplus C = \{(2, -4, 4)\}$, while $(A \oplus C) = \{(3, -4, 3), (2, -4, 4)\}$, $(B \oplus C) =$

$\{(2, -4, 4)\}$, $(A \oplus C) \cap (B \oplus C) = \{(2, -4, 4)\} \not\subset G$. The property holds, but the weak dilations result the empty set in both sides of the expression.

We note here that all the listed properties are satisfied for the strong dilation, however, in the cases when the weak dilation results the empty set, usually, there is no displayable result with the strong dilation. Thus, even the result is not the empty set of some operations, the result cannot be directly seen as an image of the triangular grid.

In the following part we analyze the properties of the variously defined erosion.

Table 2. Summary of erosion properties according to their definitions.

Properties	1	2	3	4	5	6	7	8	9	10	11
Strict	✔	Δ	t^Δ	✔	Δ	Δ	✔	✔	✔	✔	t^Δ
Weak	✔	✔	t	+	No	No	+	+	+	+	t, +
Strong	✔	✔	✔	✔	✔	✔	✔	✔	✔	✔	✔

- ✔ the property holds on the triangular grid without any additional conditions.
- Δ the expressions are defined only with the condition $A \subset G^\Delta$. With this condition the property holds.
- t^Δ the translation is defined with the condition that t is a vector with zero sum. If it is so, then the property holds.
- + the property is fulfilled, but the result could be the empty set.
- t the expressions are defined and the property holds with the condition $t \in G^\Delta$.

Similarly to strict dilation, at strict erosion the structuring element (and so, any image that is playing such a role in an expression) cannot contain vectors outside of G^Δ; also translation vectors must have zero sum coordinates.

At the case of weak erosions, translations by odd vectors are also allowed, moreover the structuring element may also contain odd pixels (vectors with coordinate sum one). In this case, it may happen that some of the expressions results the empty set. We present an example for property 7. Let $A = \{(1, 0, -1), (1, 0, 0)\}$, $B = \{(1, 0, -1), (0, 0, 1)\}$, $C = \{(1, -1, 1)\}$, $A \cap B = \{(1, 0, -1)\}$, then $A \ominus C = \{(0, 1, -2), (0, 1, -1)\}$ and $A \ominus B = \{(0, 1, -2), (-1, 1, 0)\}$, $(A \ominus C) \cap (B \ominus C) = \{(0, 1, -2)\} = (A \cap B) \ominus C$, but $(0, 1, -2) \notin G$. Thus, the expressions of both sides of property 7 give the empty set for the given sets A, B and C.

By related examples, one can easily characterize the cases when the weak erosion results empty set. In a similar manner, at the case of strong erosion, it can be characterized when there is nothing to display, i.e., the resulted set is completely outside of the triangular grid G. The next subsection gives some related ideas.

6.2   A Note on Usable Vectors Outside of the Grid

In case of strong dilation, we may use vectors both for the image and the structuring element, which are outside of the grid, e.g., have coordinate sum -1.

Example 9. Let $A = \{(1, 2, -2), (0, 2, -2), (1, 2, -3), (0, 3, -2), (0, 3, -3), (1, 3, -3)\}$ and let $B = \{(0, 0, -1), (0, -1, 0)\}$, then $A \oplus_s B = \{(0, 3, -3), (1, 1, -2), (0, 2, -2), (0, 2, -3), (1, 2, -4), (1, 1, -3), (1, 2, -3), (0, 3, -4), (0, 1, -2), (1, 3, -4)\}$. The strong erosion $A \ominus_s B = \{(1, 3, -2), (0, 3, -2)\}$.

The result of the strong dilation and erosion may contain some vectors outside the grid, thus we may keep this information for using in some next operations, e.g., in a dilation of an odd point with this vector yield to an even point of the grid. Thus in strong operations it is allowed for both the image and the structuring element to contain vectors outside of the grid. However, only those vectors can be displayed as pixels that have sum 0 or 1.

6.3   Using the Traditional Neighbors at Structuring Elements

In this subsection, we briefly analyze how we can use the traditional neighborhood structure in morphological operations. It is well-known that in the most cases, at applications the neighborhood of the image is used for these operations, i.e., on the square grid, the structuring element contains the origin with its city-block or chessboard neighbors.

Based on the restricted definitions not allowing to generate or obtain any vectors outside of the triangular grid G, the structuring element can contain only even vectors. Observing the possible neighborhood relations, this implies that only strict 2-neighborhood can be applied for structural element, that is, the structural element $C_2 = \{(0, 0, 0), (1, -1, 0), (-1, 1, 0), (1, 0, -1), (-1, 0, 1), (0, 1, -1), (0, -1, 1)\}$ is used including the origin itself. We note that these vectors as neighbors correspond exactly the neighborhood in the hexagonal grid.

In the weak case, since we can use only vectors with coordinate sum zero or one (after each operation we lose the resulted vectors not having this property), we can use only strict 2-neighborhood again (set C_2). For odd points to define 1-neighbors or strict 3-neighbors one need vectors with coordinate sum -1. Actually, this is the point why the strong case could be the most useful. For 1-neighborhood, as structuring element, one can use the set $C_1 = \{(0, 0, 0), (1, 0, 0), (-1, 0, 0), (0, 1, 0), (0, -1, 0), (0, 0, 1), (0, 0, -1)\}$. Also, strict 2-neighborhood, C_2 is allowed, however we may use all the 2-neighbors including the 1-neighbors by the structuring element $C_1 \cup C_2$. The strong definitions also allow to use strict 3-neighborhood $C_3 = \{(0, 0, 0), (1, 1, -1), (1, -1, 1), (1, -1, -1), (-1, 1, 1), (-1, 1, -1), (-1, -1, 1)\}$; and 3-neighborhood $C_1 \cup C_2 \cup C_3$ as structuring element.

We present an example in Fig. 8. In Fig. 8(a) an image of a bone implant is shown. This implant was used to insert in a leg of a rabbit. Similar picture was also used in [19] for a binary tomography problem. Here, strict dilation and erosion of this figure are obtained by the structural element C_2 on the triangular grid as they are shown in Fig. 8 (b) and (c), respectively. Those parts of the strong dilations by the 1-, 2- and 3-neighborhood (i.e., by the sets C_1, $C_1 \cup C_2$ and $C_1 \cup C_2 \cup C_3$) that belong to the grid G are shown in Fig. 8(d), (f) and (h), respectively. Similarly, strong erosions obtained by the same sets of structuring elements are shown in Fig. 8(e), (g) and (i), respectively.

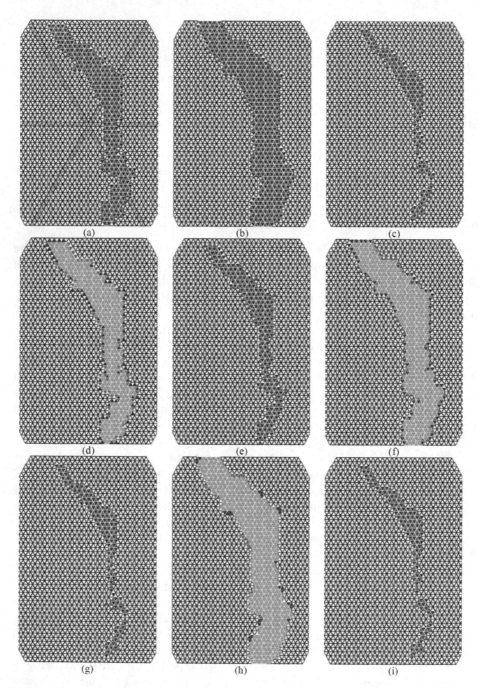

Fig. 8. (a) resampled image of a bone implant inserted in a leg of a rabbit, (b) strict dilation by structuring element C_2, (c) strict erosion by C_2. (d)–(i) the displayable result of strong dilations and erosions, by C_1 (d) and (e), by $C_1 \cup C_2$ (f) and (g), and by $C_1 \cup C_2 \cup C_3$ (h) and (i), respectively.

7   Conclusions

Digital binary image processing has frequently been applied in various places. The most applications, so far, connected to the square grid. In this paper, we have investigated/generalized related concepts to the triangular grid. This grid has interesting symmetric properties [16], since it is not a lattice, but has more symmetry axes than the square grid has. The basic concepts are defined by three different approaches, and fundamental basic properties of them are discussed. In strict operations, a condition is applied to the structuring element. In the strong case, the triangular grid is seen, as a subspace of the cubic grid [14, 15], and we also keep information that cannot be displayed (somehow in a similar manner as topological coordinates can be used [18]). While in the weak case, we may simply lose the information that is shifted outside of the grid.

References

1. Deutsch, E.S.: Thinning algorithms on rectangular, hexagonal, and triangular arrays. Commun. ACM **15**(9), 827–837 (1972)
2. Dineen, G.P.: Programming pattern recognition. In: Proceedings of Western Joint Computer Conference, Los Angeles, CA, pp. 94–100, 1–3 March 1955
3. Ghosh, P.K., Deguchi, K.: Mathematics of Shape Description. Wiley, New York (2009)
4. Golay, M.J.E.: Hexagonal parallel pattern transformations. IEEE Trans. Comput. **18**(8), 733–740 (1969)
5. Gonzalez, R.C., Woods, R.E.: Digital Image Processing, 3rd edn. Prentice-Hall Inc., Upper Saddle River (2006)
6. Her, I.: Geometric transformations on the hexagonal grid. IEEE Trans. Image Process. **4**(9), 1213–1222 (1995)
7. Kardos, P., Palágyi, K.: Topology preservation on the triangular grid. Ann. Math. Artif. Intell. **75**, 53–68 (2015)
8. Kirsch, R.A.: Experiments in processing life motion with a digital computer. In: Proceedings of Eastern Joint Computer Conference, pp. 221–229 (1957)
9. Klette, R., Rosenfeld, A.: Digital Geometry. Geometric Methods for Digital Picture Analysis. Morgan Kaufmann, San Francisco (2004)
10. Luczak, E., Rosenfeld, A.: Distance on a hexagonal grid. IEEE Trans. Comput. **C-25**(5), 532–533 (1976)
11. Matheron, G.: Random Sets and Integral Geometry. Wiley, New York (1975)
12. Minkowski, H.: Volumen und Oberfläche. Math. Ann. **57**, 447–495 (1903)
13. Nagy, B.: Finding shortest path with neighborhood sequences in triangular grids. In: ITI-ISPA 2001, 2nd IEEE R8-EURASIP International Symposium on Image and Signal Processing and Analysis, Pula, Croatia, pp. 55–60 (2001)
14. Nagy, B.: A family of triangular grids in digital geometry. In: ISPA 2003, 3rd International Symposium on Image and Signal Processing and Analysis, Rome, Italy, pp. 101–106 (2003)
15. Nagy, B.: Generalized triangular grids in digital geometry. Acta Math. Acad. Paedagogicae Nyíregyháziensis **20**, 63–78 (2004)

16. Nagy, B.: Isometric transformations of the dual of the hexagonal lattice. In: ISPA 2009 – 6th International Symposium on Image and Signal Processing and Analysis, Salzburg, Austria, pp. 432–437 (2009)

17. Nagy, B.: Weighted distances on a triangular grid. In: Barneva, R.P., Brimkov, V.E., Šlapal, J. (eds.) IWCIA 2014. LNCS, vol. 8466, pp. 37–50. Springer, Heidelberg (2014). doi:10.1007/978-3-319-07148-0_5

18. Nagy, B.: Cellular topology and topological coordinate systems on the hexagonal and on the triangular grids. Ann. Math. Artif. Intell. **75**, 117–134 (2015)

19. Nagy, B., Lukić, T.: Dense projection tomography on the triangular tiling. Fundam. Informaticae **145**, 125–141 (2016)

20. Serra, J.: Image Analysis and Mathematical Morphology. Academic Press, New York (1982)

21. Shih, F.: Binary Morphology. Image Processing and Mathematical Morphology. CRC Press, Boca Raton (2009)

22. Soille, P., Rivest, J.F.: Principles and Applications of Morphological Image Analysis. Springer, Berlin (1992)

Boundary and Shape Complexity of a Digital Object

Mousumi Dutt[1]($\boxtimes$) and Arindam Biswas[2]

[1] Department of Computer Science and Engineering,
St. Thomas' College of Engineering and Technology, Kolkata, India
duttmousumi@gmail.com
[2] Department of Information Technology,
Indian Institute of Engineering Science and Technology,
Shibpur, India
barindam@gmail.com

Abstract. The orthogonal convex hull is the minimal area convex polygon covering a digital object whereas an orthogonal convex skull is the maximal area convex polygon inscribing the digital object. A quantitative approach to analyse the complexity of a given hole-free digital object is presented in this paper. The orthogonal convex hull and an orthogonal convex skull are used together to derive the complexity of an object. The analysis is performed based on the regions added while deriving the orthogonal convex hull and the regions deleted while obtaining an orthogonal convex skull. Another measure for shape complexity using convexity tree derived from the orthogonal convex skull is also presented. The simple and novel approach presented in this paper is useful to derive several shape features of a digital object.

Keywords: Outer isothetic cover · Inner isothetic cover · Orthogonal convex hull · Orthogonal convex skull · Concavity · Convexity · Convexity tree

1 Introduction

Shape analysis of digital objects has various applications in diversifying fields. Similarly, boundary complexity analysis is also important in this context. The detailed patterns of boundary give an idea about the shape of the object. The description of the detailed pattern of the contour can be analysed using wavelet local extrema, which are obtained through wavelet transform [9]. The similarities between two patterns, examined by this method uses the detailed features of contours and their arrangement. From 2D contours of an image, 3D shape can be inferred, which is an important problem in machine vision. Two kinds of symmetries, i.e., parallel and mirror symmetries, give significant information about the surface shape for a variety of objects [15]. Image retrieval can be performed based on color, shape, and spatial properties of an image. Such a technique is proposed in [8], in which a prototype has been implemented to retrieve a particular image

© Springer International Publishing AG 2017
R.P. Barneva et al. (Eds.): CompIMAGE 2016, LNCS 10149, pp. 105–117, 2017.
DOI: 10.1007/978-3-319-54609-4_8

from an image database, and to calculate the symmetry between two images. The boundary analysis can be accomplished using features, like shape and texture, which represents the objects [11].

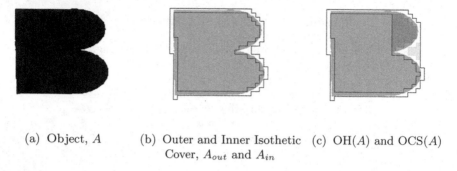

(a) Object, A

(b) Outer and Inner Isothetic Cover, A_{out} and A_{in}

(c) OH(A) and OCS(A)

Fig. 1. A sample 2D object and its orthogonal convex hull and skull (*grid size* = 10).

The applications of shape analysis and retrieval are not bounded in image processing domain. It has various real-life applications. Shape analysis can be used for road-sign detection in conjunction with image segmentation [10]. Shape based features can also be applied to construct OCR for Bengali script [12]. The working conditions of content-based retrieval, patterns of use, types of pictures, the role of semantics, the sensory gap, and the computational steps for image retrieval systems are presented in [14].

In this paper, measures for the boundary complexity and the shape complexity of a digital object are proposed. The boundary of a digital object has a close relation with outer isothetic cover (orthogonal polygon that tightly covers the digital object) and the inner isothetic cover (orthogonal polygon which tightly inscribes the digital object) [3,4]. Here, the complexity of the boundary of a digital object is analyzed by using the orthogonal hull and skull in conjunction with each other. The boundary is analyzed based on the portions added while deriving the orthogonal convex hull [5,6] and the portions deleted while deriving an orthogonal convex skull [1,7] from inner isothetic cover. The added (discarded) portions imply concave (convex) region in the contour of the object. The shape complexity gives a measure of the global shape of the object.

A digital object has been shown in Fig. 1(a) and its outer isothetic cover (black line), A_{out}, and inner isothetic cover (blue line), A_{in}, are shown Fig. 1(b). The yellow portion is added to outer cover to obtain the orthogonal convex hull (OH(A)) and the red portion is deleted from inner cover to extract an orthogonal convex skull (OCS(A)) (Fig. 1(c)). The resulting skull is shown in blue line and the hull in black line (Fig. 1(c)). The object has a concavity on its contour, which can be identified from one added portion and one deleted portion. These information are useful for analysing the boundary complexity and shape complexity of the digital object.

The paper is organized as follows. Section 2 includes definitions and an overview of constructing orthogonal hull and skull. The boundary complexity and shape complexity are discussed in Sect. 3. A discussion of shape complexity from convexity tree is proposed in Sect. 4. Section 5 contains experimental results along with the data obtained from digital objects which are related to complexity analysis. Concluding remarks are presented in Sect. 6.

2   Definitions and Preliminaries

A (finite) subset of $\mathbb{Z}^2$ in which every pair of points is k-connected[1] is called a k-connected set. A digital object A is defined to be an 8-connected subset of $\mathbb{Z}^2$ whose complement $\mathbb{Z}^2 \setminus A$ is a 4-connected set [13]. The background grid is given by $\mathcal{G} = (\mathcal{H}, \mathcal{V})$, where $\mathcal{H}$ and $\mathcal{V}$ represent two sets of equi-spaced horizontal and vertical grid lines respectively. The grid size, g, is defined as the (integer) distance between two consecutive horizontal/vertical grid lines. A grid point is the point (with integer coordinates) of intersection of a horizontal and a vertical grid line.

P is said to be an orthogonal polygon if and only if each of its vertices is a grid point and each of its edges lie on a grid line. P is an orthogonal convex polygon if and only if its intersection with any horizontal or vertical line is either a single line segment or empty. The orthogonal convex hull, or simply orthogonal hull, of a digital object A, denoted by $OH(A)$, is the smallest area orthogonal polygon such that (i) no point $p \in A$ lies on or outside $OH(A)$ and (ii) intersection of $OH(A)$ with any horizontal or vertical line is either empty or a line segment. An orthogonal convex skull, or simply orthogonal skull, of a digital object A, denoted by $OCS(A)$, is a maximal-area orthogonal polygon such that (i) no point $p \in \mathbb{Z}^2 \setminus A$ lies on or inside $OCS(A)$ and (ii) $OCS(A)$ is orthogonally convex.

The construction of outer isothetic cover and inner isothetic cover are depicted in Sects. 2.1 and 2.2. The description of the method of deriving orthogonal convex hull is presented in Sect. 2.3 and that of orthogonal convex skull in Sect. 2.4.

2.1   Deriving the Outer Isothetic Cover, (OIC)

The outer isothetic cover, A_{out}, is the minimum-area orthogonal polygon that covers the digital object, A, imposed on background grid, $\mathcal{G}$. The algorithm in [3, 4] computes the ordered set of vertices of A_{out} using a combinatorial classification of the grid points lying on/inside/outside the object boundary. The characteristics of a grid point p in $\mathcal{G}$ is determined by object containments of the four neighboring cells of size $g \times g$ incident at p. If the number of cells occupied by the object, incident at p is $i \in [0, 4]$, then p is classified to class C_i as shown in Fig. 2. The significance of a class is as follows. (i) C_0: p is not a vertex since none

---

[1] Two points p and q are said to be k-connected ($k = 4$ or 8) in a set S if and only if there exists a sequence $\langle p = p_0, p_1, \ldots, p_n = q \rangle \subseteq S$ such that $p_i \in N_k(p_{i-1})$ for $1 \leq i \leq n$. The 4-neighborhood of a point (x, y) is given by $N_4(x, y) = \{(x', y') : |x - x'| + |y - y'| = 1\}$ and its 8−neighborhood by $N_8(x, y) = \{(x', y') : \max(|x - x'|, |y - y'|) = 1\}$.

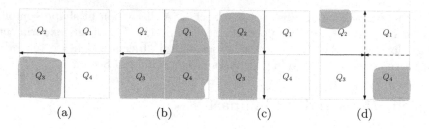

Fig. 2. Different vertex types.

of Q_is has object containment; (ii) C_1: Q_i is a 90° vertex of A_{out} (Fig. 2(a));
(iii) C_2: (a) if two adjacent cell has object containment, then p is an edge point
(Fig. 2(c)); (b) if diagonally opposite cells contain object, then p is a 270° vertex
of A_{out} (Fig. 2(d)); (iv) C_3: p is classified as a 270° vertex (Fig. 2(b)); (v) C_4: p
is not a vertex of A_{out} and lies inside A_{out}.

The start point can be derived from the given top-left point of the digital
object. The traversal is made in the anticlockwise direction such that the background lies right of the traversal. During the traversal, from each grid point,
v_{i-1}, the next direction, d_i, of traversal is decided by the type, t_i, of the grid
point, v_i, and the previous direction, d_{i-1}. The next direction, d_i, is given by
$d_i = (d_{i-1} + t_i) \bmod 4$, where $d_i \in [0, 3]$, indicating the direction towards right,
top, left, and bottom respectively. Once d_i is computed, the next grid point is
determined, and its class is evaluated. Henceforth, in this paper, a 90° vertex is
referred as a Type 1 vertex and a 270° vertex as a Type 3 vertex.

2.2 Deriving the Inner Isothetic Cover, (*IIC*)

The inner isothetic cover, namely A_{in}, is the maximum-area orthogonal polygon
inscribing the digital object A, which is imposed on the background grid $\mathcal{G}$.
The construction of inner isothetic cover is similar to the outer isothetic cover
(Sect. 2.1), except for the consideration of the grid point, q. A grid point, q, is
classified based on the number of fully occupied cells incident at q.

2.3 Construction of Orthogonal Convex Hull, (*OH*)

The orthogonal hull [5,6] can be detected without any prior knowledge of outer
isothetic cover. The concavities in the outer cover are detected and removed
when the object boundary is traversed along the grid lines, as mentioned in
Sect. 2.1. A region is said to be concave if it has two or more consecutive Type 3
vertices. If the intersection of a horizontal or vertical grid line with the orthogonal polygon has more than one line segment, then it can be said that the object
contains one or more concavities. The goal is to identify such regions and derive
the edges of the orthogonal hull such that the properties of orthogonal convexity
are maintained. In this incremental algorithm, the part of the orthogonal hull
obtained upto a point does not contain two consecutive Type 3 vertices which

acts as the invariant of the algorithm. Whenever a concavity is detected, necessary combinatorial rules are applied to maintain the algorithm invariant and to ensure the orthogonal convexity, thereof. However, rest of the patterns, **13**, **31**, and **11**, are in conformance with the algorithm invariant and hence do not violate the properties of orthogonal convexity.

2.4   Construction of Orthogonal Convex Skull (*OCS*)

The algorithm to determine an orthogonal convex skull [1,7] first obtains an inner isothetic cover (Sect. 2.2). An orthogonal convex skull is computed by applying rules on the obtained inner isothetic cover. The resulting skull is not unique and it is dependent on the direction of traversal and the starting point of traversal.

Two or more consecutive Type 3 vertices imply a concave region, which defies the properties of orthogonal convexity, as the intersection of a vertical or horizontal grid line with the orthogonal polygon (here, A_{in}) has more than one line segment. The concavity line passing through two consecutive Type 3 vertices, divides the polygon into three different parts: two separate sub-polygons lying on one side of l, and the rest of the polygon at the other side. To achieve orthogonal convexity, one of these two sub-polygons has to be dropped. Hence, first we check whether dropping of any sub-polygon divides (disconnects) the polygon into two parts. If so, then it cannot be dropped. If both the sub-polygons do not affect the connectivity, then the sub-polygon having larger area is included in the skull, thus maximizing the area of the skull. As a result, the convex skull, obtained thereof, also contains no two consecutive vertices of Type 3. Here, in this paper, this algorithm is slightly modified, to obtain OCS for each of the isothetic polygons when the object is disconnected.

3   Boundary Complexity and Shape Complexity

The boundary complexity can be analysed from the boundary of orthogonal convex hull and skull. Orthogonal convex hull is a superset of the outer isothetic cover and orthogonal convex skull is a subset of the inner isothetic cover. Several analyses are presented in this paper using A_{out} with $OH(A)$, A_{in} with $OCS(A)$, A_{out} with A_{in}, and $OCS(A)$ with $OH(A)$.

Let us consider A_{out} and $OH(A)$ together for complexity analysis. The number of portions included in $OH(A)$ provides one measure for the complexity of A. The number of regions added may be zero or one or more. If it is zero, we can certainly say that A_{out} is convex. A shape with more concavities is considered to be more complex. Wherever there is a concavity in the object, a region is included in $OH(A)$ to make it convex. When only one region is added, there are two scenarios depending on the area of the added regions. If the added area is small w.r.t. A_{out}, then the complexity is negligible. If the added area is large w.r.t. A_{out}, then it can be inferred that the complexity lies in the portion of the object where the region has been added. Thus, it can be said that the shape is complex. On the other hand, if the area of the added portion is small enough, then it implies that the object is less complex. If more than one region is added, then there are more than one concavity in A_{out}. Depending on the area of each such portions, the complexity can be determined.

Now considering $OCS(A)$ in conjunction with A_{in}, while determining $OCS(A)$ some convex portions are discarded. Based on the number of portions discarded and the area of those portions, the complexity of the object can be analysed. Similarly, if the number of discarded region is zero, we can certainly say that A_{in} is convex. However, it cannot be inferred that the object is not complex. Because it may happen that the object contains a narrow portion, where A_{in} has not entered. If one region is discarded and the area is large w.r.t. A_{in}, it implies that the complexity lies in that portion of the object. It generally happens with spiral shaped object. On the other hand, if the area is small enough, then it implies that the object is less complex. If the number of discarded portion is greater than one, then there are more than one concavities in A_{in}. Depending on the area of each such portion, the complexity can be determined as said above.

Also, by considering A_{out} and A_{in}, some analyses can be done. If there is a long neck in A, A_{in} may not enter the portion (depending on the grid size and the imposition of the object on the background grid). On the other hand, if the object contains long neck like protrusion in it, A_{out} may not be able to detect it for the same reason. Several measures for boundary complexity and shape complexity are discussed below (Sects. 3.1 and 3.2).

3.1   Boundary Complexity

Three measures for boundary complexity are defined here. Let p, p_i, p_o, p_s, and p_h be the perimeter of A, A_{in}, A_{out}, $OCS(A)$, and $OH(A)$ respectively. The boundary complexity can be defined as follows.

$$\zeta_1 = \frac{(p_h - p_o) + (p_i - p_s)}{p} \times 100 \tag{1}$$

The range is $0 < \zeta_1 < 100$. It is the percentage of the boundary for which the object becomes non-convex. When the object is convex, the value of ζ_1 is equal to zero. High value of ζ_1 implies more concavities (i.e., more complexity) along the boundary. When the grid size decreases, the complexity in the boundary increases, e.g., more concavities are detected along the contour.

During the construction of orthogonal hull, some portions of the outer isothetic cover are included inside it. For the construction of orthogonal convex skull, some portions are removed from the inner isothetic cover. Let $\triangle\,p_o$ be the total perimeter included to construct the orthogonal hull. Similarly, $\triangle\,p_i$ denotes the total perimeter excluded from the inner isothetic cover to construct orthogonal convex skull. The two other measures for boundary complexity can be defined based on these two parameters as follows.

$$\zeta_2 = \frac{\triangle\,p_o}{p_o} + \frac{\triangle\,p_i}{p_i} \tag{2}$$

$$\zeta_3 = \frac{\left| \frac{\Delta p_o}{p_o} - \frac{\Delta p_i}{p_i} \right|}{max\{ \frac{\Delta p_o}{p_o}, \frac{\Delta p_i}{p_i} \}} \tag{3}$$

ζ_2 is the sum of the complexities in the inner and the outer boundary of the object. The value of ζ_2 is higher in the lower grid sizes compared to the higher grid sizes. The range of ζ_2 is $0 \leqslant \zeta_2 < 2$. When the object is convex then the value of ζ_2 is zero. When the object is too complex most of the portions of outer cover will be included in the hull and most of the portions of the inner cover will be discarded. The value of each fraction will be almost equal to 1. ζ_3 gives a measure on the difference between the inner and the outer complexity. Its range is $0 \leqslant \zeta_3 \leqslant 1$. If the value is zero then the complexity of the inner and outer boundaries are almost same. Otherwise, it means the inner (outer) boundary contains some complexities, which are not detected in the outer (inner) boundary. It is maximum when either inner cover or outer cover is convex. ζ_3 is not dependent on ζ_2.

3.2 Shape Complexity

The two measures for shape complexity are discussed here. Let a, a_i, a_o, a_s, and a_h be the area of A, A_{in}, A_{out}, $OCS(A)$, and $OH(A)$ respectively. The shape complexity can be proposed w.r.t. the area (or regions) of the object which makes the object non-convex. It can be determined as follows.

$$\zeta_4 = \frac{(a_h - a_o) + (a_i - a_s)}{a} \times 100 \tag{4}$$

The range is $0 < \zeta_4 < 100$. This measure determines the percentage of the boundary for which the object becomes non-convex. When the object is convex, the value of ζ_4 is almost equal to zero. High value of ζ_4 implies more complex shape. For lower grid sizes the value of shape complexity increases.

The second measure of the shape complexity can be defined as follows.

$$\zeta_5 = \frac{a_s}{a_h} \tag{5}$$

Its range is $0 < \zeta_5 < 1$. When the object is very complex then $\zeta_5 \simeq 0$ as the area of an orthogonal skull will be much lesser and that of the orthogonal hull will be much higher. The area of the orthogonal convex skull can never be zero, so lower bound of ζ_5 can never be equal to zero. When the object is convex both areas will be almost equal. In case of a convex object, the orthogonal hull is the outer isothetic cover and the orthogonal skull is inner isothetic cover. Always the area of the outer isothetic cover of an object will be greater than that of inner isothetic cover. Thus, the upper bound of ζ_5 cannot be equal to one. If the value of ζ_5 is near to one, it means that the object is less complex. Otherwise, the object is very complex.

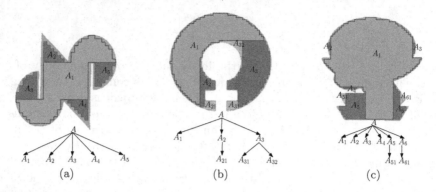

Fig. 3. Convexity tree for three digital objects: (a) Logo 211 for $g = 8$; (b) Logo 844 for $g = 7$; (c) Logo 347 for $g = 5$.

4   Shape Complexity from Orthogonal Convexity Tree

The orthogonal convexity tree can be formed from the orthogonal convex skull. A_{in} consists of the orthogonal convex skull and discarded portions which are actually orthogonal polygons. Some of these are convex and some are not. If orthogonal convex skull can be obtained recursively on the non-convex components till all the components are orthogonally convex, then an orthogonal convexity tree can be determined as shown in Fig. 3. The concept of orthogonal concavity tree is already proposed in [2]. These are useful for determining the shape complexity of a digital object.

Numerous information can be found from the convexity tree, e.g., the depth of convexity tree, the number of children, the number of children per level, etc. If the convexity tree branches out uniformly, then the corresponding object may be symmetric (provided the area of a pair of convex components are same). The object in Fig. 3(a) is symmetric. Here, all the leaves are at same depth. Each pair (A_2, A_4 and A_3, A_5) has almost similar area and perimeter. It is to be noted here that area and perimeter comparison are not the only criteria to check whether the object is symmetric or not. If the depth of the tree is higher and there are only one child at each level of the tree, then the shape of the object is very complex and the convexity tree will be unbalanced (e.g., spiral-shaped objects). The depth of the convexity tree is two for the objects shown in Fig. 3(a) and (b). These are more complex objects compared to Fig. 3(a) whose depth is one.

The total number of concavities in the object boundary can be determined from the convexity tree. It is the total number of nodes in the tree (internal nodes and leaves except root) minus one. The number of convex regions which are discarded is the number of concavities in the IIC. All the leaves in the convexity tree correspond to convex regions of the object. If the area is very small w.r.t. the area of the object, then the concavity is insignificant. Let a_x be the area of a convex region. If $\frac{a_x}{a} \simeq 0$, then the corresponding concavity is insignificant. It may not be detected when the object is imposed on higher grid size.

5   Experimental Results

Some of the experimental results are included to explain the complexity analysis in Figs. 4, 5 and 6 and the corresponding data are shown in Table 1. The algorithm of orthogonal hull and skull were implemented in C Ubuntu 14.04. The orthogonal convex skull is shown in blue border whereas orthogonal convex hull with black border. The regions which are included in the orthogonal convex hull are shown in yellow color. The discarded regions to construct orthogonal convex skull are shown in red color. ζ_1 and ζ_2 decreases when grid size increases. For less complex boundary the value is lower. The inner isothetic cover of Logo 220 for $g = 20$ is convex and its outer isothetic cover has only one concavity. The value of ζ_1 and ζ_2 are very low for this object, whereas $\zeta_3 = 1$ (maximum value).

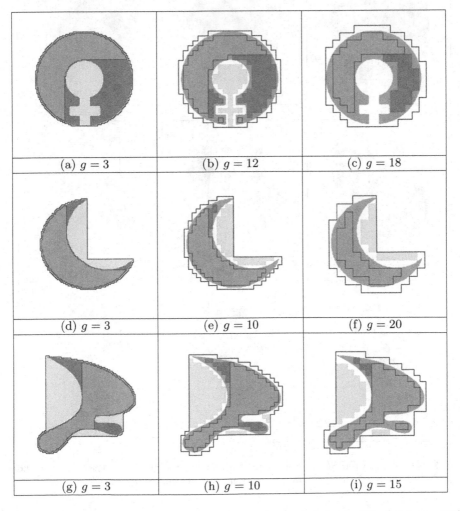

Fig. 4. $OH(A)$ and $OCS(A)$ on a set of objects (Logo 844, Logo 181, and Logo 426) which contains circular portion (g refers to grid size). (Color figure online)

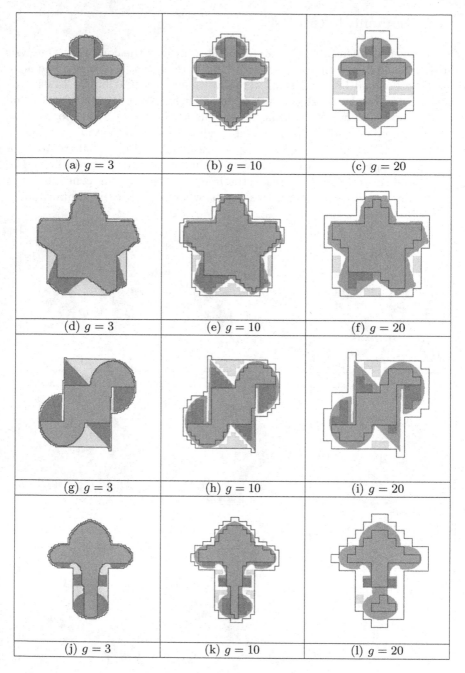

Fig. 5. $OH(A)$ and $OCS(A)$ on a set of symmetric objects (Logo 1287, Logo 5, Logo 211, and Logo 257). Here, g refers to grid size. (Color figure online)

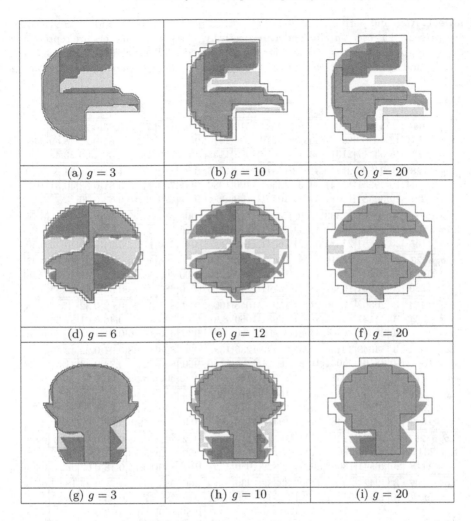

(a) $g = 3$	(b) $g = 10$	(c) $g = 20$
(d) $g = 6$	(e) $g = 12$	(f) $g = 20$
(g) $g = 3$	(h) $g = 10$	(i) $g = 20$

Fig. 6. $OH(A)$ and $OCS(A)$ on another set of objects (Logo 23, Logo 220, and Logo 347). Here, g refers to grid size. (Color figure online)

From these three measures, it can be said that either A_{in} or A_{out} is convex and concavities in the boundary are very less. If Logo 5 is considered for the three grid sizes, it is seen that the value ζ_3 increases with the increase of grid size but ζ_1 and ζ_2 decreases for the same. It implies that the difference in the boundary complexity for A_{in} and A_{out} is very less for the lower grid sizes which detects almost all the concavities. For higher grid sizes, the difference increases as some of the concavities are not being detected in both. The nature of ζ_4 is same with ζ_1. The value of ζ_5 remains high for the three grid sizes in Logo 347. It means that there is less difference in shape while considering orthogonal convex hull and skull. If the range of ζ_5 is high for lower and higher grid sizes then there is a huge difference in shape for orthogonal convex hull and skull (consider Logo 181).

Table 1. Area and perimeter of digital object, A_{out}, A_{in}, $OH(A)$, and $OCS(A)$ (areas are represented as a, a_o, a_i, a_h, and a_s respectively, whereas perimeters are represented as p, p_o, p_i, p_h, and p_s respectively) for the mentioned grid size shown in Figs. 4, 5 and 6. ζ_1, ζ_2, and ζ_3 are boundary complexity measures. Shape complexity measures are ζ_4 and ζ_5.

	g	a_o	a_i	a_h	a_s	p_o	p_i	p_h	p_s	ζ_1	ζ_2	ζ_3	ζ_4	ζ_5
Logo 844	3	37189	32662	45700	19531	1512	1494	948	882	78.4	0.93	0.28	63.39	0.43
$a = 34141$	12	43969	25911	50029	15279	1536	1416	984	888	72	0.84	0.21	48.89	0.31
$p = 1500$	18	53317	21313	53317	13663	1008	1152	1008	756	26.4	0.39	1.00	22.41	0.26
Logo 1287	3	24826	21169	31273	15310	1158	1206	840	810	58.24	1.04	0.38	54.93	0.49
$a = 22404$	10	29251	17741	34131	13591	1100	1080	860	780	44.05	0.81	0.38	40.31	0.4
$p = 1226$	20	36581	13681	39261	10761	1160	960	920	720	39.15	0.63	0.30	25	0.27
Logo 5	3	40603	37333	44299	33820	1176	1152	1008	984	29.02	0.73	0.08	18.9	0.76
$a = 38148$	10	44371	33051	46811	29871	1140	1100	1020	940	24.18	0.53	0.11	14.73	0.64
$p = 1158$	20	49781	27681	52121	26041	1160	960	1040	880	17.27	0.34	0.30	10.43	0.50
Logo 23	3	32638	28810	37882	17671	1284	1260	972	906	52.03	0.84	0.21	54.54	0.47
$a = 30040$	10	37631	24811	41401	14731	1260	1220	1000	860	48.44	0.79	0.32	46.11	0.36
$p = 1280$	20	45021	19361	48521	10741	1240	1120	1040	680	50	0.69	0.40	40.35	0.22
Logo 181	3	20302	17146	27421	15985	1056	1008	912	828	31.21	0.62	0.32	45.7	0.58
$a = 18120$	10	24131	13711	30171	12981	1060	820	940	760	17.34	0.46	0.61	37.36	0.43
$p = 1038$	20	28541	9541	34081	9541	1080	680	960	680	11.56	0.31	1.00	30.57	0.28
Logo 211	3	34648	30466	39232	21442	1236	1386	936	888	56.28	0.91	0.22	42.75	0.55
$a = 31832$	10	38161	25041	41281	18121	1120	1280	960	840	42.31	0.74	0.43	31.54	0.44
$p = 1418$	20	44981	20581	46921	15201	1160	1160	1040	800	33.85	0.60	0.60	23	0.32
Logo 220	6	37759	28099	46075	16699	1572	1004	996	924	40.95	1.13	0.36	60.77	0.36
$a = 32442$	12	42517	24793	49177	15109	1512	1200	1008	840	53.93	0.84	0.17	50.38	0.31
$p = 1602$	20	53361	18482	54121	18482	1120	960	1040	960	4.99	0.09	1.00	2.34	0.34
Logo 257	3	25594	22249	27241	18532	1110	1098	912	882	37.03	0.69	0.14	22.9	0.68
$a = 23425$	10	30271	18631	31381	15121	1140	1060	960	840	35.78	0.61	0.44	19.72	0.48
$p = 1118$	20	36541	13662	36921	12822	1080	920	1040	840	10.73	0.18	0.62	5.21	0.35
Logo 347	3	42421	38704	44503	33382	1230	1212	966	936	43.62	0.78	0.27	18.54	0.75
$a = 39946$	10	45851	33951	47391	30041	1100	1100	980	880	27.46	0.53	0.34	13.64	0.63
$p = 1238$	20	52121	28841	51741	27201	1080	880	1040	800	9.69	0.19	0.64	3.15	0.53
Logo 426	3	26821	22897	39835	19387	1314	1260	972	918	52.78	0.84	0.32	68.37	0.49
$a = 24168$	10	31540	18601	43101	16151	1259	1199	1000	900	43.06	0.68	0.21	57.97	0.37
$p = 1296$	15	35926	16967	44836	16007	1200	1079	1020	960	23.07	0.45	0.55	40.84	0.36

6 Conclusions

A simple and novel approach to analyse the complexity of a digital object is presented here. It gives a measure of the complexity of the global shape of the object. The concept of convexity tree to determine shape complexity is completely a novel technique. Various experimental results are given to show the effectiveness of the proposed scheme. As a future direction, some more features of the object can be derived and metrics can be formulated to classify digital objects based on shapes.

References

1. Biswas, A., Dutt, M., Bhowmick, P., Bhattacharya, B.B.: On finding the orthogonal convex skull of a digital object. In: Proceedings of 13th International Workshop on Combinatorial Image Analysis: IWCIA 2009, pp. 25–36. Progress in Combinatorial Image Analysis, Research Publishing Services, Playa del Carmen, Mexico, November 2009

2. Biswas, A., Sarkar, A., Bhowmick, P., Bhattacharya, B.B.: Combinatorial construction of the orthogonal concavity tree of a digital object. In: 2011 2nd International Conference on Emerging Applications of Information Technology (EAIT), pp. 210–213 (2011)

3. Biswas, A., Bhowmick, P., Bhattacharya, B.B.: TIPS: on finding a tight isothetic polygonal shape covering a 2D object. In: Kalviainen, H., Parkkinen, J., Kaarna, A. (eds.) SCIA 2005. LNCS, vol. 3540, pp. 930–939. Springer, Heidelberg (2005). doi:10.1007/11499145_94

4. Biswas, A., Bhowmick, P., Bhattacharya, B.B.: Construction of isothetic covers of a digital object: a combinatorial approach. J. Vis. Commun. Image Represent. **21**(4), 295–310 (2010)

5. Biswas, A., Bhowmick, P., Sarkar, M., Bhattacharya, B.B.: A linear-time combinatorial algorithm to find the orthogonal hull of an object on the digital plane. Inf. Sci. **216**, 176–195 (2012)

6. Biswas, A., Bhowmick, P., Sarkar, M., Bhattacharya, B.B.: Finding the orthogonal hull of a digital object: a combinatorial approach. In: Brimkov, V.E., Barneva, R.P., Hauptman, H.A. (eds.) IWCIA 2008. LNCS, vol. 4958, pp. 124–135. Springer, Heidelberg (2008). doi:10.1007/978-3-540-78275-9_11

7. Dutt, M., Biswas, A., Bhowmick, P., Bhattacharya, B.B.: On finding an orthogonal convex skull of a digital object. Int. J. Imaging Syst. Technol. **21**(1), 14–27 (2011)

8. Hung, K., Aw-Yong, M.: A content-based image retrieval system integrating color, shape and spatial analysis. In: Proceedings of IEEE International Conference on Systems, Man, and Cybernetics, vol. 2, pp. 1484–1488 (2000)

9. Hussein, E., Nakamura, Y., Ohta, Y.: Analysis of detailed patterns of contour shapes using wavelet local extrema. In: Proceedings of 13th International Conference on Pattern Recognition: ICPR 1996, vol. 2, pp. 335–339 (1996)

10. Khan, J.F., Bhuiyan, S.M.A., Adhami, R.R.: Image segmentation and shape analysis for road-sign detection. IEEE Trans. Intell. Transp. Syst. **12**(1), 83–96 (2011)

11. Liu, W., Srivastava, A., Klassen, E.: Joint shape and texture analysis of objects boundaries in images using a Riemannian approach. In: Proceedings of International Conference on Digital Image Processing, vol. 14, pp. 833–837 (2008)

12. Naser, M.A., Hasnat, M., Latif, T., Nizamuddin, S., Islam, T.: Analysis and representation of character images for extracting shape based features towards building an OCR for Bangla script. In: Proceedings of International Conference on Digital Image Processing, pp. 330–334 (2009)

13. Rosenfeld, A., Kak, A.C. (eds.): Digital Picture Processing, 2nd edn. Academic Press, Cambridge (1982)

14. Smeulders, A.W.M., Worring, M., Santini, S., Gupta, A., Jain, R.: Content-based image retrieval at the end of the early years. IEEE Trans. Pattern Anal. Mach. Intell. **22**(1), 1349–1380 (2001)

15. Ulupanar, F., Nevatia, R.: Using symmetries for analysis of shape from contour. In: Proceedings of 2nd International Conference of Computer Vision: ICCV 1998, pp. 414–426 (1998)

Interior and Exterior Shape Representations Using the Screened Poisson Equation

Laura A. Rolston and Nathan D. Cahill[✉]

Image Computing and Analysis Lab (ICAL), School of Mathematical Sciences,
Rochester Institute of Technology, Rochester, NY, USA
nathan.cahill@rit.edu

Abstract. Shape classification is a required task in many systems for image and video understanding. Implicit shape representations, such as the solutions to the Eikonal or Poisson equations defined on the shape, have been shown to be particularly effective for generating features that are useful for classification. The Poisson-based shape representation can be derived at each point inside the shape as the expected time for a particle undergoing Brownian motion to hit the shape boundary. This representation has no natural generalization when considering points *outside* of a shape, however, because the corresponding Brownian motion would have infinite expected hitting time. In this article, we modify the Brownian motion model by introducing an exponential lifetime for the particle, yielding a random variable whose expected value satisfies a screened Poisson equation that can be solved at points *both interior and exterior* to the shape. We then show how moments of this new random variable can be used to improve classification results on experiments with natural silhouettes and handwritten numerals.

Keywords: Implicit shape representation · Shape classification

1 Introduction

Humans have a remarkable ability to easily discern the classification of an object simply from its binary silhouette. In the computer vision literature, various explicit and implicit *shape representations* have been proposed for use in feature extraction and shape classification systems.

We focus in this article on implicit shape representations, which typically involve solving boundary value problems on the shape. The solution to the Eikonal equation yields the Euclidean Distance Transform of a shape, and the solution to Poisson's equation yields the expected time for Brownian motion to hit the boundary of the shape [7]. While the Poisson representation has been shown to yield superior classification results to the Euclidean Distance Transform, it has so far been limited to represent points on the interior of a shape, as it is not well defined for points exterior to the shape.

To overcome this limitation of the Poisson representation, we show how a slight modification of the Brownian motion model can lead to a well-defined

© Springer International Publishing AG 2017
R.P. Barneva et al. (Eds.): CompIMAGE 2016, LNCS 10149, pp. 118–131, 2017.
DOI: 10.1007/978-3-319-54609-4_9

implicit shape representation for both interior and exterior points. By introducing an exponential dying time to the Brownian motion, we define a new random variable called the *stopping time* to be the minimum of the boundary hitting time and the dying time of the Brownian motion. The expected stopping time satisfies a *screened* Poisson equation and is well defined both inside and outside the shape. In addition, the entire stopping time distribution is straightforward to derive, enabling shape representations based on any of its standard moments.

2 Background

Shape representations have a long history in the fields of visual psychology and computer vision. Early representations are explicit ones, including the medial axis function and the symmetric axis transform [3,4,9,12,14–16], which represent a shape in terms of its medial axis or skeleton, and part-based representations [1,11,13,17], where describe shapes by a collection of geometric primitives.

Implicit shape representations have also been developed, the simplest of which are based on the Euclidean distance transform [5,19], which describes the minimum distance from each point inside a shape to the boundary of the shape, and which satisfies the Eikonal equation with homogeneous Dirichlet boundary conditions. The use of distance transforms as implicit shape representations was extended to the *Poisson transform* in [7], which describes the expected number of steps required for a symmetric random walk to hit the boundary of a shape given that it starts at a point inside the shape. The Poisson transform involves solving a nonhomogeneous Poisson equation on the shape, and it yields a shape representation that is smoother and more robust to boundary noise than the Euclidean distance transform. It has been applied to a variety of problems including gait recognition [18], general human action recognition [2], and 3-D face recognition [10].

While the Euclidean distance transform can easily be computed for points both inside and outside a shape, it is unclear that the Poisson transform can be directly extended to exterior points due to the fact that the symmetric random walk on $\mathbb{Z}^n$ is not positive recurrent for $n \geq 2$. In the following chapters, we show that it *is* possible to define a shape representation that is similar to, but not exactly the same as the Poisson transform, and that can be defined for both interior and exterior points. In order define the new shape representation, however, we first show how to arrive at the Poisson transform in a different manner than described in [7]: by considering Brownian motion hitting times.

3 Brownian Motion Hitting Times

To formalize the concept of shape, we define a shape as a compact set $\bar{\Omega} \subset \mathbb{R}^n$, and we denote the interior and boundary of the shape by Ω and $\partial\Omega$, respectively. To formalize the concept of hitting time, let $\{X(t) \in \mathbb{R}^n, t \geq 0\}$ denote the position of a particle undergoing Brownian motion, and define the hitting time T_B as the first time that the particle hits the boundary of the shape:

$$T_B = \begin{cases} \inf_{t \geq 0} \{t | X(t) \in \mathbb{R}^n \backslash \bar{\Omega}\}, & X(0) \in \bar{\Omega}, \\ \inf_{t \geq 0} \{t | X(t) \in \bar{\Omega}\}, & X(0) \in \mathbb{R}^n \backslash \bar{\Omega}. \end{cases} \tag{1}$$

We begin this section by considering only particles that originate inside the shape (i.e., $X(0) \in \bar{\Omega}$), in which case T_B can be considered the first exit time of the particle from $\bar{\Omega}$.

3.1 Survival Function Initial Boundary Value Problem

For any point $\mathbf{x} \in \mathbb{R}^n$, we define the *survival function* $S(\mathbf{x}, t)$ to be the probability that the hitting time exceeds t given that the particle is initially located at $\mathbf{x}$; that is,

$$S(\mathbf{x}, t) = P(T_B > t | X(0) = \mathbf{x}). \tag{2}$$

As shown in Appendix A, for points $\mathbf{x} \in \bar{\Omega}$, $S(\mathbf{x}, t)$ satisfies the initial boundary value problem (IBVP):

$$\begin{aligned} \mathcal{L}\{S(\mathbf{x}, t)\} = 0, & \quad \forall \mathbf{x} \in \Omega, \quad t \geq 0, \\ S(\mathbf{x}, t) = 0, & \quad \forall \mathbf{x} \in \partial\Omega, \quad t \geq 0, \\ S(\mathbf{x}, 0) = 1, & \quad \forall \mathbf{x} \in \Omega, \end{aligned} \tag{3}$$

where the operator $\mathcal{L}$ is defined by:

$$\mathcal{L}\{S(\mathbf{x}, t)\} = \frac{\partial}{\partial t} S(\mathbf{x}, t) - \frac{1}{2n} \Delta S(\mathbf{x}, t). \tag{4}$$

3.2 Moments

The kth moment about zero of the hitting time T_B of the Brownian motion starting at point $\mathbf{x}$ inside the shape can be computed from the survival function by:

$$U_k(\mathbf{x}) = k \int_0^\infty t^{k-1} S(\mathbf{x}, t) \, dt. \tag{5}$$

Multiplying each side of (3) by kt^{k-1}, integrating, and employing (5) yields the recursive set of boundary value problems:

$$\begin{aligned} -\frac{1}{2n} \Delta U_k(\mathbf{x}) = k U_{k-1}(\mathbf{x}), & \quad \forall \mathbf{x} \in \Omega, \\ U_k(\mathbf{x}) = 0, & \quad \forall \mathbf{x} \in \partial\Omega, \end{aligned} \tag{6}$$

for $k = 1, 2, \ldots$, with $U_0(\mathbf{x}) = E\left[T_B^0\right] = 1$. Central moments, denoted $V_k(\mathbf{x})$ can be computed from the moments about zero by:

$$V_k(\mathbf{x}) = \sum_{m=0}^k \binom{k}{m} (-1)^{k-m} U_m(\mathbf{x}) U_1(\mathbf{x})^{k-m}, \tag{7}$$

and then standardized moments (skewness, kurtosis) can then be computed as $W_k(\mathbf{x}) = V_k(\mathbf{x})/(V_2(\mathbf{x}))^{k/2}$.

3.3   Discretization of Moment BVPs

Suppose $\bar{\Gamma} \subset \bar{\Omega}$ is a n-dimensional uniform grid of points. Each grid point $\mathbf{x}$ has neighborhood $\mathcal{N}(\mathbf{x}) = \bar{\Gamma} \cap \{\mathbf{x} \pm h\mathbf{e}_j | j = 1, \ldots, n\}$, where $\mathbf{e}_j$ is the jth column of the n-dimensional identity matrix, and h is the grid spacing. The interior Γ contains all points in $\bar{\Gamma}$ that have $2n$ neighbors (i.e., points for which $|\mathcal{N}(\mathbf{x})| = 2n$), and the discrete boundary $\partial\Gamma = \bar{\Gamma} \backslash \Gamma$.

Using centered differences to approximate the Laplacian of the kth hitting time moment yields:

$$\Delta U_k(\mathbf{x}) \approx \sum_{i=1}^{n} \frac{U_k(\mathbf{x} - h\mathbf{e}_i) - 2U_k(\mathbf{x}) + U_k(\mathbf{x} + h\mathbf{e}_i)}{h^2}, \tag{8}$$

and, hence, discrete versions of the boundary value problems in (6) are given by:

$$-\sum_{i=1}^{n} \frac{U_k(\mathbf{x} - h\mathbf{e}_i) - 2U_k(\mathbf{x}) + U_k(\mathbf{x} + h\mathbf{e}_i)}{2nh^2} = kU_{k-1}(\mathbf{x}), \quad \forall \mathbf{x} \in \Gamma,$$

$$U_k(\mathbf{x}) = 0, \quad \forall \mathbf{x} \in \partial\Gamma, \tag{9}$$

for $k = 1, 2, \ldots$, with $U_0(\mathbf{x}) = 1$.

3.4   Computational Examples of Moments

Figure 1 shows a shape of a horse along with various moments of Brownian motion hitting time computed for points inside the shape. For points outside the shape, all moments are zero. As can be seen in these figures, the level sets of the mean $U_1(\mathbf{x})$, standard deviation $\sqrt{V_2(\mathbf{x})}$, and variance $V_2(\mathbf{x})$ are smoothed versions of the boundary of the shape. The head, leg and tail regions of the horse exhibit much smaller values than the torso region for these first and second-order moments. Hitting time skewness $W_3(\mathbf{x})$ and kurtosis $W_4(\mathbf{x})$ achieve maximum values in the ear, tail and leg regions, while the values in the torso region are relatively small. (Log-scaled versions of skewness and kurtosis are shown in the figure in order to provide better contrast than the linear-scaled versions.)

4   From Hitting Times to Stopping Times

A difficulty immediately arises if we attempt to compute expected Brownian motion hitting times for particles that originate outside of a shape: in two or more dimensions, Brownian motion is not recurrent, so the expected hitting time T_B is infinite.

If, however, we equip the particle with a (random) natural lifetime, T_L, that is independent of position and of hitting time, then the *stopping time* $T = \min(T_B, T_L)$ (the minimum of the hitting time and the natural lifetime) *does* have finite moments no matter where the Brownian motion originates. If we assume that T_L has an exponential distribution with mean β, so that $P(T > t) = e^{-t/\beta}$ for $t \geq 0$ and 1 otherwise, then the mean stopping time will always be bounded above by β.

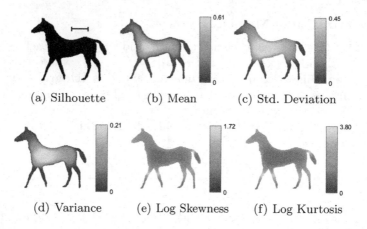

(a) Silhouette   (b) Mean   (c) Std. Deviation

(d) Variance   (e) Log Skewness   (f) Log Kurtosis

Fig. 1. Example shape (a) containing a unit-length hashmark for scale. Brownian motion hitting time moments inside the shape are shown, including the (b) mean: $U_1(\mathbf{x})$, (c) standard deviation: $\sqrt{V_2(\mathbf{x})}$, (d) variance: $V_2(\mathbf{x})$, (e) log skewness: $\log_{10}(1 + W_3(\mathbf{x}))$, (f) log kurtosis: $\log_{10}(1 + W_4(\mathbf{x}))$.

4.1   Stopping Time Survival Function

The survival function of stopping time given the initial position of the particle is denoted $S_\beta(\mathbf{x}, t)$ and is given by:

$$S_\beta(\mathbf{x}, t) = P(T > t | X(0) = \mathbf{x}). \tag{10}$$

The β subscript is added to show the dependence of T on the expected value of the natural lifetime T_L. As $\beta \to \infty$, the stopping time survival function (10) approaches the hitting time survival function (2).

As shown in Appendix B, $S_\beta(\mathbf{x}, t)$ satisfies the initial boundary value problem:

$$
\begin{aligned}
\left(\beta^{-1}\mathcal{I} + \mathcal{L}\right)\{S_\beta(\mathbf{x}, t)\} &= 0, &\quad \forall \mathbf{x} \in \mathbb{R}^n \backslash \partial\Omega, \quad t \geq 0, \\
S_\beta(\mathbf{x}, t) &= 0, &\quad \forall \mathbf{x} \in \partial\Omega, \quad t \geq 0, \\
\lim_{\|\mathbf{x}\| \to \infty} S_\beta(\mathbf{x}, t) &= e^{-t/\beta}, &\quad t \geq 0, \\
S_\beta(\mathbf{x}, 0) &= 1, &\quad \forall \mathbf{x} \in \mathbb{R}^n \backslash \partial\Omega,
\end{aligned}
\tag{11}
$$

where $\mathcal{L}$ is the diffusion operator (4).

4.2   Moments

If we define $U_{k,\beta}(\mathbf{x})$ to be the kth moment of T given $X(0) = \mathbf{x}$ and given $E[T_L] = \beta$, then $U_{k,\beta}(\mathbf{x})$ can be computed according to:

$$U_{k,\beta}(\mathbf{x}) = k \int_0^\infty t^{k-1} S_\beta(\mathbf{x}, t)\, dt. \tag{12}$$

Multiplying each side of (11) by kt^{k-1}, integrating, and employing (12) yields the recursive set of boundary value problems:

$$\left(\beta^{-1}\mathcal{I} - \frac{1}{2n}\Delta\right)\{U_{k,\beta}(\mathbf{x})\} = kU_{k-1,\beta}(\mathbf{x}), \quad \forall\mathbf{x} \in \mathbb{R}^n\backslash\partial\Omega,$$

$$U_{k,\beta}(\mathbf{x}) = 0, \qquad\qquad \forall\mathbf{x} \in \partial\Omega,$$

$$\lim_{\|\mathbf{x}\|\to\infty} U_{k,\beta}(\mathbf{x}) = k!\beta^k, \tag{13}$$

for $k = 1, 2, \ldots$, with $U_{0,\beta}(\mathbf{x}) = E[T^0] = 1$. Noting that $U_{1,\beta}(\mathbf{x})$ satisfies a screened Poisson equation, we can refer to the mean Brownian motion stopping time as a *screened Poisson transform* of the shape.

As in (6), central moments $V_{k,\beta}(\mathbf{x})$ can then be computed as:

$$V_{k,\beta}(\mathbf{x}) = \sum_{m=0}^{k}\binom{k}{m}(-1)^{k-m}U_{m,\beta}(\mathbf{x})U_{1,\beta}(\mathbf{x})^{k-m}, \tag{14}$$

and standardized moments as $W_{k,\beta}(\mathbf{x}) = V_{k,\beta}(\mathbf{x})/(V_{2,\beta}(\mathbf{x}))^{k/2}$.

4.3   Discretization of Moment BVPs

If Γ is extended to include a grid of points outside of the shape as well as inside, we can use centered difference approximations to express discretized versions of the boundary value problems in (13). For $\mathbf{x} \in \Gamma$, we have:

$$\left(\beta^{-1} + \frac{1}{h^2}\right)U_{k,\beta}(\mathbf{x}) - \sum_{i=1}^{n}\frac{U_{k,\beta}(\mathbf{x} - h\mathbf{e}_i) + U_{k,\beta}(\mathbf{x} + h\mathbf{e}_i)}{2nh^2} = kU_{k-1,\beta}(\mathbf{x}).$$

$$\tag{15}$$

For $\mathbf{x} \in \partial\Gamma$, we have $U_{k,\beta} = 0$. To handle the far-field Dirichlet boundary condition, we must end the grid some large but finite distance from the shape and specify that $U_{k,\beta} = k!\beta^k$ at these points.

4.4   Computational Examples of Moments

As we can see in Fig. 2, the mean stopping time $U_{1,\beta}(\mathbf{x})$ exhibits different properties as β varies. (Log mean stopping time is shown to provide better contrast across the interior and exterior of the shape). For small values of β, $U_{1,\beta}(\mathbf{x})$ is approximately $E[T_L] = \beta$ for all spatial positions except for those close to the shape boundary. As β grows, $U_{1,\beta}(\mathbf{x})$ approaches $U_1(\mathbf{x}) = E[T_B]$ on the inside of the shape, and on the outside of the shape, it grows from 0 to β as $\mathbf{x}$ moves away from the shape boundary.

In Fig. 3, we see higher order central and standardized stopping time moments for various choices of β. The skewness $W_{3,\beta}(\mathbf{x})$ and kurtosis $W_{4,\beta}(\mathbf{x})$ achieve maximum values near the shape boundary for all values of β, whereas the standard deviation behaves in a similar manner to the mean.

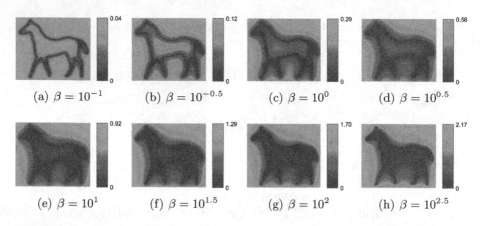

Fig. 2. Log mean stopping time, $\log_{10}(1 + U_{1,\beta}(\mathbf{x}))$, for various values of β.

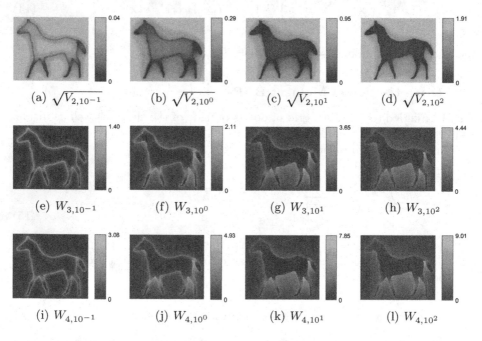

Fig. 3. Higher order Brownian motion stopping time moments for $\beta = 10^{-1}$, 10^0, 10^1, 10^2: (a)–(d) standard deviation, (e)–(h) skewness, (i)–(l) kurtosis. All figures show $\log_{10}(1 + \Phi)$ for each quantity Φ.

5   Shape Classification Experiments

In order to determine the efficacy of utilizing features derived from stopping time moments for shape classification, we perform experiments on two datasets. The first dataset is the collection of natural silhouettes used in [7]. This set

Fig. 4. Sample images from natural silhouettes dataset.

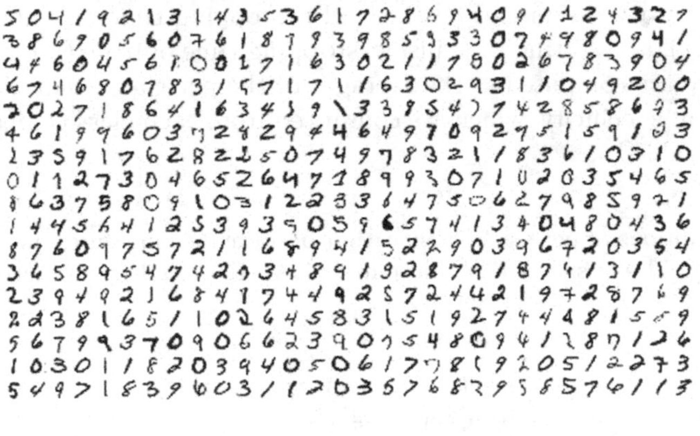

5 0 4 1 9 2 1 3 1 4 3 5 3 6 1 7 2 8 6 9 4 0 9 1 1 2 4 3 2 7
3 8 6 9 0 5 6 0 7 6 1 8 7 9 3 9 8 5 9 3 3 0 7 4 9 8 0 9 4 1
4 4 6 0 4 5 6 1 0 0 1 7 1 6 3 0 2 1 1 7 8 0 2 6 7 8 3 9 0 4
6 7 4 6 8 0 7 8 3 1 5 7 1 7 1 1 6 3 0 2 9 3 1 1 0 4 9 2 0 0
7 0 2 7 1 8 6 4 1 6 3 4 5 9 1 3 3 8 5 4 7 7 4 2 8 5 8 6 9 3
4 6 1 9 9 6 0 3 1 2 4 2 9 4 4 6 4 9 7 0 9 2 9 5 1 5 9 1 9 3
1 3 5 9 1 7 6 2 8 2 2 5 0 7 4 9 7 8 3 2 1 1 8 3 6 1 0 3 1 0
0 1 1 2 7 3 0 4 6 5 2 6 4 7 1 8 9 9 3 0 7 1 0 2 0 3 5 4 6 5
8 6 3 7 5 8 0 9 1 0 3 1 2 2 3 3 6 4 7 5 0 6 2 7 9 8 5 9 2 1
1 4 4 5 6 4 1 2 5 3 9 3 5 0 5 9 6 5 7 4 1 3 4 0 4 8 0 4 3 6
8 7 6 0 7 5 7 2 1 1 6 8 9 4 1 5 2 2 9 0 3 9 6 7 2 0 3 5 4
3 6 5 8 9 5 4 7 4 2 9 3 4 8 9 1 9 2 8 7 9 1 8 7 4 1 3 1 1 0
2 3 9 4 9 2 1 6 8 4 1 7 4 4 9 2 5 7 2 4 4 2 1 9 7 2 8 7 6 9
2 2 3 8 1 6 5 1 1 0 2 6 4 5 8 3 1 5 1 9 2 7 4 4 4 8 1 5 6 9
5 6 7 9 9 3 7 0 9 0 6 6 2 3 9 0 1 5 4 8 0 9 4 1 1 8 7 1 2 6
1 0 3 0 1 1 8 2 0 3 9 4 0 5 0 6 1 7 9 8 1 9 2 0 5 1 2 2 7 3
5 4 9 7 1 8 3 9 6 0 3 1 1 2 0 3 5 7 6 8 7 9 5 8 5 7 6 1 1 3

Fig. 5. Sample images from MNIST database.

includes 490 silhouettes comprising 12 classes of images (cups, hands, humans, horses, birds, fish, rays, cats, dogs and elephants), examples of which are shown in Fig. 4. The second dataset is the MNIST database of handwritten digits [8], which comprises a training set of 60,000 examples and a test set of 10,000 examples, a subset of which are shown in Fig. 5.

For each shape in each dataset, we first computed the hitting time mean $U_1(\mathbf{x})$, standard deviation $W_2(\mathbf{x})$, skewness $W_3(\mathbf{x})$, and kurtosis $W_4(\mathbf{x})$. Then, we defined a canonical choice for β for each shape: $\hat{\beta} = \max_{\mathbf{x} \in \Omega} U_1(\mathbf{x})$, and we computed the stopping time mean $U_{1,\hat{\beta}}(\mathbf{x})$, standard deviation $W_{2,\hat{\beta}}(\mathbf{x})$, skewness $W_{3,\hat{\beta}}(\mathbf{x})$, and kurtosis $W_{4,\hat{\beta}}(\mathbf{x})$. This $\hat{\beta}$ is selected to be large enough so that $U_{1,\hat{\beta}}$ has properties similar to U_1 on the interior of the shape, but small enough so that $U_{1,\hat{\beta}}$ does not grow too large outside the shape.

To extract features from each of these functions, we first compute the gradient magnitude, and then we extract the second, third, and fourth-order Flusser invariants [6], yielding a set of 11 rotation invariant features. Computing and concatenating these features from the first four moments of hitting/stopping time yields a 44-element feature descriptor for the shape.

Feature vectors are provided as input to a classifier, through which shape labels will be predicted and compared to the ground truth. The classifier we use is based on one-versus-all SVMs with Gaussian RBF kernels as implemented in MATLAB's `fitcsvm` function. We employ the heuristic procedure available in `fitcsvm` in order to automatically determine the kernel scaling. For the natural silhouette dataset, 20% of the shapes are used as training data; for the MNIST dataset, we use the provided training set of 60,000 digits.

Tables 2 and 3 show the classification results for each dataset. The average accuracy (AA), average precision (AP), average sensitivity (ASe) and average specificity (ASp) are shown in the first four rows for each set of features. The remaining rows show the average accuracy for each method across all classes. The columns show various combinations of hitting/stopping times inside and outside the shape that were tested: hitting time on inside/zero on outside (HT), stopping time both inside and outside (ST), stopping time on outside/zero on inside (STOut), and hitting time inside/stopping time outside (Fused). AA, AP, ASe, and ASp are computed as the averages of the per-class accuracy, precision, sensitivity, and specificity, which are computed from each class confusion matrix as shown in Table 1.

Table 1. Definitions of per-class classification performance measures, in terms of true positives (TP), false positives (FP), false negatives (FN), and true negatives (TN).

Measure	Definition
Accuracy	$\frac{TP+TN}{TP+FP+FN+TN}$
Precision	$\frac{TP}{TP+FP}$
Sensitivity	$\frac{TP}{TP+FN}$
Specificity	$\frac{TN}{FP+TN}$

We see from Table 2 that classification based on stopping time moments computed outside the shape fused with hitting time moments computed inside the shape outperformed the other methods in accuracy, precision, sensitivity and specificity over all twelve classes of natural silhouettes. This is also consistent with the results for classification of MNIST digits reported in Table 3.

Table 2. Natural silhouettes classification results. Class rows show per-class accuracy. All quantities are percentages, with the exception of the number of samples.

	No. of samples	HT	ST	STOut	Fused
Average accuracy	-	94.14	92.03	91.99	**95.39**
Average precision	-	64.46	48.89	49.73	**74.22**
Average sensitivity	-	67.39	49.46	48.76	**72.26**
Average specificity	-	96.82	95.64	95.62	**97.47**
Class 1: Mug	26	98.71	95.09	96.64	**99.22**
Class 2: Hand	26	**99.74**	96.90	95.87	99.48
Class 3: Person	55	97.67	97.67	**97.93**	97.67
Class 4: Horse	80	83.46	91.21	91.47	**91.73**
Class 5: Bird	50	**89.92**	84.75	83.98	89.41
Class 6: Shark	43	96.64	94.83	94.57	**96.90**
Class 7: Fish	28	95.87	91.47	91.47	**97.93**
Class 8: Stingray	47	93.54	86.56	87.60	**94.32**
Class 9: Cat	42	89.66	86.30	85.79	**92.51**
Class 10: Dog	49	91.73	88.11	88.63	**91.99**
Class 11: Hawk	17	96.64	96.38	**97.16**	96.90
Class 12: Elephant	27	96.12	95.09	92.76	**96.64**

Table 3. Digits classification results. Class rows show per-class accuracy. All quantities are percentages, with the exception of the number of samples.

	No. of samples	HT	ST	STOut	Fused
Average accuracy	-	88.46	88.26	87.94	**90.99**
Average precision	-	41.29	40.44	38.97	**54.21**
Average sensitivity	-	39.29	37.57	36.53	**52.37**
Average specificity	-	93.66	93.56	93.37	**95.07**
Class 0	1,000	86.99	85.84	85.75	**93.51**
Class 1	1,000	97.34	96.18	95.77	**98.08**
Class 2	1,000	87.36	85.94	86.56	**88.26**
Class 3	1,000	86.22	87.47	87.18	**88.82**
Class 4	1,000	86.78	87.82	86.59	**89.34**
Class 5	1,000	88.21	88.66	88.24	**89.58**
Class 6	1,000	87.57	**91.60**	90.64	91.18
Class 7	1,000	90.93	87.88	87.80	**92.02**
Class 8	1,000	85.85	85.50	84.00	**88.87**
Class 9	1,000	87.33	85.67	86.89	**90.28**

6   Conclusion

In this article, we showed how implicit shape representations based on Brownian motion hitting times can be generalized to be meaningful both inside and outside a shape. We accomplish this generalization by introducing a natural lifetime for the Brownian motion and then defining the *stopping time* to be the minimum of the hitting time and the natural lifetime. The expected stopping time, both inside and outside of a shape, satisfies a screened Poisson equation. Using features extracted from both hitting time and stopping time moments, we show that classification of both natural silhouettes and handwritten digits can be improved when compared to using features extracted from hitting time moments alone.

Acknowledgments. The authors would like to thank Lena Gorelick for helpful discussions.

A   Proof That (2) Satisfies (3)

Suppose $\bar{\Gamma} \subset \bar{\Omega}$ is the n-dimensional uniform grid of points defined in Sect. 3.3. Consider a particle that undergoes a symmetric random walk on $\bar{\Gamma}$. Let τ be the finite time between steps of the random walk. The position of the particle is given by $X(t)$, where t represents continuous time; hence, $X(t)$ is piecewise constant with steps at $t = k\tau$, $k = 0, 1, 2, \ldots$. The hitting time T_B corresponds to the first time the particle lands in $\partial\Gamma$; i.e. $T_B = \inf_{t \geq 0}\{t | X(t) \in \partial\Gamma\}$.

From (2), we have $S(\mathbf{x}, t) = P(T_B > t | X(0) = \mathbf{x})$. If we condition on the first step of the random walk, for points in Γ we have:

$$
\begin{aligned}
S(\mathbf{x}, t+\tau) &= \Pr\{T_B > t + \tau \,|\, X(0) = \mathbf{x}\} \\
&= \sum_{j=1}^{n} \left[\begin{array}{l} \Pr\{T_B > t + \tau | X(0) = \mathbf{x}, X(1) = \mathbf{x} + h\mathbf{e}_j\} \\ \quad \cdot \Pr\{X(1) = \mathbf{x} + h\mathbf{e}_j | X(0) = \mathbf{x}\} \\ + \Pr\{T_B > t + \tau | X(0) = \mathbf{x}, X(1) = \mathbf{x} - h\mathbf{e}_j\} \\ \quad \cdot \Pr\{X(1) = \mathbf{x} - h\mathbf{e}_j | X(0) = \mathbf{x}\} \end{array} \right] \\
&= \frac{1}{2n} \sum_{j=1}^{n} \left[\begin{array}{l} \Pr\{T_B > t \,|\, X(0) = \mathbf{x} + h\mathbf{e}_j\} \\ + \Pr\{T_B > t \,|\, X(0) = \mathbf{x} - h\mathbf{e}_j\} \end{array} \right] \\
&= \frac{1}{2n} \sum_{j=1}^{n} [S(\mathbf{x} + h\mathbf{e}_j, t) + S(\mathbf{x} - h\mathbf{e}_j, t)].
\end{aligned}
\tag{16}
$$

Alternatively, expanding $S(\mathbf{x}, t)$ about t yields:

$$
S(\mathbf{x}, t + \tau) = S(\mathbf{x}, t) + \tau \frac{\partial}{\partial t} S(\mathbf{x}, t) + O(\tau^2),
\tag{17}
$$

and so combining (16) and (17) gives:

$$
S(\mathbf{x}, t) + \tau \frac{\partial}{\partial t} S(\mathbf{x}, t) = \frac{1}{2n} \sum_{j=1}^{n} [S(\mathbf{x} + h\mathbf{e}_j, t) + S(\mathbf{x} - h\mathbf{e}_j, t)] + O(\tau^2).
\tag{18}
$$

Expanding $S(\mathbf{x}, t)$ about $\mathbf{x}$ yields:

$$S(\mathbf{x} \pm h\mathbf{e}_j, t) = S(\mathbf{x}, t) \pm hS_{x_j}(\mathbf{x}, t) + \frac{h^2}{2}\frac{\partial^2}{\partial x_j^2}S(\mathbf{x}, t) + O(h^3), \qquad (19)$$

so

$$S(\mathbf{x} + h\mathbf{e}_j, t) + S(\mathbf{x} - h\mathbf{e}_j, t) = 2S(\mathbf{x}, t) + h^2\frac{\partial^2}{\partial x_j^2}S(\mathbf{x}, t) + O(h^3), \qquad (20)$$

and (18) can be written as:

$$S(\mathbf{x}, t) + \tau\frac{\partial}{\partial t}S(\mathbf{x}, t) = S(\mathbf{x}, t) + \frac{h^2}{2n}\sum_{j=1}^{n}\left[\frac{\partial^2}{\partial x_j^2}S(\mathbf{x}, t)\right] + O(\tau^2) + O(h^3).$$
$$(21)$$

Rearranging terms and dividing both sides of (21) by τ yields:

$$\frac{\partial}{\partial t}S(\mathbf{x}, t) - \frac{1}{2n}\frac{h^2}{\tau}\sum_{j=1}^{n}\left[\frac{\partial^2}{\partial x_j^2}S(\mathbf{x}, t)\right] = O(\tau) + O\left(\frac{h^3}{\tau}\right). \qquad (22)$$

Taking the limit of both sides of (22) as h and τ approach zero while h^2/τ is fixed at 1 yields:

$$\frac{\partial}{\partial t}S(\mathbf{x}, t) - \frac{1}{2n}\Delta S(\mathbf{x}, t) = 0, \qquad (23)$$

where Δ is the spatial Laplacian. (23) now applies for all $\mathbf{x} \in \Omega$ and $t \geq 0$. Note that $T_B = 0$ on the boundary, so $S(\mathbf{x}, t) = 0$ for $\mathbf{x} \in \partial\Omega$. Furthermore, $S(\mathbf{x}, 0) = 1$ for any point on the interior, and hence, $S(\mathbf{x}, t)$ satisfies (3).   □

B   Proof That (10) Satisfies (11)

From (10), we have $S_\beta(\mathbf{x}, t) = P(T > t | X(0) = \mathbf{x})$. In addition, the density function of T_L is:

$$f_{T_L}(s) = \frac{d}{ds}P(T_L < s) = \frac{d}{ds}\left[1 - e^{-s/\beta}\right] = \beta^{-1}e^{-s/\beta}. \qquad (24)$$

Conditioning on T_L allows us to write $S_\beta(\mathbf{x}, t)$ as:

$$S_\beta(\mathbf{x}, t) = \int_0^\infty P(\min(T_L, T_B) > t | X(0) = \mathbf{x}, T_L = s)f_{T_L|X(0)}(s)\,ds$$

$$= \int_0^\infty P(\min(s, T_B) > t | X(0) = \mathbf{x})f_{T_L}(s)\,ds$$

$$= \int_0^\infty P(\min(s, T_B) > t | X(0) = \mathbf{x})\beta^{-1}e^{-s/\beta}\,ds. \qquad (25)$$

For $s \in [0, t]$, $P(\min(s, T_B) > t | X(0) = \mathbf{x}) = 0$, so (25) simplifies to:

$$
\begin{aligned}
S_\beta(\mathbf{x}, t) &= \int_t^\infty P(\min(s, T_B) > t | X(0) = \mathbf{x}) \beta^{-1} e^{-s/\beta} ds \\
&= \int_t^\infty P(T_B > t | X(0) = \mathbf{x}) \beta^{-1} e^{-s/\beta} ds \\
&= \left(\int_t^\infty \beta^{-1} e^{-s/\beta} ds \right) P(T_B > t | X(0) = \mathbf{x}) \\
&= e^{-t/\beta} S(\mathbf{x}, t).
\end{aligned}
\tag{26}
$$

Now, consider applying the operator $\mathcal{L}$ to both sides of (26):

$$
\begin{aligned}
\mathcal{L}\{S_\beta(\mathbf{x}, t)\} &= \frac{\partial}{\partial t} S_\beta(\mathbf{x}, t) - \frac{1}{2n} \Delta S_\beta(\mathbf{x}, t) \\
&= \frac{\partial}{\partial t} e^{-t/\beta} S(\mathbf{x}, t) - \frac{1}{2n} \Delta e^{-t/\beta} S(\mathbf{x}, t) \\
&= -e^{-t/\beta} \beta^{-1} S(\mathbf{x}, t) + e^{-t/\beta} \frac{\partial}{\partial t} S(\mathbf{x}, t) - e^{-t/\beta} \frac{1}{2n} \Delta S(\mathbf{x}, t) \\
&= e^{-t/\beta} \left(\mathcal{L} - \beta^{-1} \mathcal{I} \right) \{S(\mathbf{x}, t)\}.
\end{aligned}
\tag{27}
$$

For points that originate inside the shape, (3) tells us that $\mathcal{L}\{S(\mathbf{x}, t)\} = 0$; hence, (27) reduces to

$$
\mathcal{L}\{S_\beta(\mathbf{x}, t)\} = -\beta^{-1} e^{-t/\beta} S(\mathbf{x}, t) = -\beta^{-1} S_\beta(\mathbf{x}, t),
\tag{28}
$$

and so $\left(\beta^{-1} \mathcal{I} + \mathcal{L} \right) \{S_\beta(\mathbf{x}, t)\} = 0$ for all $\mathbf{x} \in \Omega$ and $t \geq 0$. For particles that originate on the shape boundary, it must be true that $S_\beta(\mathbf{x}, t) = 0$.

For particles that originate outside the shape boundary, an argument similar to that of Appendix A establishes that $\mathcal{L}\{S(\mathbf{x}, t)\} = 0$, and so (28) is also satisfied. For points sufficiently far from the shape boundary, $S_\beta(\mathbf{x}, t)$ is equal to the survival function of the natural lifetime, T_L. That is,

$$
\lim_{\|\mathbf{x}\| \to \infty} S_\beta(\mathbf{x}, t) = P(T_L > t | X(0) = \mathbf{x}) = P(T_L > t) = e^{-t/\beta}.
\tag{29}
$$

For all particles that originate at positions not on the shape boundary, it must be true that $S_\beta(\mathbf{x}, 0) = 0$. Combining all of these results yields (11).

C   Code

Prototype implementations of the algorithms for computing various moments of hitting time and stopping time are available for download at MATLAB Central (http://www.mathworks.com/matlabcentral/) under File ID #58754.

References

1. Biederman, I.: Recognition-by-components: a theory of human image understanding. Psychol. Rev. **94**(2), 115 (1987)
2. Blank, M., Gorelick, L., Shechtman, E., Irani, M., Basri, R.: Actions as space-time shapes. In: Proceedings of IEEE International Conference on Computer Vision, vol. 2, pp. 1395–1402 (2005)
3. Blum, H.: A transformation for extracting new descriptions of shape. In: Proceedings of Symposium on Models for the Perception of Speech and Visual Form (1964)
4. Blum, H., Nagel, R.N.: Shape description using weighted symmetric axis features. Pattern Recogn. **10**(3), 167–180 (1978)
5. Fabbri, R., Costa, L.D.F., Torelli, J.C., Bruno, O.M.: 2D Euclidean distance transform algorithms: a comparative survey. ACM Comput. Surv. **40**(1), 2 (2008)
6. Flusser, J.: On the independence of rotation moment invariants. Pattern Recogn. **33**, 1405–1410 (2000)
7. Gorelick, L., Galun, M., Sharon, E., Basri, R., Brandt, A.: Shape representation and classification using the Poisson equation. IEEE Trans. Pattern Anal. Mach. Intell. **28**(12), 1991–2005 (2006)
8. LeCun, Y., Cortes, C., Burges, C.J.C.: The MNIST database of handwritten digits. http://yann.lecun.com/exdb/mnist/ (1998)
9. Ma, W.C., Wu, F.C., Ouhyoung, M.: Skeleton extraction of 3D objects with radial basis functions. In: Proceedings of IEEE Shape Modeling International, pp. 207–215 (2003)
10. Maes, C., Fabry, T., Keustermans, J., Smeets, D., Suetens, P., Vandermeulen, D.: Feature detection on 3D face surfaces for pose normalisation and recognition. In: Proceedings of IEEE International Conference on Biometrics: Theory Applications and Systems, pp. 1–6 (2010)
11. Marr, D., Nishihara, H.K.: Representation and recognition of the spatial organization of three-dimensional shapes. Proc. R. Soc. Lond. B: Biol. Sci. **200**(1140), 269–294 (1978)
12. Palágyi, K., Kuba, A.: A 3D 6-subiteration thinning algorithm for extracting medial lines. Pattern Recogn. Lett. **19**(7), 613–627 (1998)
13. Pentland, A.P.: Recognition by parts. Technical report, DTIC Document (1987)
14. Pizer, S.M., Fletcher, P.T., Joshi, S., Thall, A., Chen, J.Z., Fridman, Y., Fritsch, D.S., Gash, A.G., Glotzer, J.M., Jiroutek, M.R., et al.: Deformable m-reps for 3D medical image segmentation. Int. J. Comput. Vis. **55**(2–3), 85–106 (2003)
15. Pizer, S.M., Fritsch, D.S., Yushkevich, P.A., Johnson, V.E., Chaney, E.L.: Segmentation, registration, and measurement of shape variation via image object shape. IEEE Trans. Med. Imaging **18**(10), 851–865 (1999)
16. Shaked, D., Bruckstein, A.M.: Pruning medial axes. Comput. Vis. Image Underst. **69**(2), 156–169 (1998)
17. Siddiqi, K., Kimia, B.B.: Parts of visual form: computational aspects. IEEE Trans. Pattern Anal. Mach. Intell. **17**(3), 239–251 (1995)
18. Yogarajah, P., Condell, J.V., Prasad, G.: PRWGEI: poisson random walk based gait recognition. In: Proceedings of IEEE International Symposium on Image and Signal Processing and Analysis, pp. 662–667 (2011)
19. Zhou, Y., Kaufman, A., Toga, A.W.: Three-dimensional skeleton and centerline generation based on an approximate minimum distance field. Vis. Comput. **14**(7), 303–314 (1998)

Picture Scanning Automata

Henning Fernau[1], Meenakshi Paramasivan[1(✉)], and D. Gnanaraj Thomas[2]

[1] Fachbereich 4 - Abteilung Informatik, Universität Trier, 54286 Trier, Germany
{fernau,paramasivan}@uni-trier.de
[2] Department of Mathematics, Madras Christian College, Chennai 600059, India
dgthomasmcc@yahoo.com

Abstract. We are systematically discussing finite automata working on rectangular-shaped arrays (i.e., pictures), reading them with different scanning strategies. We show that all 32 different variants only describe two different classes of array languages. We also introduce and discuss Mealy machines working on pictures and show how these can be used in a modular design of picture processing devices.

Keywords: Picture languages · Image processing · Finite automata

1 Introduction

Syntactic considerations of digital images have a tradition of about five decades. They should (somehow) reflect methods applied to picture processing. However, two basic methods of scanning pictures in practice have not been thoroughly investigated from a more theoretical point of view: that of scanning pictures line by line and that of using space-filling curves. Here, we continue such an investigation with what can be considered as the most simple way of defining space-filling curves: scanning line after line of an image, alternating the direction of movement every time when the image boundary is encountered, as well as by scanning line by line following the same direction each time. We consider finite automata that work the described ways, extending previous studies [2]. This shows tighter connections between finite automata working on pictures and array grammars of different types; for overviews on this topic, we refer to [4,5].

Our main contributions can be summarized as follows: (a) We show that an amazing variety of picture scanning devices basically only describe two different array language families. (b) This result is obtained by making use of connections to the theory of dihedral groups. (c) Further closure properties of array language families are derived. (d) We also introduce Mealy picture machines and show how they could be useful in a modular design of picture processing automata.

General Definitions. In the following, we briefly recall some standard definitions and notations regarding two-dimensional words and languages. A *two-dimensional word* (also called *picture, matrix* or *array*) *over* Σ is a tuple

$$W := ((a_{1,1}, a_{1,2}, \ldots, a_{1,n}), (a_{2,1}, a_{2,2}, \ldots, a_{2,n}), \ldots, (a_{m,1}, a_{m,2}, \ldots, a_{m,n})),$$

© Springer International Publishing AG 2017
R.P. Barneva et al. (Eds.): CompIMAGE 2016, LNCS 10149, pp. 132–147, 2017.
DOI: 10.1007/978-3-319-54609-4_10

where $m, n \in \mathbb{N}$ and, for every i, $1 \leq i \leq m$, and j, $1 \leq j \leq n$, $a_{i,j} \in \Sigma$. We define the *number of columns* (or *width*) and *number of rows* (or *height*) of W by $|W|_c := n$ and $|W|_r := m$, respectively. We also denote W by $[a_{i,j}]_{m,n}$ or by a matrix in a more pictorial form. If we want to refer to the jth symbol in row i of the picture W, then we use $W[i,j] = a_{i,j}$. By Σ^{++}, we denote the set of all (non-empty) pictures over Σ. Every subset $L \subseteq \Sigma^{++}$ is a *picture language*.

2   Operations on Pictures and Picture Languages

2.1   Binary Operations

Let $W := [a_{i,j}]_{m,n}$ and $W' := [b_{i,j}]_{m',n'}$ be two pictures over Σ. The *column concatenation* of W and W', denoted by $W \oplus W'$, is undefined if $m \neq m'$ and is otherwise obtained by writing W to the left of W', yielding the picture

$$
\begin{matrix}
a_{1,1} & a_{1,2} & \cdots & a_{1,n} & b_{1,1} & b_{1,2} & \cdots & b_{1,n'} \\
a_{2,1} & a_{2,2} & \cdots & a_{2,n} & b_{2,1} & b_{2,2} & \cdots & b_{2,n'} \\
\vdots & \vdots & \ddots & \vdots & \vdots & \vdots & \ddots & \vdots \\
a_{m,1} & a_{m,2} & \cdots & a_{m,n} & b_{m',1} & b_{m',2} & \cdots & b_{m',n'}
\end{matrix}
$$

The *row concatenation* of W and W', denoted by $W \ominus W'$, is undefined if $n \neq n'$ and is otherwise obtained by writing W above of W'. The row and column catenation operations can be also viewed as operations on languages. Also, we can define n-fold iterations (powers) of column catenation as W^n and n-fold iterations (powers) of row catenation as W_n. Accordingly, Σ_m^n is understood, as well as $\Sigma_m^+ = \bigcup_{n \geq 1} \Sigma_m^n$ and similarly Σ_+^n. In this sense, $\Sigma^{++} = \Sigma_+^+$.

2.2   Unary Operations and Connections to Group Theory

As pictures are (also) geometrical objects, several further unary operations can be introduced [7]: *quarter-turn* (rotate clockwise by 90°) Q, *half-turn* (rotate by 180°) H, *anti-quarter-turn* (rotate anti-clockwise by 90° (or rotate clockwise by 270°)) Q^{-1}, *transpose* T (reflection along the main diagonal), *anti-transpose* T' (reflection along the anti-diagonal), R_b (reflection along a horizontal (base) line), R_v (reflection along a vertical line). Together with the *identity* I, these (now eight) operators form a non-commutative group (with respect to composition), the well-known dihedral group D_4 [1]; see Table 1a.

In Table 1a, $\circ$ is the function composition. So, if $f : X \to Y$ and $g : Y \to Z$ are two functions, then $g \circ f : X \to Z$ is defined by $(g \circ f)(x) = g(f(x))$ for all $x \in X$. How Table 1a works is shown below for $W := [a_{i,j}]_{m,n}$ in the following.

$$
(T \circ R_v)(W) = T(R_v(W)) = T\begin{pmatrix} a_{1,n} & \cdots & a_{1,2} & a_{1,1} \\ a_{2,n} & \cdots & a_{2,2} & a_{2,1} \\ \vdots & \vdots & \ddots & \vdots \\ a_{m,n} & \cdots & a_{m,2} & a_{m,1} \end{pmatrix} = \begin{matrix} a_{1,n} & a_{2,n} & \cdots & a_{m,n} \\ \vdots & \ddots & \vdots & \vdots \\ a_{1,2} & a_{2,2} & \cdots & a_{m,2} \\ a_{1,1} & a_{2,1} & \cdots & a_{m,1} \end{matrix} = Q^{-1}(W).
$$

Table 1. Unary operators

∘	I	Q^{-1}	H	Q	R_v	R_b	T	T'
I	I	Q^{-1}	H	Q	R_v	R_b	T	T'
Q^{-1}	Q^{-1}	H	Q	I	T	T'	R_b	R_v
H	H	Q	I	Q^{-1}	R_b	R_v	T'	T
Q	Q	I	Q^{-1}	H	T'	T	R_v	R_b
R_v	R_v	T'	R_b	T	I	H	Q	Q^{-1}
R_b	R_b	T	R_v	T'	H	I	Q^{-1}	Q
T	T	R_v	T'	R_b	Q^{-1}	Q	I	H
T'	T'	R_b	T	R_v	Q	Q^{-1}	H	I

(a) Composition table of unary operators.

	Normal Form
I	$(Q \circ Q) \circ (Q \circ Q)$
Q^{-1}	$Q \circ (Q \circ Q)$
H	$Q \circ Q$
R_v	$Q \circ T$
R_b	$T \circ Q$
T'	$Q \circ (Q \circ T)$

(b) Normal form for the unary operators with Q and T.

Let $\mathcal{O} = \{I, Q^{-1}, H, Q, R_v, R_b, T, T'\}$ be the set of these 8 unary operators comprising D_4. The operators in D_4 are usually partitioned into the four rotations (including the identity) $\{I, Q^{-1}, H, Q\}$, which form the subgroup D_2 of D_4, and four reflections $\{R_v, R_b, T, T'\}$. These operations can be also applied (picture-wise) to picture languages and (language-wise) to families of picture languages. It is interesting to add the fact that one single rotation Q generates all rotations (as a subgroup of D_4) and that all of D_4 are generated by one rotation Q and one reflection T. In Table 1b we make explicit how any operator can be written using the composition of the two mentioned operators. Table 1b can be deduced from Table 1a, for instance $Q^{-1} = Q \circ (Q \circ Q)$ since $Q^{-1} = Q \circ H$ and $H = Q \circ Q$. This simple observation helps simplify several of our arguments.

For instance, we can combine sequences of catenation and unary operations from D_4 to obtain Table 2, starting out from the four simple observations
(1) $Q(W_1 \ominus W_2) = Q(W_2) \oplus Q(W_1)$ (however, mind the sequence of arguments),
(2) $Q(W_1 \oplus W_2) = Q(W_2) \ominus Q(W_1)$, (3) $T(W_1 \ominus W_2) = T(W_1) \oplus T(W_2)$ and
(4) $T(W_1 \oplus W_2) = T(W_1) \ominus T(W_2)$. For example,

$$
\begin{aligned}
H(W_1 \ominus W_2) &= (Q \circ Q)(W_1 \ominus W_2) \text{ (By Table 1 b)} \\
&= Q(Q(W_1 \ominus W_2)) \text{ (By the definition of } \circ) \\
&= Q(Q(W_2) \oplus Q(W_1)) \text{ (Apply Observation (1))} \\
&= Q(Q(W_1)) \ominus Q(Q(W_2)) \text{ (Apply Observation (2))} \\
&= H(W_1) \ominus H(W_2) \text{ (By Table 1 b)}
\end{aligned}
$$

3   General Boustrophedon Finite Automata

We now give one of the main definitions of this paper, introducing a new, parameterized automaton model for picture processing.

A *general boustrophedon finite automaton*, or GBFA for short, can be specified as an 8-tuple $M = (Q, \Sigma, R, s, F, \#, \square, D)$, where Q is a finite set of states, where Q is partitioned into Q_f and Q_b, Σ is an input alphabet, $R \subseteq Q \times (\Sigma \cup \{\#\}) \times Q$ is a finite set of rules. A rule $(q, a, p) \in R$ is usually written as $qa \rightarrow p$. We

Table 2. Table of operators.

	$W_1 \ominus W_2$	$W_1 \oplus W_2$
I	$I(W_1) \ominus I(W_2)$	$I(W_1) \oplus I(W_2)$
Q	$Q(W_2) \oplus Q(W_1)$	$Q(W_1) \ominus Q(W_2)$
Q^{-1}	$Q^{-1}(W_1) \oplus Q^{-1}(W_2)$	$Q^{-1}(W_2) \ominus Q^{-1}(W_1)$
H	$H(W_2) \ominus H(W_1)$	$H(W_2) \oplus H(W_1)$
T	$T(W_1) \oplus T(W_2)$	$T(W_1) \ominus T(W_2)$
T'	$T'(W_2) \oplus T'(W_1)$	$T'(W_2) \ominus T'(W_1)$
R_v	$R_v(W_1) \ominus R_v(W_2)$	$R_v(W_2) \oplus R_v(W_1)$
R_b	$R_b(W_2) \ominus R_b(W_1)$	$R_b(W_1) \oplus R_b(W_2)$

impose some additional restrictions. If $q \in Q_f$ and $a \in \Sigma$, then $qa \to p \in R$ is only possible if $p \in Q_f$. Such transitions are also called *forward transitions* and collected within R_f. Similarly, if $q \in Q_b$ and $a \in \Sigma$, $qa \to p \in R$ is only possible if $p \in Q_b$ (*backward transitions*, collected in R_b). Finally, *border transitions* (collected in $R_\#$) are of the form $q\# \to p$ with $q \in Q_f$ iff $p \in Q_b$. Namely, the special symbol $\# \notin \Sigma$ indicates the border of the rectangular picture that is processed, $s \in Q_f$ is the initial state, F is the set of final states, and $D \in \mathcal{D}$ indicates the move directions. Here,

$$\mathcal{D} = \left\{ \begin{pmatrix} s \to & \downarrow \\ \downarrow & \leftarrow \end{pmatrix}, \begin{pmatrix} s \downarrow & \to \\ \to & \uparrow \end{pmatrix}, \begin{pmatrix} \downarrow & \leftarrow s \\ \to & \downarrow \end{pmatrix}, \begin{pmatrix} \leftarrow & \downarrow s \\ \uparrow & \leftarrow \end{pmatrix}, \begin{pmatrix} \to & \downarrow \\ s\uparrow & \to \end{pmatrix}, \begin{pmatrix} \uparrow & \leftarrow \\ s \to & \uparrow \end{pmatrix}, \begin{pmatrix} \downarrow & \leftarrow \\ \leftarrow & \uparrow s \end{pmatrix}, \begin{pmatrix} \to & \uparrow \\ \uparrow & \leftarrow s \end{pmatrix} \right\}$$

We are now going to discuss the notions of configurations, valid configurations and an according configuration transition to formalize the work of GBFAs, based on snapshots of their work.

Let $\square$ be a new symbol indicating an *erased* position and let $\Sigma_{\#,\square} := \Sigma \cup \{\#, \square\}$. Then $C_M := Q \times (\Sigma_{\#,\square}^{++} \cap (\{\#\}^+ \ominus (\{\#\}_+ \oplus (\Sigma \cup \{\square\})^{++} \oplus \{\#\}_+) \ominus \{\#\}^+)) \times \{f, d\}$ is the set of configurations of M. Hence, the first and last columns and the first and last rows are completely filled with $\#$, and these are the only positions that contain $\#$.

The *initial configuration* is determined by the input array $A \in \Sigma^{++}$. More precisely, if A has m rows and n columns, then

$$\left(s, \#^{n+2} \ominus (\#_m \oplus A \oplus \#_m) \ominus \#^{n+2}, f \right)$$

shows the according initial configuration $c_{init}(A)$. Similarly, a *final configuration* is then given by

$$\left(q_f, \#^{n+2} \ominus (\#_m \oplus \square_m^n \oplus \#_m) \ominus \#^{n+2}, d \right)$$

for some $q_f \in F$ and $d \in \{b, f\}$.

The processing of the automaton is then crucially depending on $D \in \mathcal{D}$. The arrow that appears together with s indicates the direction of the forward processing of the first, third, fifth etc. line. For instance, the first listed direction contains $s \to$, determining that the odd-numbered rows of the input array are scanned left to right. Similarly, the second listed direction contains $s \downarrow$, determining that the odd-numbered columns of the input array are scanned top to bottom. When the automaton encounters a border symbol, it processes the next line in the reversed way (backward processing). This is also indicated in the little pictures that describe $D \in \mathcal{D}$. For instance, in the first case, the $\downarrow$ in the first row indicates that after hitting the border, the automaton moves downwards, processing (as indicated by the $\leftarrow$ in the last row) now from right to left, until the border is hit again, which means to move downwards one more row, as indicated by the $\downarrow$ in the last row. The other $D \in \mathcal{D}$ can be interpreted in a similar fashion, as explained below. Let us now formalize this description. Notice that an odd-numbered row of the input array corresponds to an even-numbered row if we consider the input array bordered by a #-layer.

- If (p, A, f) and (q, A', f) are two configurations such that A and A' are identical but for one position (i, j), $1 \le i \le m+2$, $1 \le j \le n+2$, where $A'[i,j] = \square$ while $A[i,j] \in \Sigma$, then $(p, A, f) \vdash_M (q, A', f)$ if $pA[i,j] \to q \in R_f$. Moreover, i is even.
- Conversely, if (p, A, f) and (q, A', f) are two configurations such that A and A' are identical but for one position (i, j), $1 \le i \le m+2$, $1 \le j \le n+2$, where $A'[i, j] = \square$ while $A[i, j] \in \Sigma$, then $(p, A, b) \vdash_M (q, A', b)$ if $pA[i,j] \to q \in R_b$. Moreover, i is odd.
- If (p, A, f) and (q, A, b) are two configurations, then $(p, A, f) \vdash_M (q, A, b)$ or $(p, A, b) \vdash_M (q, A, f)$ if $p\# \to q \in R_\#$.

The reflexive transitive closure of the relation $\vdash_M$ is denoted by $\vdash_M^*$. $A \in \Sigma^{++}$ is accepted by a GBFA M with direction $D_{BFA} := \left(\begin{smallmatrix} s\to & \downarrow \\ \downarrow & \leftarrow \end{smallmatrix} \right)$ if $c_{init}(A) \vdash_M^* c$ such that c is a final configuration.

The following illustrates how such a GBFA scans some input picture and how a picture of a valid configuration looks like; it can be seen that the sequence of $\square$ only indicates how far the input has been processed:

It should be also clear that the representation on the right-hand side of the previous picture contains all information necessary to describe a configuration apart from the state. Now, we define the other modes by applying transformations according to the following table.

$D=$	$\begin{pmatrix} s\downarrow & \to \\ \to & \uparrow \end{pmatrix}$	$\begin{pmatrix} \downarrow & \leftarrow s \\ \to & \downarrow \end{pmatrix}$	$\begin{pmatrix} \leftarrow & \downarrow s \\ \uparrow & \to \end{pmatrix}$	$\begin{pmatrix} \to & \downarrow \\ s\uparrow & \to \end{pmatrix}$	$\begin{pmatrix} \uparrow & \leftarrow \\ s\to & \uparrow \end{pmatrix}$	$\begin{pmatrix} \downarrow & \leftarrow \\ \leftarrow & \uparrow s \end{pmatrix}$	$\begin{pmatrix} \to & \uparrow \\ \uparrow & \leftarrow s \end{pmatrix}$
$f_D(A)=$	$T(A)$	$R_v(A)$	$Q^{-1}(A)$	$Q(A)$	$R_b(A)$	$T'(A)$	$H(A)$

A is accepted by a GBFA M with a different direction D if $f_D(A)$ is accepted by the GBFA M_{BFA} that coincides with M in every detail except for the direction, which is now D_{BFA}. This means, for instance, in the case of $D = \begin{pmatrix} s\downarrow & \to \\ \to & \uparrow \end{pmatrix}$, that instead of scanning the input array A column by column, the first column top-down, the second bottom-up, and so forth, we rather transpose A and then scan the transposed array row by row, the first row left-right, the second right-left, and so forth.

The GBFA M is deterministic, or a GBDFA for short, if for each $p \in Q$ and $a \in \Sigma \cup \{\#\}$, there is at most one $q \in Q$ with $pa \to q \in R$.

This way, we define language classes like $\mathcal{L}_D(\text{GBFA})$ of those array languages accepted by GBFAs working with direction D, as well as

$$\mathcal{L}(\text{GBFA}) := \bigcup_{D \in \mathcal{D}} \mathcal{L}_D(\text{GBFA}).$$

Of course, the interesting question is if the eight language families that we can obtain in these ways are really different from each other or not. Also, the situation of $\mathcal{L}(\text{GBFA})$ needs to be investigated, as well as the role of determinism. We will see that the group-theoretic excursion from above will simplify our reasoning a lot. From the definitions themselves, we can immediately derive the following characterization result.

Theorem 1. *The class* $\mathcal{L}_{D_{BFA}}(\text{GBFA})$ *coincides with the following classes of picture languages:* $T\left(\mathcal{L}_{\left(\begin{smallmatrix} s\downarrow & \to \\ \to & \uparrow \end{smallmatrix}\right)}(\text{GBFA})\right)$, $R_v\left(\mathcal{L}_{\left(\begin{smallmatrix} \downarrow & \leftarrow s \\ \to & \downarrow \end{smallmatrix}\right)}(\text{GBFA})\right)$,
$Q^{-1}\left(\mathcal{L}_{\left(\begin{smallmatrix} \leftarrow & \downarrow s \\ \uparrow & \to \end{smallmatrix}\right)}(\text{GBFA})\right)$, $Q\left(\mathcal{L}_{\left(\begin{smallmatrix} \to & \downarrow \\ s\uparrow & \to \end{smallmatrix}\right)}(\text{GBFA})\right)$, $R_b\left(\mathcal{L}_{\left(\begin{smallmatrix} \uparrow & \leftarrow \\ s\to & \uparrow \end{smallmatrix}\right)}(\text{GBFA})\right)$,
$T'\left(\mathcal{L}_{\left(\begin{smallmatrix} \downarrow & \leftarrow \\ \leftarrow & \uparrow s \end{smallmatrix}\right)}(\text{GBFA})\right)$, *and* $H\left(\mathcal{L}_{\left(\begin{smallmatrix} \to & \uparrow \\ \uparrow & \leftarrow s \end{smallmatrix}\right)}(\text{GBFA})\right)$.

Due to the connections to group theory described above, we can easily infer from the previous theorem characterizations of the seven other classes, referring back to $\mathcal{L}_{D_{BFA}}(\text{GBFA})$. We collect these results in the following theorem.

Theorem 2. *We obtain the following list of characterizations.*

- $\mathcal{L}_{\left(\begin{smallmatrix} s\downarrow & \rightarrow \\ \rightarrow & \uparrow \end{smallmatrix}\right)}(\text{GBFA}) = T\left(\mathcal{L}_{D_{BFA}}(\text{GBFA})\right).$
- $\mathcal{L}_{\left(\begin{smallmatrix} \downarrow & \leftarrow s \\ \rightarrow & \downarrow \end{smallmatrix}\right)}(\text{GBFA}) = (Q \circ T)\left(\mathcal{L}_{D_{BFA}}(\text{GBFA})\right).$
- $\mathcal{L}_{\left(\begin{smallmatrix} \rightarrow & \downarrow \\ s\uparrow & \rightarrow \end{smallmatrix}\right)}(\text{GBFA}) = (Q \circ (Q \circ Q))\left(\mathcal{L}_{D_{BFA}}(\text{GBFA})\right).$
- $\mathcal{L}_{\left(\begin{smallmatrix} \leftarrow & \downarrow s \\ \uparrow & \leftarrow \end{smallmatrix}\right)}(\text{GBFA}) = Q\left(\mathcal{L}_{D_{BFA}}(\text{GBFA})\right).$
- $\mathcal{L}_{\left(\begin{smallmatrix} \uparrow & \leftarrow \\ s\rightarrow & \uparrow \end{smallmatrix}\right)}(\text{GBFA}) = (T \circ Q)\left(\mathcal{L}_{D_{BFA}}(\text{GBFA})\right).$
- $\mathcal{L}_{\left(\begin{smallmatrix} \downarrow & \leftarrow \\ \leftarrow & \uparrow s \end{smallmatrix}\right)}(\text{GBFA}) = (Q \circ (Q \circ T))\left(\mathcal{L}_{D_{BFA}}(\text{GBFA})\right).$
- $\mathcal{L}_{\left(\begin{smallmatrix} \rightarrow & \uparrow \\ \uparrow & \leftarrow s \end{smallmatrix}\right)}(\text{GBFA}) = (Q \circ Q)\left(\mathcal{L}_{D_{BFA}}(\text{GBFA})\right).$

These characterizations are also valid for the corresponding deterministic classes.

In [2], *boustrophedon finite automata* have been introduced that basically work as GBFAs do when working in mode D_{BFA}, apart from the fact that we have integrated in our definition of GBFA that the automaton 'knows' if it processes an even- or odd-numbered row. In other words, they are aware of the direction of their movement (left to right or right to left). However, as the basic model is finite automata, it is not difficult to show BFAs can be turned into one that is direction-aware, by counting the number of reads of #-symbols modulo two. Hence, we can easily profit from results in [2]. We give one example now.

Theorem 3. *For each direction mode D, we know:* $\mathcal{L}_D(\text{GBFA}) = \mathcal{L}_D(\text{GBDFA})$.

Proof. By the previous reasoning, we can derive from [2] that $\mathcal{L}_D(\text{GBFA}) = \mathcal{L}_D(\text{GBDFA})$ is true for $D = D_{BFA}$. By Theorem 2, we have characterizations of $\mathcal{L}_D(\text{GBFA})$ in terms of $\mathcal{L}_{D_{BFA}}(\text{GBFA})$. These characterizations are also valid for the corresponding deterministic classes. □

4   Generalized Returning Finite Automata

A useful auxiliary model introduced in [2] was so-called *returning finite automata* (RFA). We are going to generalize their work in the following, again by introducing working modes for them. Now, a pair of directions like $D = (\,s\rightarrow\downarrow\,)$ is sufficient, indicating that an input array is always processed row by row, top down, where each row is scanned from left to right; moreover, the procedure is (here) started at the upper left corner of the array, as indicated by the position of s. The processing mode just describes coincide with that of RFAs from [2]. Leaving out the formal definition for now, we arrive at language families like $\mathcal{L}_D(\text{GRFA})$. It is sufficient to understand that there is no need to give any direction information. There are (again) eight natural processing modes D:

$$(\,s\rightarrow\downarrow\,),\ (\,s\downarrow\rightarrow\,),\ (\,\downarrow\leftarrow s\,),\ (\,\leftarrow\downarrow s\,),\ (\,s\rightarrow\uparrow\,),\ (\,s\uparrow\rightarrow\,),\ (\,\uparrow\leftarrow s\,),\ (\,\leftarrow\uparrow s\,).$$

These can be naturally partioned into the row-first modes $\mathcal{D}_{row-f}\{(s\rightarrow\downarrow)$, $(\downarrow\leftarrow s)$, $(s\rightarrow\uparrow)$, $(\uparrow\leftarrow s)\}$ and the four other column-first modes in $\mathcal{D}_{col-f}$. Again, we have deterministic variants, and we can consider the union of all languages pertaining to these GRFA-variants.

Example 1. The set of all arrays over $\{a, b\}$ such that each array in the set has exactly one row completely filled with b's and a's everywhere else is accepted by a GRFA M as shown in Fig. 1.

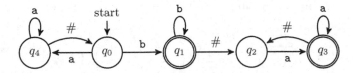

Fig. 1. GRFA M that accepts the language in Example 1.

As with GBFAs, we can alternatively describe the work of a GRFA working in mode D by first performing a unary operation on the image and then processing the image in the mode $D_{RFA} = (s\rightarrow\downarrow)$ that corresponds to RFAs as introduced in [2]. Therefore, we obtain the following characterizations for these modes.

Theorem 4. *The class* $\mathcal{L}_{D_{RFA}}(GRFA)$ *coincides with the following classes:*
$T\left(\mathcal{L}_{(s\downarrow\rightarrow)}(GRFA)\right)$, $\qquad R_v\left(\mathcal{L}_{(\downarrow\leftarrow s)}(GRFA)\right)$, $\qquad Q^{-1}\left(\mathcal{L}_{(\leftarrow\downarrow s)}(GRFA)\right)$,
$Q\left(\mathcal{L}_{(s\uparrow\rightarrow)}(GRFA)\right)$, $\quad R_b\left(\mathcal{L}_{(s\rightarrow\uparrow)}(GRFA)\right)$, $\quad T'\left(\mathcal{L}_{(\leftarrow\uparrow s)}(GRFA)\right)$, *and*
$H\left(\mathcal{L}_{(\uparrow\leftarrow s)}(GRFA)\right)$.

Likewise, we can use the previous theorem to obtain a representation like Theorem 2 also for GRFAs.

Theorem 5. *We obtain the following list of characterizations.*

- $\mathcal{L}_{(s\downarrow\rightarrow)}(GRFA) = T\left(\mathcal{L}_{D_{RFA}}(GRFA)\right)$.
- $\mathcal{L}_{(\downarrow\leftarrow s)}(GRFA) = (Q\circ T)\left(\mathcal{L}_{D_{RFA}}(GRFA)\right)$.
- $\mathcal{L}_{(s\uparrow\rightarrow)}(GRFA) = (Q\circ(Q\circ Q))\left(\mathcal{L}_{D_{RFA}}(GRFA)\right)$.
- $\mathcal{L}_{(\leftarrow\downarrow s)}(GRFA) = Q\left(\mathcal{L}_{D_{RFA}}(GRFA)\right)$.
- $\mathcal{L}_{(s\rightarrow\uparrow)}(GRFA) = (T\circ Q)\left(\mathcal{L}_{D_{RFA}}(GRFA)\right)$.
- $\mathcal{L}_{(\leftarrow\uparrow s)}(GRFA) = (Q\circ(Q\circ T))\left(\mathcal{L}_{D_{RFA}}(GRFA)\right)$.
- $\mathcal{L}_{(\uparrow\leftarrow s)}(GRFA) = (Q\circ Q)\left(\mathcal{L}_{D_{RFA}}(GRFA)\right)$.

We can conclude from [2, Theorem 2]:

Corollary 1. $\mathcal{L}_{D_{BFA}}(GBFA) = \mathcal{L}_{D_{RFA}}(GRFA)$.

With Theorems 2 and 5, we also get a complete list of correspondences between GBFAs and GRFAs as in Table 3. For instance, $Q(\mathcal{L}_{D_{RFA}}(GRFA)) = \mathcal{L}_D(GRFA)$ for $D = (\leftarrow\downarrow s)$ can be read off as the table entry. The previous result means that determinism does not restrict the power of GRFA in all processing modes.

5   Language Families Under the Unary Operators

We are now going to collect and prove several results relating the different language families that we introduced so far by means of the unary operators that we discussed above. The proofs will also show that it is quite valuable to have the different processing modes available.

Lemma 1. $\mathcal{L}_D(\mathrm{GRFA}) = R_v(\mathcal{L}_D(\mathrm{GRFA}))$ for $D \in \mathcal{D}_{row-f}$.

Table 3. Operators/directions for GBFAs and GRFAs

$\mathcal{O}$	T	R_v	Q^{-1}	Q	R_b	T'	H
GBFA	$\left(\begin{smallmatrix} s\downarrow & \to \\ \to & \uparrow \end{smallmatrix}\right)$	$\left(\begin{smallmatrix} \downarrow & \leftarrow s \\ \to & \downarrow \end{smallmatrix}\right)$	$\left(\begin{smallmatrix} \to & \downarrow \\ s\uparrow & \to \end{smallmatrix}\right)$	$\left(\begin{smallmatrix} \leftarrow & \downarrow s \\ \uparrow & \leftarrow \end{smallmatrix}\right)$	$\left(\begin{smallmatrix} \uparrow & \leftarrow \\ s\to & \uparrow \end{smallmatrix}\right)$	$\left(\begin{smallmatrix} \downarrow & \leftarrow \\ \leftarrow & \uparrow s \end{smallmatrix}\right)$	$\left(\begin{smallmatrix} \to & \uparrow \\ \uparrow & \leftarrow s \end{smallmatrix}\right)$
GRFA	$(s\downarrow \to)$	$(\downarrow \leftarrow s)$	$(s\uparrow \to)$	$(\leftarrow \downarrow s)$	$(s\to \uparrow)$	$(\leftarrow \uparrow s)$	$(\uparrow \leftarrow s)$

Proof. Let $M = (Q, \Sigma, R, s, F, \#, \Box, (s\to \downarrow))$ be some GRFA and let $L = L(M)$. Let us construct some GRFA M^v that accepts $R_v(L)$. $M^v = (Q^v, \Sigma, R^v, Q_I, Q_F, \#, \Box, (s\to \downarrow))$ is defined by $Q^v = Q \times Q \times Q$, $Q_\# = \{q \mid pa \to q \wedge q\# \to r \in R\}$, $Q_I = \{(s, q, r)\}$ where $q \in Q$, $r \in Q_\#$,

$$R^v = \{(\ell, q, r)a \to (\ell, p, r) \mid pa \to q \in R, \ell \in Q, r \in Q_\#, a \in \Sigma\}$$
$$\cup \{(\ell, \ell, p)\# \to (q, t, r) \mid p\# \to q \in R, \ell, r, t \in Q, a \in \Sigma\}$$
$$\cup \{(\ell, q, f)a \to (\ell, p, f) \mid pa \to q \in R, \ell \in Q, f \in F, a \in \Sigma\}$$

and $Q_F \subseteq Q^v$, $Q_F = \{(\ell, \ell, f) \mid f \in F, \ell \in Q\}$. The idea of the construction is that the mirror-image construction well-known from classical formal language theory. Here, the first component ℓ of some triple $(\ell, q, r) \in Q \times Q \times Q$ memorizes the state associated to the left-most symbol of that row, q is the current state and r is associated to the right-most symbol in the row. Reading $\#$ switches to the next row until the final state is reached. We only considered the first processing mode here, the other three modes can be shown using similar constructions. $\square$

As the vertical reflection can be likewise seen as a change in the processing mode, we can immediately conclude:

Corollary 2. $\mathcal{L}_{(s\to \downarrow)}(\mathrm{GRFA}) = \mathcal{L}_{(\downarrow \leftarrow s)}(\mathrm{GRFA})$; $\mathcal{L}_{(s\to \uparrow)}(\mathrm{GRFA}) = \mathcal{L}_{(\uparrow \leftarrow s)}(\mathrm{GRFA})$.

Due to Theorem 5, we can also conclude:

Lemma 2. $\mathcal{L}_D(\mathrm{GRFA}) = R_b(\mathcal{L}_D(\mathrm{GRFA}))$ for $D \in \mathcal{D}_{row-f}$.

Lemma 3. $\mathcal{L}_D(\mathrm{GRFA}) = R_b(\mathcal{L}_D(\mathrm{GRFA}))$ for $D \in \mathcal{D}_{col-f}$.

Hence, we can immediately conclude the following characterizations.

Corollary 3. $\mathcal{L}_{(s\to\,\downarrow)}(\mathrm{GRFA}) = \mathcal{L}_{(s\to\,\uparrow)}(\mathrm{GRFA})$;
$\mathcal{L}_{(\downarrow\,\leftarrow s)}(\mathrm{GRFA}) = \mathcal{L}_{(\uparrow\,\leftarrow s)}(\mathrm{GRFA})$.

Corollary 4. $\mathcal{L}_{(s\downarrow\,\to)}(\mathrm{GRFA}) = \mathcal{L}_{(s\uparrow\,\to)}(\mathrm{GRFA})$;
$\mathcal{L}_{(\leftarrow\,\downarrow s)}(\mathrm{GRFA}) = \mathcal{L}_{(\leftarrow\,\uparrow s)}(\mathrm{GRFA})$.

Lemma 1 and Corollary 1 give the following corollary; also refer to Theorem 1; similar closure properties for the other processing modes can be easily derived by combining what we have shown so far.

Corollary 5. $\mathcal{L}_{D_{BFA}}(\mathrm{GBFA}) = R_v(\mathcal{L}_{D_{BFA}}(\mathrm{GBFA}))$.

By Corollary 1 and Theorem 4, we obtain:

Theorem 6. $\mathcal{L}_{D_{BFA}}(\mathrm{GBFA}) = Q(\mathcal{L}_{(s\uparrow\,\to)}(\mathrm{GRFA}))$.

Theorem 6 immediately yields the following result, as $Q \circ Q^{-1}$ is the identity.

Corollary 6. $Q^{-1}(\mathcal{L}_{D_{BFA}}(\mathrm{GBFA})) = \mathcal{L}_{(s\uparrow\,\to)}(\mathrm{GRFA})$.

Using Theorems 2, 5, Corollary 1, and Table 3, we can conclude the following:

Corollary 7. $\mathcal{L}_{D_{BFA}}(\mathrm{GBFA}) = Q^{-1}(\mathcal{L}_{(s\uparrow\,\to)}(\mathrm{GRFA}))$.

As $Q^{-1} \circ Q^{-1} = H$, Corollaries 6 and 7 yield:

Corollary 8. $\mathcal{L}_{D_{BFA}}(\mathrm{GBFA}) = H(\mathcal{L}_{D_{BFA}}(\mathrm{GBFA}))$.

As $R_b = H \circ R_v$, Corollaries 5 and 8 give:

Corollary 9. $\mathcal{L}_{D_{BFA}}(\mathrm{GBFA}) = R_b(\mathcal{L}_{D_{BFA}}(\mathrm{GBFA}))$.

We can now summarize our characterization of picture language classes:

Theorem 7. *The picture language family* $\mathcal{L}_{D_{BFA}}(\mathrm{GBFA})$ *equals*

- $\mathcal{L}_D(\mathrm{GBFA})$ *for* $D \in \left\{ D_{BFA}, \left(\begin{smallmatrix} & \downarrow & \leftarrow s \\ \to & & \downarrow \end{smallmatrix}\right), \left(\begin{smallmatrix} & \uparrow & \leftarrow \\ s\to & & \uparrow \end{smallmatrix}\right), \left(\begin{smallmatrix} \to & & \uparrow \\ & \uparrow & \leftarrow s \end{smallmatrix}\right) \right\}$;
- $\mathcal{L}_D(\mathrm{GRFA})$ *for* $D \in \mathcal{D}_{row-f}$.

The picture language family $T(\mathcal{L}_{D_{BFA}}(\mathrm{GBFA}))$ *equals*

- $\mathcal{L}_D(\mathrm{GBFA})$ *for* $D \in \left\{ \left(\begin{smallmatrix} s\downarrow & & \to \\ & \to & \uparrow \end{smallmatrix}\right), \left(\begin{smallmatrix} \leftarrow & & \downarrow s \\ \uparrow & & \leftarrow \end{smallmatrix}\right), \left(\begin{smallmatrix} & \to & \downarrow \\ s\uparrow & & \to \end{smallmatrix}\right), \left(\begin{smallmatrix} \downarrow & & \leftarrow \\ & \leftarrow & \uparrow s \end{smallmatrix}\right) \right\}$;
- $\mathcal{L}_D(\mathrm{GRFA})$ *for* $D \in \mathcal{D}_{col-f}$.

Proof
(a) $\mathcal{L}_{D_{BFA}}(\mathrm{GBFA}) =_{\text{Cor. 5}} R_v(\mathcal{L}_{D_{BFA}}(\mathrm{GBFA})) =_{\text{Thm. 2}} \mathcal{L}_{\left(\begin{smallmatrix} \downarrow & \leftarrow s \\ \to & \downarrow \end{smallmatrix}\right)}(\mathrm{GBFA})$.

$\mathcal{L}_{D_{BFA}}(\mathrm{GBFA}) =_{\text{Cor. 9}} R_b(\mathcal{L}_{D_{BFA}}(\mathrm{GBFA})) =_{\text{Thm. 2}} \mathcal{L}_{\left(\begin{smallmatrix} \uparrow & \leftarrow \\ s\to & \uparrow \end{smallmatrix}\right)}(\mathrm{GBFA})$.

$\mathcal{L}_{D_{BFA}}(\mathrm{GBFA}) =_{\text{Cor. 8}} H(\mathcal{L}_{D_{BFA}}(\mathrm{GBFA})) =_{\text{Thm. 2}} \mathcal{L}_{\left(\begin{smallmatrix} \to & \uparrow \\ \uparrow & \leftarrow s \end{smallmatrix}\right)}(\mathrm{GBFA})$.

(b) $\mathcal{L}_{D_{BFA}}(\text{GBFA}) =_{\text{Cor.1}} \mathcal{L}_{(s\rightarrow \downarrow)}(\text{GRFA}) =_{\text{Cor.2}} \mathcal{L}_{(\downarrow \leftarrow s)}(\text{GRFA})$
$=_{\text{Cor.3}} \mathcal{L}_{(\uparrow \leftarrow s)}(\text{GRFA}) =_{\text{Cor.2}} \mathcal{L}_{(s\rightarrow \uparrow)}(\text{GRFA})$.

(c) $T(\mathcal{L}_{D_{BFA}}(\text{GBFA})) =_{\text{Thm.2}} \mathcal{L}_{\left(\begin{smallmatrix} s\downarrow & \rightarrow \\ \rightarrow & \uparrow \end{smallmatrix}\right)}(\text{GBFA})$.

$T(\mathcal{L}_{D_{BFA}}(\text{GBFA})) =_{\text{Table 1a}} Q(R_b(\mathcal{L}_{D_{BFA}}(\text{GBFA})))$
$=_{\text{Cor.9}} Q(\mathcal{L}_{D_{BFA}}(\text{GBFA})) =_{\text{Thm.2}} \mathcal{L}_{\left(\begin{smallmatrix} \leftarrow & \downarrow s \\ \uparrow & \leftarrow \end{smallmatrix}\right)}(\text{GBFA})$.

$T(\mathcal{L}_{D_{BFA}}(\text{GBFA})) =_{\text{Table 1a}} Q^{-1}(R_v(\mathcal{L}_{D_{BFA}}(\text{GBFA})))$
$=_{\text{Cor.5}} Q^{-1}(\mathcal{L}_{D_{BFA}}(\text{GBFA})) =_{\text{Thm.2}} \mathcal{L}_{\left(\begin{smallmatrix} \rightarrow & \downarrow \\ s\uparrow & \rightarrow \end{smallmatrix}\right)}(\text{GBFA})$.

$T(\mathcal{L}_{D_{BFA}}(\text{GBFA})) =_{\text{Table 1a}} T'(H(\mathcal{L}_{D_{BFA}}(\text{GBFA})))$
$=_{\text{Cor.8}} T'(\mathcal{L}_{D_{BFA}}(\text{GBFA})) =_{\text{Thm.2}} \mathcal{L}_{\left(\begin{smallmatrix} \downarrow & \leftarrow \\ \leftarrow & \uparrow s \end{smallmatrix}\right)}(\text{GBFA})$.

(d) Similarly, using Corollary 1 and Theorem 5.   □

Corollary 10. $\mathcal{L}(\text{GBFA}) = \mathcal{L}_{(s\downarrow \rightarrow)}(\text{GRFA}) \cup \mathcal{L}_{(s\rightarrow \downarrow)}(\text{GRFA})$.

Remark 1. Unfortunately, as can be easily seen by applying the exchange property arguments derived in [2], the set of square-sized arrays of odd length whose middle row and middle column contain b's, while all other positions are filled with a's, is not in $\mathcal{L}(\text{GBFA})$.

Let $\mathcal{O}$ collect the eight unary operators introduced in Sect. 2.2.

Corollary 11. $\forall O \in \mathcal{O} : O(\mathcal{L}(\text{GBFA})) = \mathcal{L}(\text{GBFA})$.

To underline the special importance of the two operations Q and T from $\mathcal{O}$, we can also state:

Corollary 12. $\mathcal{L}(\text{GBFA}) = Q(\mathcal{L}_{D_{BFA}}(\text{GBFA})) \cup \mathcal{L}_{D_{BFA}}(\text{GBFA})$ and
$\mathcal{L}(\text{GBFA}) = T(\mathcal{L}_{D_{BFA}}(\text{GBFA})) \cup \mathcal{L}_{D_{BFA}}(\text{GBFA})$.

Let us now turn to the two binary catenation operations.

Theorem 8. $\forall L_1, L_2 \in \mathcal{L}_{D_{RFA}}(\text{GRFA}), L_1 \ominus L_2 \in \mathcal{L}_{D_{RFA}}(\text{GRFA})$.

Proof. Let $M_1 = (Q_1, \Sigma, R_1, s_1, F_1, \#, \square, (s \rightarrow \downarrow))$ and $M_2 = (Q_2, \Sigma, R_2, s_2, F_2, \#, \square, (s \rightarrow \downarrow))$ be two GRFAs, with $L_1 = L(M_1)$ and $L_2 = L(M_2)$. W.l.o.g., assume that $Q_1 \cap Q_2 = \emptyset$. Let us construct the GRFA $M^\ominus$ that accepts $L_1 \ominus L_2$ (i.e., $L(M^\ominus) = L_1 \ominus L_2$). $M^\ominus = (Q_\ominus, \Sigma, R_\ominus, s_1, F_2, \#, \square, (s \rightarrow \downarrow))$ is defined by $Q_\ominus = Q_1 \cup Q_2$, $R_\ominus = R_1 \cup R_2 \cup \{f\# \rightarrow s_2 \mid f \in F_1\}$. The idea of the construction is to first simulate M_1; whenever a final state is entered at the end of reading a line, then the simulation may switch to M_2.   □

Corollary 13. $\forall L_1, L_2 \in Q(\mathcal{L}_{D_{RFA}}(\text{GRFA})), L_1 \oplus L_2 \in Q(\mathcal{L}_{D_{RFA}}(\text{GRFA}))$.

Define a mapping $str : \Sigma_2^+ \rightarrow (\Sigma \cup \{\#\})^+$, $W \mapsto w$ where $w = w_1 \# w_2$, $|W|_c = n$, $n \geq 2$ and $|W|_r = 2$; moreover, $|w_1| = |w_2| = n$, so that $|w| = 2n + 1$.

Lemma 4. *If* $L \in \mathcal{L}_{D_{RFA}}(\text{GRFA})$ *with* $L \subseteq \Sigma_2^+$, *then* $str(L)$ *is context-free.*

Proof. A pushdown automaton uses its finite control as the GRFA and simply checks if the two rows have equal length by using its pushdown store.   □

Theorem 9. $\exists L_1, L_2 \in \mathcal{L}_{D_{RFA}}(\text{GRFA}) : L_1 \oplus L_2 \notin \mathcal{L}_{D_{RFA}}(\text{GRFA}).$

Proof. Let $L = \left\{ \begin{matrix} 1 \ 0^\ell \\ 1 \ 0^\ell \end{matrix} : \ell \geq 1 \right\}$. Clearly, $L \in \mathcal{L}_{D_{RFA}}(\text{GRFA})$. However, $str(L \oplus L)$ is a variant of the crossing dependency language known to be not context-free. By Lemma 4, $L \oplus L \notin \mathcal{L}_{D_{RFA}}(\text{GRFA})$.   □

Corollary 14. $\exists L_1, L_2 \in Q(\mathcal{L}_{D_{RFA}}(\text{GRFA})) : L_1 \ominus L_2 \notin Q(\mathcal{L}_{D_{RFA}}(\text{GRFA})).$

Theorem 10. $\mathcal{L}(\text{GRFA})$ *is closed neither under column catenation nor under row catenation.*

Proof. Consider L from the proof of Theorem 9. By Theorem 8, $L' = L \ominus \{0\}_+^+ \ominus \{1\}^+ \ominus \{0\}_+^+ \in \mathcal{L}_{D_{RFA}}(\text{GRFA})$. If $L' \oplus L' \in \mathcal{L}_{D_{RFA}}(\text{GRFA})$, then also $L \oplus L \in \mathcal{L}_{D_{RFA}}(\text{GRFA})$, contradicting the reasoning from Theorem 9. Exchange property arguments derived in [2] easily show that $\{L', L' \oplus L'\} \cap Q(\mathcal{L}_{D_{RFA}}(\text{GRFA})) = \emptyset$, intuitively because horizontal lines in arbitrary positions cannot be checked by finite automata working column by column.   □

So far, we have used finite automata to only accept picture (languages). If such devices should be used in practice, at least two questions show up:

– Can we design such automata in a systematic, best modular fashion?
– Can we set up these automata so that they can tolerate certain input errors?

We will work towards this in the next section.

6   Picture Transforming Automata

To answer both questions, we introduce a generalization of Mealy machines to picture processing. A *Mealy Picture Machine*, or MPM for short, can be specified as a 9-tuple $M = (Q, \Sigma, \Gamma, \delta, \lambda, q_0, \#, F, D)$, where Q is a finite set of states, Σ is a finite set of input symbols, Γ is a finite set of output symbols, $\delta : Q \times (\Sigma \cup \{\#\}) \to Q$ is the transition function, $\lambda : Q \times \Sigma \to \Gamma$ is the output function, $q_0 \in Q$ is the initial state, $\# \notin \Sigma$, $\# \notin \Gamma$, is the special symbol that indicates the borders of the picture that is processed, F is the set of final states, and $D \in \mathcal{D}$ indicates the move directions of the GRFAs. Let us now give the notion of configurations to formalize the working of MPMs, based on the snapshots of their work. Here, we assume (w.l.o.g.) that $\Sigma \cap \Gamma = \emptyset$. Let $\Upsilon := \Sigma \cup \Gamma$ and $\Upsilon_\# := \Upsilon \cup \{\#\}$. Then $C_M := Q \times (\Upsilon_\#^{++} \cap (\{\#\}^+ \ominus (\{\#\}_+ \oplus \Upsilon^{++} \oplus \{\#\}_+)) \ominus \{\#\}^+) \times \mathbb{N}$ is the set of configurations of M. Hence, the first and last columns and the first and last rows are completely filled with $\#$, and these are the only positions that contain $\#$. The *initial configuration* is determined by the input array $A \in \Sigma^{++}$. More precisely, if A has m rows and n columns, then

$$(q_0, \#^{n+2} \ominus (\#_m \oplus A \oplus \#_m) \ominus \#^{n+2}, 1)$$

shows the according initial configuration $C_{init}(A)$. Similarly, a *final configuration* $C_{fin}(A')$ is then given by

$$(q_f, \#^{n+2} \ominus (\#_m \oplus A' \oplus \#_m) \ominus \#^{n+2}, n)$$

for some $A' \in \Gamma^{++}$ with m rows and n columns and for some $q_f \in F$. The processing of the machine depends on $D \in \mathcal{D}$ of the GRFAs. Let us now formalize the description with direction $D = (s \to \downarrow)$; the remaining seven directions can be formalized similarly but we did not explicitly write those here.[1]

- If (p, A, μ) and (q, B, μ) are two configurations such that A and B are identical but for one position (i, j), $1 \le i \le m+2$, $1 \le j \le n+2$, where $B[i, j] \in \Gamma$, while $A[i, j] \in \Sigma$, then $(p, A, \mu) \vdash_M (q, B, \mu)$ if $\delta(p, A[i, j]) = q$ and $\lambda(p, A[i, j]) = B[i, j]$.
- If (p, A, μ) and $(q, A, \mu+1)$ are two configurations, then $(p, A, \mu) \vdash_M (q, A, \mu+1)$ if $\delta(p, \#) = q$.

The reflexive transitive closure of the relation $\vdash_M$ is denoted by $\vdash_M^*$. Notice that for each $A \in \Sigma^{++}$ there is at most one $A' \in \Gamma^{++}$ such that $C_{init}(A) \vdash_M^* C_{fin}(A')$. We can hence view M as a partial function $M : \Sigma^{++} \nrightarrow \Gamma^{++}$.

Theorem 11. *Given $L \in \mathcal{L}_{DRFA}(\text{GRFA})$, with $L \subseteq \Sigma^{++}$, and an MPM M: $\Sigma^{++} \nrightarrow \Gamma^{++}$, then $M(L) \in \mathcal{L}_{DRFA}(\text{GRFA})$.*

The proof is very similar to the construction in the following theorem.

Theorem 12. *Given $L \in \mathcal{L}_{DRFA}(\text{GRFA})$, with $L \subseteq \Gamma^{++}$, and an MPM M: $\Sigma^{++} \nrightarrow \Gamma^{++}$, then $M^-(L) \in \mathcal{L}_{DRFA}(\text{GRFA})$ with $M^-(L) \subseteq \Sigma^{++}$.*

Proof. Let $R = (Q, \Gamma, P, s, F, \#, \square, (s \to \downarrow))$ be some GRFA with $L(R) \subseteq \Gamma^{++}$. Let $M : \Sigma^{++} \nrightarrow \Gamma^{++}$ be an MPM, $M = (Q_M, \Sigma, \Gamma, \delta, \lambda, s_M, \#, F_M, (s \to \downarrow))$ lifted to $M : 2^{\Sigma^{++}} \to 2^{\Gamma^{++}}$ and $M^- : 2^{\Sigma^{++}} \to 2^{\Gamma^{++}}$, i.e., $M^-(\Gamma^{++}) \subseteq \Sigma^{++}$. Now our aim is to define a GRFA R' such that $L(R') = M^-(L(R))$. $R' = (Q', \Sigma, P', s', F', \#, \square, (s \to \downarrow))$ is defined by $Q' = Q \times Q_M$, $P' = \{(p, q)a \to (p', q') \mid \delta(q, a) = q' \wedge p\lambda(q, a) \to p' \in P\} \cup \{(p, q)\# \to (p', q') \mid \delta(q, \#) = q' \wedge p\# \to p' \in P\}$, $s' = (s, s_M) \in Q'$ is the start state and $F' = F \times F_M$.   □

Let us now explain the extended power of GRFA by adding MPM with it. Also, we want to describe how to use MPMs and further closure properties to design more complicated GRFAs, starting off from very simple automata. Suppose the task is to design a GRFA R' with $L(R') = L'$, see Table 4. How can we obtain such an automaton? We see that the pictures in L' can be decomposed into two subsequent rows of x and one first column of x. It looks easier to design automata for both tasks separately and then combine them later 'somehow'. This is what MPMs can do. Assume that we have designed an MPM M that takes inputs from Σ^{++}, with $\Sigma = \{\text{x}, \cdot\}$, and does the following:

---

[1] We could also use the unary operations and their inverses to formally describe the processing.

- It converts all · symbols into a if they do not appear in the first column.
- It converts all x symbols into a if they appear in the first column, but into b if they appear in any other column.

Assume we have designed M. $M(L')$ can be found in Table 4. How can we further process this? Maybe, it would be an idea to design another MPM M' that takes inputs from $\{a, b\}^{++}$ and converts them according to the following:

- It converts a symbols into · but one special to be mentioned later.
- When it first reads a b, this must be in the second column, and this b must be followed by b's only in the current row, which is then converted into a row of ·'s only. (If the first row where M' encounters any b's is not of the form ab^+, M' will not enter an accepting state when further processing the input array, as it has observed an input error.)
- In the next row, it is checked if the first symbol is a a. This will then be converted to x.
- Any further occurrences of b will be converted into x.

Table 4. Simplifying array languages with MPMs

$$
L' = \left\{ \begin{array}{l} (\text{x } (\cdot)^n)_{\ell-1} \\ \text{x} \quad (\text{x})^n \\ \text{x} \quad (\text{x})^n \\ (\text{x } (\cdot)^n)_{m-1} \end{array} : n, m, \ell \geq 1 \right\}, \qquad M(L') = \left\{ \begin{array}{l} (\text{a } (\text{a})^n)_{\ell-1} \\ \text{a} \quad (\text{b})^n \\ \text{a} \quad (\text{b})^n \\ (\text{a } (\text{a})^n)_{m-1} \end{array} : n, m, \ell \geq 1 \right\}.
$$

As a consequence, we find:

$$
M'(M(L')) = \left\{ \begin{array}{l} (\cdot \ (\cdot)^n)_{\ell-1} \\ \cdot \quad (\cdot)^n \\ \text{x} \quad (\text{x})^n \\ (\cdot \ (\cdot)^n)_{m-1} \end{array} : n, m, \ell \geq 1 \right\} = \left\{ \begin{array}{l} (\cdot \ (\cdot)^n)_{\ell} \\ \text{x} \quad (\text{x})^n \\ (\cdot \ (\cdot)^n)_{m-1} \end{array} : n, m, \ell \geq 1 \right\}.
$$

We can devise a quite similar machine R that then accepts all arrays over $\{\cdot, \text{x}\}$ that start with at least one row filled with ·, followed by one row of length at least two completely filled with x and then followed by an arbitrary number (possibly zero many) of rows filled up with ·. Now, the GRFA R' we are looking for can be obtained by first constructing $R^\dagger$ for accepting $M'^-(L(R))$, based on R and M', and then constructing R' accepting $M^-(L(R^\dagger))$, based on $R^\dagger$ and M. The drawback of this design approach could be a certain state explosion. For instance, already the smallest GRFA implementing $L(R)$ has five states and the smallest MPM for M' has six states, which gives us 30 states for $R^\dagger$, unless we interleave steps that delete useless states or even implement some state minimization procedures.

Notice that MPMs naturally generalize array homomorphisms as considered in [6]. Also, with the possibility to implement multiple passes (by applying the described form of transductions), it is easy to also implement set operations like union or intersection, as the first reading automaton could communicate with the automaton doing the second read by changing one particular part of the picture, for instance, by printing a special character at the very end, signaling acceptance. However, we also have negative results like the following one.

Theorem 13. $T(\mathcal{L}(\mathrm{GRFA}))$ *is not closed under MPM nor under inverse MPM mappings.*

Proof. By an exchange property argument [2], one can see that the language $L = \{a\}_+^+ \ominus \{b\}^+ \ominus \{a\}_+^+ \notin T(\mathcal{L}(\mathrm{GRFA}))$. However, it is easy to design an MPM M that only translates arrays from L into some arrays over the alphabet $\{0\}$. Also, $\{0\}_+^+ \in T(\mathcal{L}(\mathrm{GRFA}))$. This shows non-closure under inverse MPM mappings, as $L = M^-(\{0\}_+^+)$. Conversely, let the MPM M' work on arrays over the alphabet $\{0, 1\}$ as follows: If a row starts with 0, then M' will translate the whole row into a row of a's. If a row starts with 1, then M' will translate the whole row into a row of b's. Now, it is easy to see that $L' := (\{0\}_+\ominus\{1\}\ominus\{0\}_+)\oplus\{0\}_+^+ \in T(\mathcal{L}(\mathrm{GRFA}))$, while $M'(L') = L$ is not.   □

It would therefore be interesting to study hierarchies of array languages defined by combining array processing devices that alternate between row-wise or column-wise processing, working on multiple passes over given images. This would also enable us to implement error-correction features, like thinning out blurred lines. This also shows that combining different processing modes in subsequent passes could be useful in practice.

7   Conclusions

We hope that the simple finite automata models that we presented in this paper can form a starting point to bring these syntactic ideas back into the practice of image processing. This is also why we studied seemingly simplistic working modes of such automata, including their use in image transformations. We also like to mention that currently a student of ours is implementing the machine transformation algorithms that we showed to prove closure properties of GBFA languages. We hope to report on this in the near future. This software will be available on request.

A possible further direction of research could be to integrate these models into pattern recognition algorithms. As exhibited by Flasiński in [3], this would necessitate the development of Grammatical Inference algorithms. In this context, it looks that we have to overcome the following technical problem: mostly, learners converge to canonical hypotheses like minimum-state automata and hence, efficient learners also comprise efficient state minimization algorithms. However, we can show the following result:

Theorem 14. *It is NP-hard to decide, given some BDFA $M = (Q, \Sigma, R, s, F, \#, \square)$ on some binary input alphabet $\Sigma = \{a, b\}$, whether or not there is an equivalent BDFA M' with only one state.*

The proof is not given due to space restrictions. This result seems to pose some difficulties in applying Grammatical Inference techniques to Syntactic Pattern Recognition in the context of picture scanning automata. It also indicates some limitations to the modular design approach of picture processing automata.

References

1. Armstrong, M.A.: Groups and Symmetry. Springer, Heidelberg (1988)
2. Fernau, H., Paramasivan, M., Schmid, M.L., Thomas, D.G.: Scanning pictures the boustrophedon way. In: Barneva, R.P., Bhattacharya, B.B., Brimkov, V.E. (eds.) IWCIA 2015. LNCS, vol. 9448, pp. 202–216. Springer, Heidelberg (2015). doi:10.1007/978-3-319-26145-4_15
3. Flasiński, M.: Syntactic pattern recognition: paradigm issues and open problems. In: Chen, C.H. (ed.) Handbook of Pattern Recognition and Computer Vision, 5th edn., pp. 3–25. World Scientific (2016)
4. Inoue, K., Takanami, I.: A survey of two-dimensional automata theory. Inf. Sci. **55**(1–3), 99–121 (1991)
5. Kari, J., Salo, V.: A survey on picture-walking automata. In: Kuich, W., Rahonis, G. (eds.) Algebraic Foundations in Computer Science. LNCS, vol. 7020, pp. 183–213. Springer, Heidelberg (2011). doi:10.1007/978-3-642-24897-9_9
6. Siromoney, G., Siromoney, R., Krithivasan, K.: Abstract families of matrices and picture languages. Comput. Graph. Image Process. **1**, 284–307 (1972)
7. Siromoney, G., Siromoney, R., Krithivasan, K.: Picture languages with array rewriting rules. Inf. Control (now Inf. Comput.) **22**(5), 447–470 (1973)

Two-Dimensional Input-Revolving Automata

S. James Immanuel[1]($\boxtimes$), D.G. Thomas[1], Henning Fernau[2],
Robinson Thamburaj[1], and Atulya K. Nagar[3]

[1] Department of Mathematics, Madras Christian College, Tambaram, Chennai, India
james_imch@yahoo.co.in, dgthomasmcc@yahoo.com, robin.mcc@gmail.com
[2] Fachbereich 4 – Abteilung Informatik, Universität Trier, 54286 Trier, Germany
Fernau@uni-trier.de
[3] Department of Mathematics and Computer Science,
Liverpool Hope University, Liverpool, UK
nagara@hope.ac.uk

Abstract. A new type of two-dimensional automaton for accepting
two-dimensional languages, called as two-dimensional input-revolving
automaton is introduced in this paper. It is an extension of input-
revolving automaton for string languages. We bring out all the variants of
this automaton which are based on the various types of column-revolving
operations considered here. We compare the families of array languages
accepted by the variants of these automata along with the well known
families of Siromoney matrix languages. We discuss some of the closure
properties of the new families of array languages and give an application
in steganography.

Keywords: Picture languages · Extended finite automaton · Rectangular arrays · Input-revolving

1 Introduction

Two-dimensional languages, also called picture languages, are sets of two-
dimensional arrays of symbols chosen from a finite alphabet. Their application
and importance can be seen in image processing, pattern recognition, character
recognition and also in studies concerning cellular automaton and other models
of computing. Due to the structure handling ability of the syntactic models, syntactic techniques of generation of digital picture arrays have become one of the
major areas of theoretical studies in picture analysis and pattern recognition.
Various grammars and automata have already been proposed for the generation
and recognition of rectangular picture languages [4,8–11,13,14,16].

Automata play a vital role in the theory of picture recognizability. The generalization of finite state automata to two-dimensional languages can be attributed
to Blum and Hewitt [2] who introduced the notion of 4-way automata. Array
automata acting on scenes (two-dimensional tapes) are defined in [5]. Motivated
by certain floor designs called "Kolam" patterns, Siromoney et al. [12] introduced
Siromoney Matrix Grammars (SMG), a simple and elegant grammar model for

© Springer International Publishing AG 2017
R.P. Barneva et al. (Eds.): CompIMAGE 2016, LNCS 10149, pp. 148–164, 2017.
DOI: 10.1007/978-3-319-54609-4_11

generating rectangular arrays. Several variations of Siromoney matrix grammars have already been added to the literature, for instance [15,17].

In this paper, we extend the concept of input-revolving automata systematically investigated in [1], with several shift operations, namely left-revolving, right-revolving and circular-interchanging operations to two dimensional languages. Input-revolving automata work similar to ordinary finite automata except that they can make three special shift operations as mentioned. Two-dimensional input-revolving automata work on an input array row by row and every row computation is same as that of input-revolving automata for string languages. Here we introduce column shift transitions, namely left column-revolving, right-column revolving and circular column-interchanging transitions which shifts an entire column. We study all the variants of these automata, both deterministic and non-deterministic, based on the various types of column-revolving operations considered. We compare various classes of languages accepted by these automata with the families of Siromoney matrix languages [12].

2   Preliminaries

In this section we recall some notions related to formal language theory and array grammars (see [7,12]). Let Σ be a finite alphabet, Σ^* is the set of words over Σ including the empty word λ. $\Sigma^+ = \Sigma^* - \{\lambda\}$. For $w \in \Sigma^*$ and $a \in \Sigma$, $|w|_a$ denotes the number of occurrences of a in w. An array consists of finitely many symbols from Σ that are arranged as rows and columns in some particular order and is written in the form, $A = \begin{bmatrix} a_{11} & \cdots & a_{1n} \\ \vdots & \ddots & \vdots \\ a_{m1} & \cdots & a_{mn} \end{bmatrix}$ or in short $A = [a_{ij}]_{m \times n}$, for all $a_{ij} \in \Sigma$, $i = 1, 2, \ldots, m$ and $j = 1, 2, \ldots, n$. The set of all arrays over Σ is denoted by Σ^{**} which also includes the empty array Λ (zero rows and zero columns). $\Sigma^{++} = \Sigma^{**} - \{\Lambda\}$. For $a \in \Sigma$, $|A|_a$ denotes the number of occurrences of a in A. The column concatenation of $A = \begin{bmatrix} a_{11} & \cdots & a_{1p} \\ \vdots & \ddots & \vdots \\ a_{m1} & \cdots & a_{mp} \end{bmatrix}$, and $B = \begin{bmatrix} b_{11} & \cdots & b_{1q} \\ \vdots & \ddots & \vdots \\ b_{n1} & \cdots & b_{nq} \end{bmatrix}$, defined only when $m = n$, is given by $A \oplus B = \begin{bmatrix} a_{11} & \cdots & a_{1p} & b_{11} & \cdots & b_{1q} \\ \vdots & \ddots & \vdots & \vdots & \ddots & \vdots \\ a_{m1} & \cdots & a_{mp} & b_{n1} & \cdots & b_{nq} \end{bmatrix}$. As $1 \times n$-dimensional arrays can be easily interpreted as words of length n (and vice versa), we will then write their column catenation by juxtaposition (as usual). Similarly, the row concatenation, defined only when $p = q$, is given by $A \ominus B = \begin{bmatrix} a_{11} & \cdots & a_{1p} \\ \vdots & \ddots & \vdots \\ a_{m1} & \cdots & a_{mp} \\ b_{11} & \cdots & b_{1q} \\ \vdots & \ddots & \vdots \\ b_{n1} & \cdots & b_{nq} \end{bmatrix}$. The empty array acts as the identity for column and row catenation of arrays of arbitrary dimensions. If $A = \begin{bmatrix} a_{11} & \cdots & a_{1n} \\ \vdots & \ddots & \vdots \\ a_{m1} & \cdots & a_{mn} \end{bmatrix}$, then the transpose of A is $A^T = \begin{bmatrix} a_{11} & \cdots & a_{m1} \\ \vdots & \ddots & \vdots \\ a_{1n} & \cdots & a_{mn} \end{bmatrix}$.

Given a picture $A \in \Sigma^{**}$ of size (m, n), we define $\hat{A}$ as the picture of size $(m, n+1)$ obtained by adjoining a special symbol $\# \notin \Sigma$ to the right end of the input array. $|A|_c$ is the number of columns in picture A and $|A|_r$ is the number of rows in picture A. If every element in an $m \times n$ array is a, then we write this array as a_m^n.

Definition 1. *A* Context-sensitive matrix grammar (CSMG) (Context-free matrix grammar (CFMG), Right-linear matrix grammar (RLMG)) *is defined by a 7-tuple* $G = (V_h, V_v, \Sigma_I, \Sigma, S, R_h, R_v)$, *where:* V_h *is a finite set of* horizontal nonterminals; V_v *is a finite set of* vertical nonterminals; $\Sigma_I \subseteq V_v$ *is a finite set of* intermediates; Σ *is a finite set of* terminals; $S \in V_h$ *is a* starting symbol; R_h *is a finite set of* horizontal context-sensitive (context-free, right-linear) rules; R_v *is a finite set of* vertical right-linear rules.

There are two phases of derivation of the Siromoney Matrix Grammars. In the first phase, a horizontal string of intermediate symbols is generated by means of any type of Chomsky grammar rules in R_h. During the second phase treating each intermediate as a start symbol, vertical generation of the actual picture is done parallely, by applying R_v. Parallel application ensures that the terminating rules are all applied simultaneously in every column and that the column grows only in downward direction. The language generated by a CSMG (CFMG, RLG) is called a CSML (CFML, RML). For more information, we can refer to [12]. We denote the family of Siromoney matrix languages by $\mathfrak{L}(X)$, where $X \in \{CSML, CFML, RML\}$.

3   Two-Dimensional Input-Revolving Finite Automata

In this section we introduce the two-dimensional input-revolving finite automaton and explain its working with an example.

Definition 2. *A two-dimensional non-deterministic column revolving finite automaton, a 2-NCRFA for short, is a 6-tuple* $M = (Q, \Sigma, \delta, \Delta, q_0, F)$, *where* Q *is a finite set of states,* Σ *is an input alphabet,* $Q \cap \Sigma = \emptyset$, $\delta \subseteq Q \times (\Sigma \cup \{\lambda, \#\}) \times 2^Q$ *is the finite set of transition rules,* $\#$ *is a special symbol not in* $Q \cup \Sigma$, $\Delta \subseteq Q \times (\Sigma \cup \{\lambda\}) \times 2^Q$ *is the finite set of column revolving rules,* $q_0 \in Q$ *is the initial state, and* F *is the set of final states.*

M is said to be λ-free *if* δ *is a mapping from* $Q \times (\Sigma \cup \{\#\})$ *to* 2^Q *and* Δ *is a mapping from* $Q \times \Sigma$ *to* 2^Q.

Members of δ, Δ *are referred to as the rules of M and instead of* $(p, y, q) \in \delta$ *(or* Δ*), we write* $py \rightarrow q \in \delta$ *(or* Δ*). Based on the different interpretations of the mapping* Δ*, we formally distinguish the different operations on the input array and hence categorize various types of 2-NCRFA. For this, we consider configurations of this automaton for some input array* $A = \begin{matrix} X \\ uv \\ Z \end{matrix}$*, of size* $m \times n$ *with* $X, Z, u, v \in \Sigma^{**}$*,* uv *is* $1 \times n$ *array, to be of the form* $\begin{matrix} X \\ uqv \\ Z \end{matrix}$ *where* $q \in Q$*. This* means that q is the current state of the finite control with tape head reading the first element in v. (X, u) and (v, Z) are therefore the read and unread part of the input array, respectively. The transition of a configuration to next configuration can be induced by either δ or Δ.

1. Let $a \in \Sigma \cup \{\lambda\}$ and $A = \begin{matrix} X \\ uav \\ Z \end{matrix}$ be an $m \times n$ array in Σ^{**} with $|X|_c =$

$|uav|_c = |Z|_c$. If $qa \to p$ is in δ then, $\begin{matrix} \hat{X} \\ uqav\# \\ \hat{Z} \end{matrix} \vdash_M \begin{matrix} \hat{X} \\ uapv\# \\ \hat{Z} \end{matrix}$. These transitions

are called as the normal transitions. Let $A = \begin{matrix} X \\ u \\ Z \end{matrix}$ be a $m \times n$ array in Σ^{**} with

$|X|_c = |u|_c = |Z|_c, u$ is a $1 \times n$ array. If $q\# \to p$ is in δ then, $\begin{matrix} \hat{X} \\ u \ q\# \\ \hat{Z} \end{matrix} \vdash_M \begin{matrix} \hat{X}' \\ pu'\# \\ \hat{Z}' \end{matrix}$,

where, $X' = \dfrac{X}{u}, Z = \dfrac{u'}{Z'}$ and u' is a $1 \times n$ array.

2. Column revolving operations are performed by applying the rules from Δ.
   For $a \in \Sigma \cup \{\lambda\}, b \in \Sigma$, a $m \times n$ array, $A = \begin{matrix} X_1\ X_2\ X_3\ X_4 \\ u\quad a\quad v\quad b \\ Z_1\ Z_2\ Z_3\ Z_4 \end{matrix}$ in Σ^{**} with

   $|X_1|_c = |u|_c = |Z_1|_c, |X_3|_c = |v|_c = |Z_3|_c, |X_2|_c = |Z_2|_c = |X_4|_c = |Z_4|_c = 1$ and a rule $qa \to p$ in Δ,

   (a) a left column-revolving transition is defined by

$$
\begin{array}{|llll|} \hline X_1 & X_2\ X_3\ \mathbf{X_4} & \# \\ u\ q & a\quad v\quad \mathbf{b} & \# \\ Z_1 & Z_2\ Z_3\ \mathbf{Z_4} & \# \\ \hline \end{array} \vdash_M
\begin{array}{|llll|} \hline X_1 & \mathbf{X_4}\ X_2\ X_3 & \# \\ u\ q & \mathbf{b}\quad a\quad v & \# \\ Z_1 & \mathbf{Z_4}\ Z_2\ Z_3 & \# \\ \hline \end{array}
$$

   (b) a right column-revolving transition is defined by

$$
\begin{array}{|llll|} \hline X_1 & \mathbf{X_2}\ X_3\ X_4 & \# \\ u\ q & \mathbf{a}\quad v\quad b & \# \\ Z_1 & \mathbf{Z_2}\ Z_3\ Z_4 & \# \\ \hline \end{array} \vdash_M
\begin{array}{|llll|} \hline X_1 & X_3\ X_4\ \mathbf{X_2} & \# \\ u\ q & v\quad b\quad \mathbf{a} & \# \\ Z_1 & Z_3\ Z_4\ \mathbf{Z_2} & \# \\ \hline \end{array}
$$

   (c) a circular column-interchanging transition is defined by

$$
\begin{array}{|llll|} \hline X_1 & \mathbf{X_2}\ X_3\ \mathbf{X_4} & \# \\ u\ q & a\quad v\quad b & \# \\ Z_1 & \mathbf{Z_2}\ Z_3\ \mathbf{Z_4} & \# \\ \hline \end{array} \vdash_M
\begin{array}{|llll|} \hline X_1 & \mathbf{X_4}\ X_3\ \mathbf{X_2} & \# \\ u\ q & b\quad v\quad a & \# \\ Z_1 & \mathbf{Z_4}\ Z_3\ \mathbf{Z_2} & \# \\ \hline \end{array}
$$

If $a = \lambda$, then the column-revolving operation is carried out irrespective of what symbol is being read by the finite state control. If 'a' is being read by the state q at the end of some row of the input array then there is no column-revolving involved and simply the state will be changed to p.

Whenever there is a choice between a normal transition or a column-revolving transition, the next move is non-deterministically chosen by the automaton. A deterministic 2-dimensional column revolving automaton, a 2-DCRFA for short is defined in the same way as 2-NCRFA except that $\delta \subseteq Q \times (\Sigma \cup \{\lambda, \#\}) \times Q$, $\Delta \subseteq Q \times (\Sigma \cup \{\lambda\}) \times Q$ *and for which there is at most one choice for any possible configuration.*

The reflexive transitive closure of $\vdash_M$ *is denoted by,* $\vdash_M^*$.

Based on the column-revolving operation involved, a 2-NCRFA is referred to as two-dimensional non-deterministic left-column revolving (2-NLCRA), right column-revolving (2-NRCRA) or circular column-interchanging finite automaton (2-NCIRA). The corresponding deterministic variants are referred to as 2-DLCRA, 2-DRCRA and 2-DCIRA. We can also consider bi-column-revolving automaton, where both left column-revolving and right column-revolving transitions are involved. We can formally define this two-dimensional non-deterministic bi-column-revolving automaton or in short 2-NBCRA to be a 7-tuple, $M = (Q, \Sigma, \delta, \Delta_l, \Delta_r, q_0, F)$, *where* $M = (Q, \Sigma, \delta, \Delta_l, q_0, F)$ *is a 2-NLCRA and* $M = (Q, \Sigma, \delta, \Delta_r, q_0, F)$ *is a 2-NRCRA. Whenever we refer to an automaton as two-dimensional input-revolving finite automaton it is either left column-revolving, right column-revolving, bi-column-revolving or circular column interchanging.*

We define the language accepted by a two-dimensional input-revolving finite automaton M to be,

$$
L(M) = \left\{ A = \begin{matrix} A_1 \\ \vdots \\ A_m \end{matrix} \in \Sigma^{**} \ \middle| \ \begin{matrix} q_0 A_1 \\ \vdots \\ A_m \end{matrix} \vdash_M^* \begin{matrix} A_1 \\ \vdots \\ A_m \\ q\lambda \end{matrix} \ or \ simply \ \begin{matrix} q_0 A_1 \\ \vdots \\ A_m \end{matrix} \vdash_M^* q \ with \ q \in F \right\}.
$$

We denote the family of languages accepted by devices of type X by $\mathfrak{L}(X)$.

Example 1. Let, $M = (\{q_0, q_1, q_2\}, \{0, 1, 2\}, \delta, \Delta, q_0, \{q_0\})$ be a 2-DCRFA, where $\delta = \{q_0 0 \rightarrow q_1, q_1 1 \rightarrow q_2, q_2 2 \rightarrow q_0, q_0 \# \rightarrow q_0\}$ and $\Delta = \{q_0 1 \rightarrow q_0, q_0 2 \rightarrow q_0, q_1 0 \rightarrow q_1, q_1 2 \rightarrow q_1, q_2 0 \rightarrow q_2, q_2 1 \rightarrow q_2\}$.

The automaton M accepts the following language regarded as either 2-DLCRA or 2-DRCRA,

$$
L(M) = \left\{ X = \begin{matrix} X_1 \\ \vdots \\ X_m \end{matrix} \in \{0, 1, 2\}^{**} \ \middle| \ \begin{matrix} \text{each } X_i\text{'s are of size } 1 \times n \text{ and} \\ |X_i|_0 = |X_i|_1 = |X_i|_2, \forall \quad i = 1, 2, \ldots, m \end{matrix} \right\}
$$

Working:

The automaton works on $\hat{X}$ if X is given as the input array. Starting from state q_0 and the tape head on the first element of the first row in the input X, the automaton M tries to read 0, 1 and 2 sequentially beginning with 0. It uses the δ transitions to store the current missing symbol in its finite control in order to search for it. Being in a search state, all non-matching symbol are column-wise shifted by the transitions from Δ. If the automaton reads #, it means that a complete row is read. Now the automaton uses the rule $q_0\# \to q_0$ in δ and moves to the first element of the very next row. This computational process is repeated for the subsequent rows until the entire array is read. If after reading the entire row, the automaton is in state q_0, then the input array is accepted or else it is rejected. Thus accepted array has equal number of 0s, 1s and 2s in each of its rows.

Example 2. The language $L = \{(\bullet^n X \bullet^n)_m \mid n, m > 0\}$ is accepted by a 2-DLCRA, $M = (Q, \{\bullet, X\}, \delta, \Delta, q_0, \{q_4\})$, where $Q = \{q_0, q_1, q_2, q_3, q_4, q_5\}$, $\delta = \{q_1\bullet \to q_2, q_2\bullet \to q_0, q_0X \to q_3, q_3\# \to q_4, q_4\bullet \to q_5, q_5\bullet \to q_4, q_4X \to q_3\}$ and $\Delta = \{q_0\bullet \to q_1\}$. The working of this automaton for an input array
$$\begin{array}{l} \bullet\ \bullet\ X\ \bullet\ \bullet \\ \bullet\ \bullet\ X\ \bullet\ \bullet \\ \bullet\ \bullet\ X\ \bullet\ \bullet \\ \bullet\ \bullet\ X\ \bullet\ \bullet \end{array}$$
is given as follows (Successive configurations are listed one below the other. The configuration at the end of a column is succeeded by the configuration at the top of the next column. The last configuration at the last column of this page is succeeded by the first configuration on the first column of the next page):

Column 1

```
q0 • • X • • #
   • • X • • #
   • • X • • #
   • • X • • #
        ⊢
q1 • • • X • #
   • • • X • #
   • • • X • #
   • • • X • #
        ⊢
 • q2 • • X • #
 •    • • X • #
 •    • • X • #
 •    • • X • #
        ⊢
 • • q0 • X • #
 • •    • X • #
 • •    • X • #
 • •    • X • #
        ⊢
 • • q1 • • X #
 • •    • • X #
 • •    • • X #
 • •    • • X #
        ⊢
```

Column 2

```
 • • • q2 • X #
 • • •    • X #
 • • •    • X #
 • • •    • X #
        ⊢
 • • • • q0 X #
 • • • •    X #
 • • • •    X #
 • • • •    X #
        ⊢
 • • • • X q3 #
 • • • • X    #
 • • • • X    #
 • • • • X    #
        ⊢
 • • • • X #
q4 • • • • X #
 • • • • X #
 • • • • X #
        ⊢
 •    • • • X #
 • q5 • • • X #
 •    • • • X #
 •    • • • X #
        ⊢
```

Column 3

```
 • •    • • X #
 • • q4 • • X #
 • •    • • X #
 • •    • • X #
        ⊢
 • • •    • X #
 • • • q5 • X #
 • • •    • X #
 • • •    • X #
        ⊢
 • • • •  ' X #
 • • • • q4 X #
 • • • •    X #
 • • • •    X #
        ⊢
 • • • • X    #
 • • • • X q3 #
 • • • • X    #
 • • • • X    #
        ⊢
 • • • •   X #
 • • • •   X #
q4 • • • • X #
 • • • •   X #
        ⊢
```

Column 4

```

     
```

```
 •    • • • X #
 •    • • • X #
 • q5 • • • X #
 •    • • • X #
        ⊢
 • •    • • X #
 • •    • • X #
 • • q4 • • X #
 • •    • • X #
        ⊢
 • • •    • X #
 • • •    • X #
 • • • q5 • X #
 • • •    • X #
        ⊢
 • • • •   X #
 • • • •   X #
 • • • • q4 X #
 • • • •   X #
        ⊢
 • • • • X    #
 • • • • X    #
 • • • • X q3 #
 • • • • X    #
        ⊢
```

$$
\begin{array}{l}
\bullet\ \bullet\ \bullet\ \bullet\ X\ \# \\
\bullet\ \bullet\ \bullet\ \bullet\ X\ \# \\
\bullet\ \bullet\ \bullet\ \bullet\ X\ \# \\
q_4\ \bullet\ \bullet\ \bullet\ \bullet\ X\ \#
\end{array}
\qquad
\begin{array}{l}
\bullet\ \bullet\qquad \bullet\ \bullet\ X\ \# \\
\bullet\ \bullet\qquad \bullet\ \bullet\ X\ \# \\
\bullet\ \bullet\qquad \bullet\ \bullet\ X\ \# \\
\bullet\ \bullet\ q_4\ \bullet\ \bullet\ X\ \#
\end{array}
\qquad
\begin{array}{l}
\bullet\ \bullet\ \bullet\ \bullet\qquad X\ \# \\
\bullet\ \bullet\ \bullet\ \bullet\qquad X\ \# \\
\bullet\ \bullet\ \bullet\ \bullet\qquad X\ \# \\
\bullet\ \bullet\ \bullet\ \bullet\ q_4\ X\ \#
\end{array}
\qquad q_4
$$

$$\vdash \qquad\qquad \vdash \qquad\qquad \vdash$$

$$
\begin{array}{l}
\bullet\qquad \bullet\ \bullet\ \bullet\ X\ \# \\
\bullet\qquad \bullet\ \bullet\ \bullet\ X\ \# \\
\bullet\qquad \bullet\ \bullet\ \bullet\ X\ \# \\
\bullet\ q_5\ \bullet\ \bullet\ \bullet\ X\ \#
\end{array}
\qquad
\begin{array}{l}
\bullet\ \bullet\ \bullet\qquad \bullet\ X\ \# \\
\bullet\ \bullet\ \bullet\qquad \bullet\ X\ \# \\
\bullet\ \bullet\ \bullet\qquad \bullet\ X\ \# \\
\bullet\ \bullet\ \bullet\ q_5\ \bullet\ X\ \#
\end{array}
\qquad
\begin{array}{l}
\bullet\ \bullet\ \bullet\ \bullet\ X\qquad \# \\
\bullet\ \bullet\ \bullet\ \bullet\ X\qquad \# \\
\bullet\ \bullet\ \bullet\ \bullet\ X\qquad \# \\
\bullet\ \bullet\ \bullet\ \bullet\ X\ q_3\ \#
\end{array}
$$

$$\vdash \qquad\qquad \vdash \qquad\qquad \vdash$$

The definition of two-dimensional column revolving finite automaton allows λ-transitions of both δ and Δ. The λ moves do not increase the computational power of any such automaton. (It is so, because, for every non-deterministic column-revolving automaton involving λ transitions in normal rules or column-revolving rules, an equivalent λ-free non-deterministic column-revolving automaton can be easily constructed (same for the deterministic variants)).

4   Comparison of the Variants

In this section we consider two-dimensional deterministic and non-deterministic variants of the column-revolving finite automata in a detailed way. In particular, we give the comparison among the different column-revolving rules and it turns out that the non-deterministic variants are better than the deterministic ones.

Theorem 1. *The language $L = \{(\bullet^n X \bullet^n)_m \mid n, m > 0\}$ is not accepted by any 2-NRCRA.*

Proof. We assume that the language L is accepted by some 2-NRCRA, $M = (Q, \Sigma, \delta, \Delta, q_0, F)$ with $|Q| = t$. Let us consider an array A with $|A|_c > 2t$, which is given as an input to the automaton M. The automaton begins its computation starting from the first element of the first row. During its computation on the first row some state, say p, appears at least twice, because of Pigeon Hole Principle. Let us consider that the first appearance of state p is reached after i ordinary moves and j right column-revolving moves with $0 \leq j < t$. Therefore we have,

$$
\begin{array}{l}
q_0\ \bullet\ \bullet^{n-1}\ X\ \bullet^n\ \# \\
(\ \bullet\ \bullet^{n-1}\ X\ \bullet^n\ \#)_{m-1}
\end{array}
\vdash_M^*
\begin{array}{l}
\bullet^i\ p\ \bullet\ \bullet^{n-1-i-j}\ X\ \bullet^{n+j}\ \# \\
(\bullet^i\ \ \bullet\ \bullet^{n-1-i-j}\ X\ \bullet^{n+j}\ \#)_{m-1}
\end{array}
$$

Let us consider that the second appearance of state p is reached after k ordinary moves and l right column-revolving moves with $1 \leq l < t - j$. Therefore we have,

$$
\begin{array}{l}
\bullet^i\ p\ \bullet\ \bullet^{n-1-i-j}\ X\ \bullet^{n+j}\ \# \\
(\bullet^i\ \ \bullet\ \bullet^{n-1-i-j}\ X\ \bullet^{n+j}\ \#)_{m-1}
\end{array}
\vdash_M^*
\begin{array}{l}
\bullet^{i+k}\ p\ \bullet\ \bullet^{n-1-i-j-k-l}\ X\ \bullet^{n+j+l}\ \# \\
(\bullet^{i+k}\ \ \bullet\ \bullet^{n-1-i-j-k-l}\ X\ \bullet^{n+j+l}\ \#)_{m-1}
\end{array}
$$

Since we have assumed that the array A is accepted, the computation reaches an accepting state, say q_f, such that,

$$\begin{array}{l} \bullet^{i+k} \; p \; \bullet \; \bullet^{n-1-i-j-k-l} \; X \; \bullet^{n+j+l} \; \# \\ (\bullet^{i+k} \quad \bullet \; \bullet^{n-1-i-j-k-l} \; X \; \bullet^{n+j+l} \; \#)_{m-1} \end{array} \vdash_M^* q_f$$

When we consider the input array as $A' = (\bullet^{n-k-l} X \bullet^{n+l})_m$, the automaton accepts this input too,

$$q_0 \; \bullet \; \bullet^{n-1-k-l} \; X \; \bullet^{n+l} \; \# \atop (\; \bullet \; \bullet^{n-1-k-l} \; X \; \bullet^{n+l} \; \#)_{m-1} \vdash_M^* \; \begin{array}{l} \bullet^i \; p \; \bullet \; \bullet^{n-1-i-j-k-l} \; X \; \bullet^{n+j+l} \; \# \\ (\bullet^i \quad \bullet \; \bullet^{n-1-i-j-k-l} \; X \; \bullet^{n+j+l} \; \#)_{m-1} \end{array} \vdash_M^* q_f$$

But the array A' is not a member of the language L. This is a contradiction to the initial assumption. Hence the theorem is proved. □

Theorem 2. *Let $L = \{\{0,1\}^{**} | \exists$ at least one row in which number of 0s and 1s are not equal\}. This language is not accepted by any 2-DBCRA but is accepted by some 2-NRCRA.*

Proof. We first prove that the language L is not accepted by any 2-DBCRA, by the method of contradiction. Let us assume that the language is accepted by some 2-DBCRA, $M = (Q, \Sigma, \delta, \Delta, q_0, F)$ with $|Q| = t$. Let $A = (0^n 1^{2n-j} 0^n)_m$ with $n, m, j > 2$, $j < n$, $j \neq 0$ and $n > t$ be the input given to the automaton. Based on pigeon hole principle, there exists a state, say p, which appears at least twice during the computation on first row. Let the first occurrence of state p be reached after: i normal transition, l_1 left column-revolving transition and r_1 right column-revolving transition,

$$q_0 \; 0 \; 0^{n-1} \; 1^{2n-j} \; 0^n \; \# \atop (\; 0 \; 0^{n-1} \; 1^{2n-j} \; 0^n \; \#)_{m-1} \vdash_M^* \; \begin{array}{l} 0^i \; p \; 0 \; 0^{n-1-i+l_1-r_1} \; 1^{2n-j} \; 0^{n-l_1+r_1} \; \# \\ (0^i \quad 0 \; 0^{n-1-i+l_1-r_1} \; 1^{2n-j} \; 0^{n-l_1+r_1} \; \#)_{m-1} \end{array}$$

Let the second occurrence of state p be reached after: j normal transition, l_2 left column-revolving transition and r_2 right column-revolving transition.

$$0^i \; p \; 0 \; 0^{n-1-i+l_1-r_1} \; 1^{2n-j} \; 0^{n-l_1+r_1} \; \# \atop (0^i \quad 0 \; 0^{n-1-i+l_1-r_1} \; 1^{2n-j} \; 0^{n-l_1+r_1} \; \#)_{m-1} \vdash_M^*$$

$$\begin{array}{l} 0^{i+j} \; p \; 0 \; 0^{n-1-i-j+l_1-r_1+l_2-r_2} \; 1^{2n-j} \; 0^{n-l_1+r_1-l_2+r_2} \; \# \\ (0^{i+j} \quad 0 \; 0^{n-1-i-j+l_1-r_1+l_2-r_2} \; 1^{2n-j} \; 0^{n-l_1+r_1-l_2+r_2} \; \#)_{m-1} \end{array}$$

Since the considered array A is in the language L, the computation reaches the accepting state, say q_f i.e.,

$$\begin{array}{l} 0^{i+j} \; p \; 0 \; 0^{n-1-i-j+l_1-r_1+l_2-r_2} \; 1^{2n-j} \; 0^{n-l_1+r_1-l_2+r_2} \; \# \\ (0^{i+j} \quad 0 \; 0^{n-1-i-j+l_1-r_1+l_2-r_2} \; 1^{2n-j} \; 0^{n-l_1+r_1-l_2+r_2} \; \#)_{m-1} \end{array} \vdash_M^* q_f$$

Let us now consider the input array $A' = (0^{n-j+l_2-r_2} 1^{2n-j} 0^{n-l_2+r_2})_m$. Due to the deterministic behavior, the computation on the input array A' is,

$$q_0 \; 0 \; 0^{n-1-j+l_2-r_2} \; 1^{2n-j} \; 0^{n-l_2+r_2} \; \# \atop (\; 0 \; 0^{n-1-j+l_2-r_2} \; 1^{2n-j} \; 0^{n-l_2+r_2} \; \#)_{m-1} \vdash_M^*$$

$$\begin{array}{l} 0^i \; p \; 0 \; 0^{n-1-i-j+l_1-r_1+l_2-r_2} \; 1^{2n-j} \; 0^{n-l_1+r_1-l_2+r_2} \; \# \\ (0^i \quad 0 \; 0^{n-1-i-j+l_1-r_1+l_2-r_2} \; 1^{2n-j} \; 0^{n-l_1+r_1-l_2+r_2} \; \#)_{m-1} \end{array} \vdash_M^* q_f$$

But this is a contradiction because A' is not in L. Now we prove that there is a 2-NRCRA, M', that accepts the language L. M' works on each row by reading a 0 and a 1 alternatively. This part is similar to the working of the automaton considered in Example 1. The input array is rejected if M' finds that the numbers are equal in each and every row. If there is at least one row in which the numbers are not equal then M' accepts the input array. Also, during the computation on each and every row, M' guesses whether there are only 0's or 1's remaining on present row and if that is the case then it simply reads the remaining elements on the row and the guess is verified. Hence, M' can accept the input array for correct guess on at least one row or else reject it. Hence the proof.   □

Theorem 3. *There is a language L not accepted by any 2-NLCRA but accepted by some 2-DRCRA.*

Proof. To prove this theorem, we consider the following language $L =$

$$\left\{ Y \oplus X \mid Y = (0^{n+1}1)_m, X = \begin{matrix} X_1 \\ \vdots \\ X_m \end{matrix}, X_i \in \{0,1\}^+, n + |X_i|_0 = 1 + |X_i|_1, \forall\, n, m \geq 0, i = 1, \ldots, m \right\}$$

Let us assume that L is accepted by some 2-NLCRA, $M = (Q, \Sigma, \delta, \Delta, q_0, F)$ with $|Q| = t$. Let us consider the array $A = (0^{n+1}1^{2n}0^n)_m$ with $n, m > 0$ and $n > t$ to be the input given to the automaton. Based on assumption, there is an accepting computation such that a state, say p appears at least twice during the computation on first row. Let the first occurrence of state p be reached after i normal transition and j left column-revolving transition and its second appearance be reached after k normal transition and l left column-revolving transition with $0 \leq i, j, k, l \leq n$ and $k + l > 0$. Hence we have,

$$
\begin{matrix} q_0\, 0\, 0^n\, 1^{2n}\, 0^n\, \# \\ (\ 0\, 0^n\, 1^{2n}\, 0^n\, \#)_{m-1} \end{matrix} \vdash_M^*
\begin{matrix} 0^i\, p\, 0\, 0^{n-i+j}\, 1^{2n}\, 0^{n-j}\, \# \\ (0^i\ \ 0\, 0^{n-i+j}\, 1^{2n}\, 0^{n-j}\, \#)_{m-1} \end{matrix} \vdash_M^+
$$

$$
\begin{matrix} 0^{i+k}\, p\, 0\, 0^{n-i+j-k+l}\, 1^{2n}\, 0^{n-j-l}\, \# \\ (0^{i+k}\ \ 0\, 0^{n-i+j-k+l}\, 1^{2n}\, 0^{n-j-l}\, \#)_{m-1} \end{matrix} \vdash_M^* q_f
$$

Let us consider the computation of M for the input array $A' = (0^{n+1-k+l}1^{2n}0^{n-l})$,

$$
\begin{matrix} q_0\, 0\, 0^{n-k+l}\, 1^{2n}\, 0^{n-l}\, \# \\ (\ 0\, 0^{n-k+l}\, 1^{2n}\, 0^{n-l}\, \#)_{m-1} \end{matrix} \vdash_M^*
\begin{matrix} 0^i\, p\, 0\, 0^{n-i+j-k+l}\, 1^{2n}\, 0^{n-j-l}\, \# \\ (0^i\ \ 0\, 0^{n-i+j-k+l}\, 1^{2n}\, 0^{n-j-l}\, \#)_{m-1} \end{matrix} \vdash_M^* q_f
$$

But this implies that $A' \in L$. Since, $n + \frac{l-k}{2} + n - l = 2n - \frac{l+k}{2} \neq 2n$ implies $A' \notin L$, we arrive at a contradiction.

The 2-DRCRA, $M = (\{q_0, q_1, q_2, q_3, q_4\}, \{0,1\}, \delta, \Delta, q_0, \{q_4\})$, where $\delta = \{q_0 0 \to q_1, q_1 1 \to q_2, q_2 0 \to q_3, q_3 1 \to q_2, q_3 \# \to q_4, q_4 0 \to q_3\}$ and $\Delta = \{q_1 0 \to q_1, q_2 1 \to q_2, q_3 0 \to q_3\}$ accepts the language L.

Theorem 4. *(i) The family $\mathfrak{L}(2\text{-}DLCRA)$ is properly included in $\mathfrak{L}(2\text{-}NLCRA)$*

(ii) The family $\mathfrak{L}(2\text{-}DRCRA)$ is properly included in $\mathfrak{L}(2\text{-}NRCRA)$

(iii) The family $\mathfrak{L}(\text{2-DBCRA})$ *is properly included in* $\mathfrak{L}(\text{2-NBCRA})$
(iv) The family $\mathfrak{L}(\text{2-DRCRA})$ *is properly included in* $\mathfrak{L}(\text{2-DBCRA})$
(v) The family $\mathfrak{L}(\text{2-NRCRA})$ *is properly included in* $\mathfrak{L}(\text{2-NBCRA})$
(vi) The family $\mathfrak{L}(\text{2-DLCRA})$ *is properly included in* $\mathfrak{L}(\text{2-DBCRA})$
(vii) The family $\mathfrak{L}(\text{2-NLCRA})$ *is properly included in* $\mathfrak{L}(\text{2-NBCRA})$

Proof. All inclusions are trivial. From Theorem 1, $\exists$ a language not accepted by any 2-NRCRA and hence also by any 2-DRCRA. But, as seen in Example 2 it is accepted by some 2-DLCRA and hence by some 2-DBCRA and 2-NBCRA. This proves (iv) and (v). By Theorem 2, $\exists$ a language not accepted by any 2-DBCRA and hence by any 2-DLCRA, 2-DRCRA but is accepted by 2-NRCRA, and hence by some 2-NBCRA. This proves (ii) and (iii). By Theorem 3 $\exists$ a language not accepted by any 2-NLCRA and hence by any 2-DLCRA but is accepted by some 2-DRCRA and hence by 2-DBCRA and 2-NBCRA. This proves (vi) and (vii).

To prove strict inclusion of (i), we consider the language L from Example 2. Let $L' = L \cup (L \oplus L'')$, where $L'' = \{Y_m \oplus \{\bullet, Y\}^{**}\}$. Clearly L' is accepted by some 2-NLCRA. But L' is not accepted by any 2-DLCRA. We prove this by contradiction. Let us assume that there is a 2-DLCRA, $M = (Q, \Sigma, \delta, \Delta, q_0, F)$ with $|Q| = t$. We consider an input array $A = (\bullet^{2n} X \, \bullet^{2n} Y^n \bullet^{2n})_m \in L'$ with $n > t$. During the computation on the first row some state, say $p \in Q$ appears at least twice due to our choice of n, let the first appearance be after i normal moves and j left column-revolving moves and the second appearance be after k normal moves and l left column-revolving moves, with $0 \le i, j, k, l \le n, k+l > 0$. Since M is deterministic we have,

$$
\begin{array}{l}
q_0 \bullet \bullet^{2n-1} X \bullet^{2n} Y^n \bullet^{2n} \# \\
(\ \bullet \bullet^{2n-1} X \bullet^{2n} Y^n \bullet^{2n} \#)_{m-1}
\end{array}
\vdash_M^*
\begin{array}{l}
\bullet^i p \bullet \bullet^{2n-1-i+j} X \bullet^{2n} Y^n \bullet^{2n-j} \# \\
(\bullet^i \ \bullet \bullet^{2n-1-i+j} X \bullet^{2n} Y^n \bullet^{2n-j} \#)_{m-1}
\end{array}
$$

$$
\vdash_M^+
\begin{array}{l}
\bullet^{i+k} p \bullet \bullet^{2n-1-i+j-k+l} X \bullet^{2n} Y^n \bullet^{2n-j-l} \# \\
(\bullet^{i+k} \ \bullet \bullet^{2n-1-i+j-k+l} X \bullet^{2n} Y^n \bullet^{2n-j-l} \#)_{m-1}
\end{array}
\vdash_M^* q_f
$$

We find that there exists an accepting configuration for the input array $A' = (\bullet^{2n-k+l} X \bullet^{2n} Y^n \bullet^{2n-l})_m$ such that,

$$
\begin{array}{l}
q_0 \bullet \bullet^{2n-1-k+l} X \bullet^{2n} Y^n \bullet^{2n-l} \# \\
(\ \bullet \bullet^{2n-1-k+l} X \bullet^{2n} Y^n \bullet^{2n-l} \#)_{m-1}
\end{array}
\vdash_M^*
$$

$$
\begin{array}{l}
\bullet^i p \bullet \bullet^{2n-1-i+j-k+l} X \bullet^{2n} Y^n \bullet^{2n-j-l} \# \\
(\bullet^i \ \bullet \bullet^{2n-1-i+j-k+l} X \bullet^{2n} Y^n \bullet^{2n-j-l} \#)_{m-1}
\end{array}
\vdash_M^* q_f
$$

But this implies that $k = l$. Since $l + k > 0$, we derive at $l > 0$. Since, M is deterministic, the accepting computation for the input array $A'' = (\bullet^{2n} X \bullet^{2n})_m$ is,

$$
\begin{array}{l}
q_0 \bullet \bullet^{2n-1} X \bullet^{2n} \# \\
(\ \bullet \bullet^{2n-1} X \bullet^{2n} \#)_{m-1}
\end{array}
\vdash_M^*
\begin{array}{l}
\bullet^i p \bullet \bullet^{2n-1-i+j} X \bullet^{2n-j} \# \\
(\bullet^i \ \bullet \bullet^{2n-1-i+j} X \bullet^{2n-j} \#)_{m-1}
\end{array}
\vdash_M^+
$$

$$
\begin{array}{l}
\bullet^{i+k} p \bullet \bullet^{2n-1-i+j-k+l} X \bullet^{2n-j-l} \# \\
(\bullet^{i+k} \ \bullet \bullet^{2n-1-i+j-k+l} X \bullet^{2n-j-l} \#)_{m-1}
\end{array}
\vdash_M^* q_f'
$$

where $q'_f \in Q$. We can also obtain an accepting configuration for the input array $A''' = (\bullet^{2n-k+l} X \bullet^{2n-l})_m$ as follows,

$$
\begin{array}{c}
q_0 \bullet \bullet^{2n-1-k+l} X \bullet^{2n-l} \# \\
(\bullet \bullet^{2n-1-k+l} X \bullet^{2n-l} \#)_{m-1}
\end{array}
\vdash^*_M
\begin{array}{c}
\bullet^i p \bullet \bullet^{2n-1-i+j-k+l} X \bullet^{2n-j-l} \# \\
(\bullet^i \quad \bullet \bullet^{2n-1-i+j-k+l} X \bullet^{2n-j-l} \#)_{m-1}
\end{array}
\vdash^*_M q'_f
$$

Since we have $l = k$ and $l > 0$, the array $A''' \notin L'$, which is a contradiction. Hence this proves (i).   □

Theorem 5. *(i) The family $\mathfrak{L}(2\text{-}DLCRA)$ is incomparable with $\mathfrak{L}(2\text{-}DRCRA)$*
(ii) The family $\mathfrak{L}(2\text{-}DLCRA)$ is incomparable with $\mathfrak{L}(2\text{-}NRCRA)$
(iii) The family $\mathfrak{L}(2\text{-}NLCRA)$ is incomparable with $\mathfrak{L}(2\text{-}NRCRA)$
(iv) The family $\mathfrak{L}(2\text{-}NLCRA)$ is incomparable with $\mathfrak{L}(2\text{-}DRCRA)$
(v) The family $\mathfrak{L}(2\text{-}NLCRA)$ is incomparable with $\mathfrak{L}(2\text{-}DBCRA)$
(vi) The family $\mathfrak{L}(2\text{-}NRCRA)$ is incomparable with $\mathfrak{L}(2\text{-}DBCRA)$

Proof. Follows from the proof of Theorem 4.   □

5   Comparison with Siromoney Matrix Languages

In this section we give the comparison of the family of languages accepted by all the variants of 2-NCRFA with the family of Siromoney matrix languages. Here we consider RML^T to be the transposition of RML.

Theorem 6. $\mathfrak{L}(RML^T) = \mathfrak{L}(2\text{-}DCIRA) = \mathfrak{L}(2\text{-}NCIRA)$

Proof. The transposition of any regular matrix language is accepted by a returning finite automaton (RFA), as can be seen in [3], which can be viewed as a 2-DCIRA where $\Delta = \emptyset$ and which is also a particular 2-NCIRA. Hence it remains to prove that any 2-NCIRA accepts a regular matrix language.

While we consider a 2-NCIRA, during the application of Δ rules, the automaton performs a circular-column-interchanging operation and it interchanges the column of the currently read symbol of the input array with the last column. So, once the last letter of the current reading row is known, a simulating automaton can remember the current last symbol of the row in its finite control to act correctly. Also we see that, initially the last symbol in the current row of the input can be guessed and finally, it can also be verified. Thus during the acceptance of any input array it can be easily seen that every row as well as column is regular. This simply corresponds with the definition of right-linear matrix grammars, where the vertical string of intermediate symbols are generated by using right-linear grammar rules followed by horizontal generation in parallel using right-linear grammar rules, yielding the regular matrix language. Hence any 2-NCIRA accepts a regular matrix language.   □

Theorem 7. *1. $\mathfrak{L}(2\text{-}NLCRA)$ and $\mathfrak{L}(CFML)$ are incomparable.*
2. $\mathfrak{L}(2\text{-}NRCRA)$ and $\mathfrak{L}(CFML)$ are incomparable.

Proof. To Prove: $\mathcal{L}(\text{2-NLCRA}) - \mathcal{L}(\text{CFML}) \neq \emptyset$ and $\mathcal{L}(\text{2-NRCRA}) - \mathcal{L}(\text{CFML}) \neq \emptyset$. Consider the language,

$$L = \left\{ X = \begin{array}{c} X_1 \\ \vdots \\ X_m \end{array} \in \{0,1,2\}^{**} \middle| \begin{array}{l} \text{each } X_i\text{'s are of size } 1 \times n \text{ and} \\ |X_i|_0 = |X_i|_1 = |X_i|_2, \forall \quad i = 1, 2, \ldots, m \end{array} \right\}$$

This language can be generated by any 2-NLCRA and 2-NRCRA, as can be seen in Example 1. But $L \notin \mathcal{L}\text{CFML}$. Hence, $\mathcal{L}(\text{2-NLCRA}) - \mathcal{L}\text{CFML} \neq \emptyset$ and $\mathcal{L}(\text{2-NRCRA}) - \mathcal{L}(\text{CFML}) \neq \emptyset$.

To prove: $\mathcal{L}(\text{CFML}) - \mathcal{L}(\text{2-NLCRA}) \neq \emptyset$ and $\mathcal{L}(\text{CFML}) - \mathcal{L}(\text{2-NRCRA}) \neq \emptyset$. Consider the language, $L' = \left\{ \begin{array}{c} X^m \ Y^m \ X \ Y^n \ X^n \\ (X^m \ \bullet^m \ X \ \bullet^n \ X^n)_i \\ X^m \ Y^m \ X \ Y^n \ X^n \end{array} \middle| \ n, m, i \geq 1 \right\}$. $L' \in \mathcal{L}(\text{CFML})$, as can be seen in [12], the vertical string of intermediate symbols can be generated by using right-linear grammar rules followed vertical generation in parallel using context-free grammar rules.

However, this language cannot be generated by any 2-NBCRA. We prove this by the method of contradiction. Let us assume that the language is accepted by some 2-NBCRA, $M = (Q, \Sigma, \delta, \Delta, q_0, F)$ with $|Q| = t$ and no (useless) loops. Let $A = \begin{array}{c} X^{n+1} \ Y^{n+1} \ X \ Y^n \ X^n \\ (X^{n+1} \ \bullet^{n+1} \ X \ \bullet^n \ X^n)_i \\ X^{n+1} \ Y^{n+1} \ X \ Y^n \ X^n \end{array}$ with $n > 2$, $i > 0$ and $n > t$ be the input given to the automaton. By the pigeon hole principle, there exists a state, say p which appears at least twice during the computation on first row. Let the first occurrence of state p be reached in r steps which involves j normal transition, l_1 left column-revolving transition and r_1 right column-revolving transition,

$$\begin{array}{c} q_0 \ X \ X^n \ Y^{n+1} \ X \ Y^n \ X^n \ \# \\ (\ X \ X^n \ \bullet^{n+1} \ X \ \bullet^n \ X^n \ \#)_i \\ X \ X^n \ Y^{n+1} \ X \ Y^n \ X^n \ \# \end{array} \vdash_M^* \begin{array}{c} X^j \ p \ X \ X^{n-j+l_1-r_1} \ Y^{n+1} \ X \ Y^n \ X^{n-l_1+r_1} \ \# \\ (X^j \ X \ X^{n-j+l_1-r_1} \ \bullet^{n+1} \ X \ \bullet^n \ X^{n-l_1+r_1} \ \#)_i \\ X^j \ X \ X^{n-j+l_1-r_1} \ Y^{n+1} \ X \ Y^n \ X^{n-l_1+r_1} \ \# \end{array}$$

Let the second occurrence of state p be reached in s steps which involves k normal transition, l_2 left column-revolving transition and r_2 right column-revolving transition,

$$\begin{array}{c} X^j \ p \ X \ X^{n-j+l_1-r_1} \ Y^{n+1} \ X \ Y^n \ X^{n-l_1+r_1} \ \# \\ (X^j \ X \ X^{n-j+l_1-r_1} \ \bullet^{n+1} \ X \ \bullet^n \ X^{n-l_1+r_1} \ \#)_i \\ X^j \ X \ X^{n-j+l_1-r_1} \ Y^{n+1} \ X \ Y^n \ X^{n-l_1+r_1} \ \# \end{array} \vdash_M^*$$

$$\begin{array}{c} X^{j+k} \ p \ X \ X^{n-j-k+l_1-r_1+l_2-r_2} \ Y^{n+1} \ X \ Y^n \ X^{n-l_1+r_1-l_2+r_2} \ \# \\ (X^{j+k} \ X \ X^{n-j-k+l_1-r_1+l_2-r_2} \ \bullet^{n+1} \ X \ \bullet^n \ X^{n-l_1+r_1-l_2+r_2} \ \#)_i \\ X^{j+k} \ X \ X^{n-j-k+l_1-r_1+l_2-r_2} \ Y^{n+1} \ X \ Y^n \ X^{n-l_1+r_1-l_2-r_2} \ \# \end{array}$$

Since the considered array A is in the language L, the computation reaches the accepting state, say $q_f \in Q$ i.e.,

$$\begin{array}{c} X^{j+k} \ p \ X \ X^{n-j-k+l_1-r_1+l_2-r_2} \ Y^{n+1} \ X \ Y^n \ X^{n-l_1+r_1-l_2+r_2} \ \# \\ (X^{j+k} \ X \ X^{n-j-k+l_1-r_1+l_2-r_2} \ \bullet^{n+1} \ X \ \bullet^n \ X^{n-l_1+r_1-l_2+r_2} \ \#)_i \vdash_M^* q_f \\ X^{j+k} \ X \ X^{n-j-k+l_1-r_1+l_2-r_2} \ Y^{n+1} \ X \ Y^n \ X^{n-l_1+r_1-l_2-r_2} \ \# \end{array}$$

We find that $l_2 - r_2 = 0$. Otherwise, the computation

$$q_0 \ X \ X^{n-k+l_2-r_2} \ Y^{n+1} \ X \ Y^n \ X^{n-l_2+r_2} \ \#$$
$$(\ X \ X^{n-k+l_2-r_1} \ \bullet^{n+1} \ X \ \bullet^n \ X^{n-l_2+r_2} \ \#)_i \ \vdash_M^*$$
$$X \ X^{n-j+l_1-r_1} \ Y^{n+1} \ X \ Y^n \ X^{n-l_1+r_1} \ \#$$

$$X^j \ p \ X \ X^{n-j-k+l_1-r_1+l_2-r_2} \ Y^{n+1} \ X \ Y^n \ X^{n-l_1+r_1-l_2+r_2} \ \#$$
$$(X^j \quad X \ X^{n-j-k+l_1-r_1+l_2-r_2} \ \bullet^{n+1} \ X \ \bullet^n \ X^{n-l_1+r_1-l_2+r_2} \ \#)_i \ \vdash_M^* \ q_f$$
$$X^j \quad X \ X^{n-j-k+l_1-r_1+l_2-r_2} \ Y^{n+1} \ X \ Y^n \ X^{n-l_1+r_1-l_2-r_2} \ \#$$

accepts the array $A' = \begin{pmatrix} X^{n+1-k+l_2-r_2} \ Y^{n+1} \ X \ Y^n \ X^{n-l_2+r_2} \\ X^{n+1-k+l_2-r_2} \ \bullet^{n+1} \ X \ \bullet^n \ X^{n-l_2+r_2} \\ X^{n+1-k+l_2-r_2} \ Y^{n+1} \ X \ Y^n \ X^{n-l_2+r_2} \end{pmatrix}_i \notin L'$. For same

reason we also find that $l_2 - r_2 - k = 0$ which implies $k = 0$. But this is a contradiction to our assumption because it follows either that $r = s$ or M loops. This implies L' is not accepted by any 2-NBCRA and hence by any 2-NLCRA and 2-NRCRA. Hence, $\mathcal{L}(CFML) - \mathcal{L}(2\text{-NLCRA}) \neq \emptyset$ and $\mathcal{L}(CFML) - \mathcal{L}(2\text{-NRCRA}) \neq \emptyset$. □

6   Closure Properties

In this section we discuss the closure properties of the family of languages accepted by any 2-DCIRA, 2-DLCRA, 2-DRCRA and 2-DBCRA.

Theorem 8. *$\mathcal{L}$(2-DCIRA) is closed under complementation and union.*

Proof. From [14], RML is closed under complementation and union. By Theorem 6, this theorem is proved. □

Theorem 9. *$\mathcal{L}$(2-DLCRA), $\mathcal{L}$(2-DRCRA) and $\mathcal{L}$(2-DBCRA) are not closed under complementation.*

Proof. Consider the language,

$$L = \left\{ X = \begin{matrix} X_1 \\ \vdots \\ X_m \end{matrix} \in \{0,1\}^{**} \middle| \begin{matrix} \text{each } X_i\text{'s are of size } 1 \times n \text{ and} \\ |X_i|_0 = |X_i|_1, \forall \quad i = 1, 2, \ldots, m \end{matrix} \right\}$$

By Theorem 2, the language $\bar{L} = \{\{0,1\}^{**} | \exists$ at least one row in which no. of 0s and 1s are not equal$\}$ is not accepted by any 2-DBCRA and hence by any 2-DLCRA and 2-DRCRA. But by Example 1, we see that the language L is accepted by some 2-DBCRA and hence by some 2-DLCRA and 2-DRCRA. □

Theorem 10. *$\mathcal{L}$(2-DLCRA), $\mathcal{L}$(2-DRCRA) and $\mathcal{L}$(2-DBCRA) are not closed under union.*

Proof. Consider the languages,

$$L_1 = \left\{ X = \begin{matrix} X_1 \\ \vdots \\ X_m \end{matrix} \in \{0,1\}^{**} \; \middle| \; \begin{matrix} \text{each } X_i\text{'s are of size } 1 \times n \text{ and} \\ |X_i|_0 = |X_i|_1, \forall \quad i = 1,2,\ldots,m \end{matrix} \right\}$$

and $L_2 = \{0\}^{**}$. The language L_1 can be accepted by a 2-DBCRA, as can be seen in Example 1. L_2 can be accepted by a 2-DBCRA, $M = (\{q_0\}, \{0\}, \{q_0 0 \to q_0, q_0\# \to q_0\}, \emptyset, \emptyset, q_0, \{q_0\})$. By the same reasoning as in the proof of Theorem 9 we can show that $L_1 \cup L_2$ is not accepted by any 2-DBCRA and hence by any 2-DLCRA and 2-DRCRA. □

Theorem 11. $\mathcal{L}$(*2-DCIRA*) *is closed under column catenation, but not under row catenation.*

Proof. From [14], RML is closed under column catenation, but not under row catenation. By Theorem 6, this theorem is proved. □

Theorem 12. $\mathcal{L}$(*2-DLCRA*), $\mathcal{L}$(*2-DRCRA*) *and* $\mathcal{L}$(*2-DBCRA*) *are neither closed under column catenation nor row catenation.*

Proof. Consider the languages, $L_1 = \left\{ \begin{matrix} X^m \, Y^m \, X \\ (X^m \;.^m \, X)_i \\ X^m \, Y^m \, X \end{matrix} \; \middle| \; m, i \geq 1 \right\}$ and $L_2 =$

$\left\{ \begin{matrix} Y^n \, X^n \\ (.^n \, X^n)_i \\ Y^n \, X^n \end{matrix} \; \middle| \; n, i \geq 1 \right\}$. Both the languages L_1 and L_2 belong to $\mathcal{L}$(2-DLCRA).

But the column catenation, $L_1 \oplus L_2 = \left\{ \begin{matrix} X^m \, Y^m \, X \, Y^n \, X^n \\ (X^m \;.^m \, X \;.^n \, X^n)_i \\ X^m \, Y^m \, X \, Y^n \, X^n \end{matrix} \; \middle| \; n, m, i \geq 1 \right\}$

does not belong to $\mathcal{L}$(2-DBCRA), as seen in proof of Theorem 7. This proves that $\mathcal{L}$(2-DLCRA) and $\mathcal{L}$(2-DBCRA) are not closed under column catenation. If we consider the language,

$$L_3 = \left\{ X = \begin{matrix} X_1 \\ \vdots \\ X_m \end{matrix} \in \{0,1\}^{**} \; \middle| \; \begin{matrix} \text{each } X_i\text{'s are of size } 1 \times n \text{ and} \\ |X_i|_0 = |X_i|_1, \forall \quad i = 1,2,\ldots,m \end{matrix} \right\}$$

belonging to $\mathcal{L}$(2-DLCRA) and also $\mathcal{L}$(2-DRCRA), it is clear that $L_3 \ominus L_2$ does not even belong to $\mathcal{L}$(2-DBCRA), because the columns shuffling involved during the simulation of the automaton accepting L_2 makes the simulation of the automaton accepting L_3 impossible. This proves that $\mathcal{L}$(2-DLCRA), $\mathcal{L}$(2-DRCRA) and $\mathcal{L}$(2-DBCRA) is not closed under row catenation. Now we consider the language,

$$L_4 = \left\{ X = \begin{matrix} 2X_1 \\ \vdots \\ 2X_m \end{matrix} \in \{0,1,2\}^{**} \; \middle| \; \begin{matrix} \text{each } X_i\text{'s are of size } 1 \times n \text{ and} \\ |X_i|_0 = |X_i|_1, \forall \quad i = 1,2,\ldots,m \end{matrix} \right\}$$

Clearly, the languages L_3 and L_4 belong to 2-DRCRA, but the language $L_1 \oplus L_2$ does not belong to 2-DRCRA (can be proved by a similar argument as in Theorem 1). Hence $\mathcal{L}$(2-DRCRA) are not closed under column catenation.

Theorem 13. $\mathcal{L}$(2-DCIRA) is closed under column Kleene star, but not under row Kleene star.

Proof. The proof follows from Theorem 11.   □

Theorem 14. $\mathcal{L}$(2-DLCRA), $\mathcal{L}$(2-DRCRA) and $\mathcal{L}$(2-DBCRA) are neither closed under column Kleene star nor row Kleene star.

Proof. The proof follows from Theorem 12.   □

Theorem 15. $\mathcal{L}$(2-DCRFA) is not closed under quarter turn and transpose.

Proof. For $\mathcal{L}$(2-DCIRA) the properties are true, from [14] and Theorem 6.
Consider the language,

$$L(M) = \left\{ X = \begin{matrix} X_1 \\ \vdots \\ X_m \end{matrix} \in \{0,1\}^{**} \middle| \begin{matrix} X_i = a_{i1}a_{i2}\dots a_{in} \\ |X_i|_0 = |X_i|_1, \forall \quad i = 1,2,\dots,m \end{matrix} \right\}. \text{ This lan-}$$

guage can be generated by some 2-DBCRA, and hence by some 2-DLCRA and 2-DRCRA as can be seen in Example 1. Let us represent the quarter turn of this language by L^{QT} and we have,

$$L^{QT} = \left\{ X = X_m \dots X_1 \in \{0,1\}^{**} \middle| X_i = \begin{matrix} a_{i1} \\ \vdots \\ a_{in} \end{matrix}, |X_i|_0 = |X_i|_1, \forall\, i = 1,2,\dots,m \right\}.$$

This language cannot be generated by any 2-DCRFA, because no such automaton can accept the input X from L^{QT} with the condition $|X_i|_0 = |X_i|_1, \forall\, i = 1,2,\dots,k$. This proves that $\mathcal{L}$(2-DBCRA) is not closed under quarter turn.
Let us represent the transpose of the language L by L^T and we have, $L^T =$
$$\left\{ X = X_1 \dots X_m \in \{0,1\}^{**} \middle| X_i = \begin{matrix} a_{i1} \\ \vdots \\ a_{in} \end{matrix}, |X_i|_0 = |X_i|_1, \forall\, i = 1,2,\dots,m \right\}. \text{ By the}$$
same argument as discussed for L^{QT}, we see that there is no 2-DCRFA to accept L^T. This proves that $\mathcal{L}$(2-DBCRA) is not closed under transpose.   □

Theorem 16. $\mathcal{L}$(2-DCRFA) is closed under half turn, reflection on rightmost vertical and reflection on base.

Proof. For $\mathcal{L}$(2-DCIRA) the property is true, from [14] and Theorem 6. We shall give the proof for $\mathcal{L}$(2-DRCRA) and reflection about the rightmost vertical, proof being similar for the rest of the cases. Let $M = (Q, \Sigma, \delta, \Delta, q_0, F)$ be a 2-DRCRA that accepts the language $L = L(M)$. We now construct a 2-DRCRA, M' that accepts the language obtained by taking reflection on the rightmost vertical of L represented by L^{RB}. $M' = (Q \cup \{q_0'\}, \Sigma, \delta', \Delta, q_0', F')$, where $\delta' = \delta \cup \{q_0' \to q \mid p\# \to q \in \delta\} \cup \{p \to q_0 \mid p \in Q\} \cup \{q \to p \mid q \in F, p \in Q\}$ and $F' = \{q \mid q_0 \to q_1, q_1 \to q_2, \dots, q_{n-1} \to q_n, q_n\# \to q \in \delta'\}$. Clearly, $L(M') = L^{RB}$.   □

7 Application

One of the applications of this newly constructed automata is in the field of steganography [6]. Steganography is the art and science of hiding information by embedding messages within others. This hidden information can be plain text, cipher text, or even images. Steganographic communications hide often in plain sight, whereas encrypted communications in Cryptography are very obvious of the fact that they are sending secrets.

Here we consider hiding text messages behind some pictures which can be retrieved only by giving them as input in the newly constructed automata and arriving at the final configuration. For example, let us consider the 2-DRCRA automaton from Example 1 and an array $2\,1\,0\,1\,0\,2$. Let $\begin{smallmatrix}\end{smallmatrix}$

$$\begin{matrix} 2\,2\,1\,0\,1\,0 \\ 2\,1\,0\,1\,0\,2 \\ 0\,1\,2\,1\,0\,2 \end{matrix} \qquad \begin{matrix} T\ \square\ \square\ S\ H\ I \\ I\ \square\ \square\ A\ S\ \square \\ E\ S\ T\ E\ C\ R \end{matrix}$$

automaton from Example 1 and an array $2\,1\,0\,1\,0\,2$. Let $I\ \square\ \square\ A\ S\ \square$ be the scrambled hidden message behind the array A. Now to get the proper hidden

$$\begin{matrix} \square\ T\ H\ I\ S\ \square \\ \square\ I\ S\ \square\ A\ \square \\ S\ E\ C\ R\ E\ T \end{matrix}$$

message $\square\ I\ S\ \square\ A\ \square$, the array A is given as the input to the automaton and an accepting configuration has to be reached. The working of this automaton on array A along with the hidden message is as follows:

$$\begin{matrix} q_0 2\,2\,1\,0\,1\,0\,\# & & I\ T\ H\ S\ \square\ \square \\ 2\,1\,0\,1\,0\,2\,\# & & \square\ I\ S\ A\ \square\ \square \\ 0\,1\,2\,1\,0\,2\,\# & & R\ E\ C\ E\ S\ T \\ & \vdash^* & \\ 0\,1\,2\,0\,1\,2\,\# & & S\ \square\ I\ \square\ H\ T \\ q_0 1\,0\,2\,2\,0\,1\,\# & & A\ \square\ \square\ \square\ S\ I \\ 1\,0\,0\,2\,2\,1\,\# & & E\ S\ R\ T\ C\ E \\ & \vdash^* & \\ 1\,2\,2\,1\,0\,0\,\# & & \square\ T\ I\ H\ S\ \square \\ 0\,1\,2\,0\,1\,2\,\# & & \square\ I\ \square\ S\ A\ \square \\ q_0 0\,1\,0\,2\,1\,2\,\# & & S\ E\ R\ C\ E\ T \\ & \vdash^* & \\ & & T\ H\ I\ S \\ q_0 & & I\ S\ \quad A \\ & & S\ E\ C\ R\ E\ T \end{matrix}$$

8 Conclusion

It is worth examining the decidability results (non-emptiness, membership, finiteness, etc.) of these automata and explore their further applications in character recognition and floor designs. Construction of two-dimensional automaton which involves both column-revolving and row-revolving can also be studied.

References

1. Bensch, S., Bordihn, H., Holzer, M., Kutrib, M.: On input-revolving deterministic and nondeterministic finite automata. Inf. Comput. **207**, 1140–1155 (2009)
2. Blum, M., Hewitt, C.: Automata on a 2-dimensional tape. In: SWAT 1967, 8th Annual Symposium on Switching and Automata Theory, pp. 155–160 (1967)
3. Fernau, H., Paramasivan, M., Schmid, M.L., Thomas, D.G.: Scanning pictures the boustrophedon way. In: Barneva, R.P., Bhattacharya, B.B., Brimkov, V.E. (eds.) IWCIA 2015. LNCS, vol. 9448, pp. 202–216. Springer, Heidelberg (2015). doi:10. 1007/978-3-319-26145-4sps15
4. Giammarresi, D., Restivo, A.: Two-dimensional languages. In: Rozenberg, G., Salomaa, A. (eds.) Handbook of Formal Languages, vol. 3, pp. 215–267. Springer, Heidelberg (1997)
5. Kamala, K., Siromoney, R.: Array automata and operations on array languages. Int. J. Comput. Appl. **4**(1–4), 3–30 (1974)
6. Kipper, G.: Investigator's Guide to Steganography. Auerbach Publications, Boca Raton (2003)
7. Meduna, A.: Automata and Languages: Theory and Applications. Springer, Heidelberg (2000)
8. Pradella, M., Cherubini, A., Crespi-Reghizzi, S.: A unifying approach to picture grammars. Inf. Comput. **209**, 1246–1267 (2011)
9. Rosenfeld, A.: Picture Languages: Formal Models for Picture Recognition. Acadamic Press, Cambridge (1979)
10. Rosenfeld, A., Siromoney, R.: Picture languages - a survey. Lang. Des. **1**(3), 229–245 (1993)
11. Rozenberg, G., Salomaa, A. (eds.): Handbook of Formal Languages, vols. 1–3. Springer, Heidelberg (1997)
12. Siromoney, G., Siromoney, R., Krithivasan, K.: Abstract families of matrices and picture languages. Comput. Graph. Image Proc. **1**, 284–307 (1972)
13. Siromoney, R.: Advances in array languages. In: Ehrig, H., Nagl, M., Rozenberg, G., Rosenfeld, A. (eds.) Graph Grammars 1986. LNCS, vol. 291, pp. 549–563. Springer, Heidelberg (1987). doi:10.1007/3-540-18771-5_75
14. Siromoney, G., Siromoney, R., Krithivasan, K.: Picture languages with array rewriting rules. Inf. Control **22**, 447–470 (1973)
15. Siromoney, R., Subramanian, K.G., Rangarajan, K.: Parallel/sequential rectangular arrays with tables. Int. J. Comput. Math. **6A**, 143–158 (1977)
16. Wang, P.S.P.: Array Grammars, Patterns and Recognizers. World Scientific Series in Computer Science, vol. 18. World Scientific Publishing, Singapore (1989)
17. Wang, P.S.P.: Sequential/parallel matrix array languages. J. Cybern. **5**, 19–36 (1975)

Application-driven Contributions

Direct Phasing of Crystalline Materials from X-ray Powder Diffraction

Hongliang Xu[1,2(✉)]

[1] Mathematics Department, SUNY Buffalo State,
1300 Elmwood Avenue, Buffalo, NY 14222, USA
xuh@buffalostate.edu
[2] Hauptman-Woodward Medical Research Institute,
700 Ellicott Street, Buffalo, NY 14203, USA

Abstract. The direct-methods phasing program, *Shake-and-Bake* for single crystal structure determination, has been adapted and modified to solve microcrystal structures from X-ray powder diffraction data.

Keywords: X-ray powder diffraction · Direct phasing methods · Structure determination

1 Introduction

No microscope has sufficient resolution to let us observe molecules and their interactions directly. Instead, scientists use tools provided by the science of crystallography. In a crystallographic experiment, a single crystal of a purified substance is irradiated, typically with X-rays. The incident radiation is scattered in many directions to produce a so-called diffraction pattern, and a complex mathematical analysis of the diffraction data (a process known as "solving" the structure) will determine the shape and atomic arrangement of the molecules comprising the crystals. Once the structure has been solved, molecular models can be constructed and examined for insight. For example, scientists can learn how the protein molecules function, what might be happening when disease occurs, and what compounds might be designed as drugs to modify activities. In chemical and petrochemical industry, scientists can study crystalline structures of microporous materials, such as zeolites and silicoaluminophosphates. They can analyze three-dimensional frameworks that contain uniform pore openings, channels, and internal cages, study their unique properties and discover new microporous sieves to improve the performance of these materials as catalysts, absorbents, and ion exchangers.

The best and most detailed structural information is obtained when the diffraction pattern of a single crystal a few tenths of a millimeter in each dimension is analyzed, but growing high-quality crystals of this size is often difficult, sometimes impossible. However, many crystallization experiments, including those for pharmaceutically important compounds and microporous materials that do not

ⓒ Springer International Publishing AG 2017
R.P. Barneva et al. (Eds.): CompIMAGE 2016, LNCS 10149, pp. 167–177, 2017.
DOI: 10.1007/978-3-319-54609-4_12

yield single crystals, produce showers of randomly oriented microcrystals that can be exposed to X-rays simultaneously to provide a powder diffraction pattern. Although single-crystal diffraction data consists of discrete spots or X-ray reflections, the diffraction of microcrystals in a powder forms rings so that the reflections overlap. Thus, the analysis is more challenging because the individual reflection amplitudes are needed before the phases of the structure factors can be found and the structure determined. In this paper, we applied a new algorithm, which is adapted and modified from a well-known program for determining single crystal structures, to solve microcrystal structures.

2   Single-Crystal Structure Determination

When X-rays are diffracted by a single crystal, each scattered wave or reflection has associated with it an amplitude or magnitude $|F|$, which is experimentally accessible, and a phase angle ϕ, which is not. Together, the magnitudes and their associated phase angles constitute the complex-valued structure factors (Fs), which can be converted into normalized structure factors (Es) having average values that are independent of scattering angle (resolution). Fourier transformation of the structure factors leads directly to the real-valued unit-cell electron density distribution that, after suitable interpretation, reveals the atomic positions and describes the molecular architecture of the crystalline material responsible for the scattering.

If the 3D structure of a similar compound is known, it may be possible, by rotating and translating a homologous model, to use molecular replacement [19] to find initial values for the missing phase angles. If no model is available, solution of the "phase-angle problem" requires *ab initio* Patterson-based [15] or direct methods [9] that rely on the values of diffraction amplitudes alone. Either of these methods can be used to locate heavy or anomalously scattering atoms that comprise protein substructures and are used for phasing by multiple isomorphous replacement [17] or anomalous scattering [10] techniques. Direct methods are more robust and have been used to solve the largest full structure (~2000 independent non-hydrogen atoms) [8] and the largest heavy-atom (Se) substructures (160 sites) [21].

Direct Methods. Direct methods of structure determination dominate the field of single-crystal diffraction because of their incredible success rate, range of applicability, speed, ease of use and reliability. Direct methods exploit the facts that the electron density is everywhere nonnegative and electrons are strongly concentrated around the centers of the atoms. The electron density, $\rho(\mathbf{r})$, at any point $\mathbf{r}$ in space can be expressed in terms of diffraction quantities as:

$$\rho(\mathbf{r}) = \sum_{\mathbf{H}} F_{\mathbf{H}} \exp(-2\pi i \mathbf{H} \cdot \mathbf{r}) \tag{1}$$

where $\mathbf{H}$ and $F_{\mathbf{H}}$ are the reciprocal-lattice vector and its structure factor. The normalized structure factor, $E_{\mathbf{H}}$, which is directly obtainable from $F_{\mathbf{H}}$, can be expressed in terms of its corresponding amplitude, $|E_{\mathbf{H}}|$, and phase angle, $\phi_{\mathbf{H}}$, as:

$$E_{\mathbf{H}} = |E_{\mathbf{H}}| \exp(i\phi_{\mathbf{H}}) = \frac{1}{\sigma_2^{1/2}} \sum_{j=1}^{N} f_j \exp(2\pi i \mathbf{H} \cdot \mathbf{r}_j) \qquad (2)$$

where N is the number of non-hydrogen atoms in the asymmetric unit, f_j and $\mathbf{r}_j = (x_j, y_j, z_j)$ are the atomic scattering factor and the position vector of atom j respectively, and $\sigma_2 = \sum_{j=1}^{N} f_j^2$.

Since there are many more reflections in a diffraction pattern than there are independent atoms in the corresponding crystal, the phase-angle problem is overdetermined, and the existence of relationships among the measured magnitudes is implied. Direct methods seek to exploit these relationships through a probabilistic or statistical approach. Direct methods theory identifies certain sums of three phase angles (termed triplet structure invariants due to their independence of the choice of origins),

$$\phi_{\mathbf{HK}} = \phi_{\mathbf{H}} + \phi_{\mathbf{K}} + \phi_{-\mathbf{H}-\mathbf{K}}, \qquad (3)$$

having large associated values

$$A_{\mathbf{HK}} = 2N^{-1/2}|E_{\mathbf{H}} E_{\mathbf{K}} E_{\mathbf{H}+\mathbf{K}}|, \qquad (4)$$

and estimates probabilistic expected values of $\cos(\phi_{\mathbf{HK}})$ [6] or analyzes statistical properties of distribution functions of $\phi_{\mathbf{HK}}$ [24].

Dual-Space Optimization. Conventional reciprocal-space direct methods, implementing tangent formula refinement in computer programs such as *MULTAN* [14,28], *SHELXS* [20] and *SIR* [3], can provide solutions for structures containing fewer than some 100 to 150 unique non-hydrogen atoms. In general, however, phase-angle refinement alone is not sufficient to solve large structures. In such cases, successful applications require a dual-space optimization procedure known as *Shake-and-Bake* that was implemented in the *SnB* program [18,22,23,25]. *Shake-and-Bake* is also a powerful method for smaller structures and substructures, effectively avoiding most cases of false minima [27]. The distinctive feature of this procedure is the repeated and unconditional alternation of reciprocal-space phase-angle refinement with a complementary real-space process that seeks to improve phase angles by applying an atomicity constraint through peak picking. In each cycle, the largest peaks are used as an updated trial structure without regard to chemical constraints other than a minimum allowed distance between atoms. Figure 1 shows a small sample of successful applications of *SnB* program.

3   Microcrystalline Structure Determination

3.1   Powder Diffraction

X-ray powder diffraction is a non-destructive technique widely applied for the characterization of crystalline materials. The method has been traditionally used

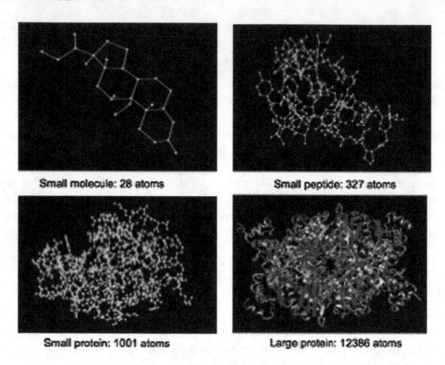

Fig. 1. Successful applications of *SnB*.

for phase identification, quantitative analysis and the determination of structure imperfections. Due to rapid development in powder diffraction theory, techniques and practices, instrumentations, computation power and analysis tools, powder diffraction applications have been extended to new area, such as the determination of crystal structures and the extraction of three-dimensional microstructural properties. Various kinds of micro- and nano-crystalline materials can be characterized from X-ray powder diffraction, including inorganics, organics, drugs, minerals, zeolites, catalysts, metals and ceramics.

As illustrated in Fig. 2, the X-ray diffraction powder method utilizes samples consisting of many randomly oriented microcrystallites (Fig. 2a) that scatter X-rays in the directions of a set of concentric cones (Fig. 2b) defined by the Bragg law $2d_{\mathbf{H}} \sin \theta_{\mathbf{H}} = \lambda$, where λ is the X-ray wavelength, $d_{\mathbf{H}}$ is the perpendicular interplanar spacing in the $\mathbf{H}$ family of crystal lattice planes, and $2\theta_{\mathbf{H}}$ is the opening angle of the cone of X-ray beams reflected from the $\mathbf{H}$ planes. If some of the crystallites are relatively large, the diffraction rings recorded by a detector or photographic film intercepting the cones of scattered X-rays will have a spotty appearance; if the crystallites are uniformly small, the rings will be continuous (Fig. 2c). In effect, a trace along the diameters of the diffraction rings gives a projection of the three-dimensional diffraction pattern from each crystallite onto a one-dimensional scan (Fig. 2d) $I(2\theta_{\mathbf{H}}) \propto |F_{\mathbf{H}}|^2$, where $F_{\mathbf{H}} = |F_{\mathbf{H}}| \exp(i\phi_{\mathbf{H}})$ is the, in general, complex-valued crystal structure factor with magnitude or amplitude $|F_{\mathbf{H}}|$, which can be extracted from the $I(2\theta_{\mathbf{H}})$ measurements, and phase $\phi_{\mathbf{H}}$, which cannot.

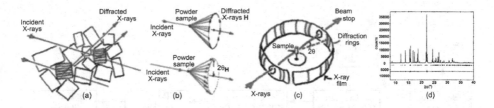

Fig. 2. Illustrations of X-ray powder diffraction (courtesy of Bryan Craven).

The collapse of the three dimensions of reciprocal space to the single dimension of a powder diffraction pattern, with the resultant Bragg-peak overlap becoming particularly severe at shorter d-spacings, results in a very substantial loss of information. When powder patterns exhibit clear discrimination between Bragg peaks and background as well as clear resolution of Bragg peaks from one another, individual structure-factor amplitudes can be reliably obtained by integrated intensity extraction methods such as the Le Bail method [13] or the Pawley method [16]. These extracted reflections are treated as *non-overlapped* or 'well determined' reflections. When powder diffraction patterns do not exhibit clear peak-background and peak-peak resolution discrimination, individual intensities obtained from existing integrated intensity extraction methods must be considered unreliable, and are treated as *overlapped* reflections. Overlap may be exact because of the equivalence of reflection d-spacings or accidental resulting from near-equivalent d-spacings that are separated from one another by an amount less than $\sim$10% of the resolving power of the instrument. The overlap can be exacerbated by sample-induced broadening effects. Exact overlap occurs in crystal systems with higher than orthorhombic symmetry, and accidental overlap mainly occurs in orthorhombic or lower systems. The overlap problem highlights most clearly the loss of information in the powder diffraction measurement and limits the applicabilities of powder diffraction as a routine tool for determining and refining crystal structures from microcrystalline powders.

3.2 Powder *Shake-and-Bake*

The *Shake-and-Bake* algorithm, specifically designed for finding the constrained global minimum from single-crystal diffraction data, has been adapted and modified for structure determination from powder diffraction data. The new algorithm, termed Powder *Shake-and-Bake* and shown by a portion (non-gray) of the flow diagram in Fig. 3, combines minimal-function phase refinement, real-space filtering and overlapped-reflection partitioning. It is an iterative process that is repeated until a solution is achieved or a designated number of cycles have been performed. For present purposes, 'solution' does not necessarily imply a complete structure but rather a set of atomic positions that can be readily refined and extended by a refinement package, such as *EXPO2004* [2]. With reference to Fig. 3, the major steps of the algorithm are described next.

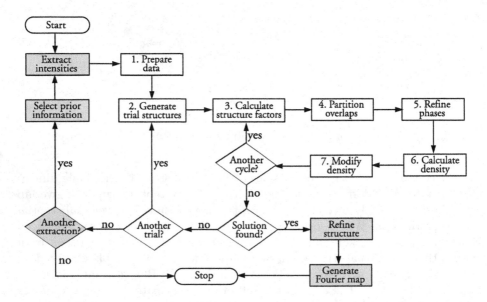

Fig. 3. Flow chart for the *Powder Shake-and-Bake* algorithm.

Step 1. Data Preparation. Reflections are divided into a non-overlapped set G_0 and overlapped sets G_j for $j = 1, 2, \ldots, M$. For reflections $\mathbf{H} \in G_0$, the structure-factor amplitudes $|F_\mathbf{H}|$ are obtained directly from the extracted integrated intensities. For reflections $\mathbf{H} \in G_j$, their initial values, $|F_\mathbf{H}|$, are obtained from the overlap constraints with an equal partition, *i.e.*,

$$|F_\mathbf{H}| = \langle |F_\mathbf{H}| \rangle_{\mathbf{H} \in G_j} = \left(\sum_{l=1}^{L_j} m_l \right)^{-1/2} |F_{G_j}|.$$

Normalized structure-factor magnitudes ($|E|$'s) are generated from the structure-factor magnitudes ($|F|$'s) by standard scaling methods such as a Wilson plot. Reflections are then sorted in descending order of $|E|$ values, and a predetermined number of top reflections is selected to generate a predetermined number of triplet invariants $\phi_{\mathbf{HK}}$ having the largest $A_{\mathbf{HK}}$ values.

Step 2. Trial-Structure Generation. A trial structure or model is generated that is comprised of a number of randomly positioned atoms and their symmetry-related mates sufficient to specify the origin and enantiomorph for the space group in question. The starting coordinate sets are subject to the restrictions that no two atoms are closer than a specified distance.

Step 3. Structure-Factor Calculation. A normalized structure-factor calculation based on the trial coordinates is used to compute initial values for all the desired phase angles and structure-factor magnitudes simultaneously. In subsequent cycles, peaks selected from the most recent Fourier series are used as atoms to generate new phase-angle values and structure-factor magnitudes.

Step 4. Overlapped-Reflection Partition. Let

$$\sum_{l=1}^{L_j} m_l |E_{\mathbf{H}_l}|^2 = |E_{G_j}|^2 \tag{5}$$

be the jth ($1 \leq j \leq M$) overlap constraint, and $|E_{\mathbf{H}_l}^{cal}|$ ($1 \leq l \leq L_j$) be the calculated structure-factor magnitudes from the previous step. The new partition for the overlapped reflections is given by

$$|E_{\mathbf{H}_l}| = \frac{|E_{\mathbf{H}_l}^{cal}|}{\left(\sum_{l=1}^{L_j} m_l |E_{\mathbf{H}_l}^{cal}|^2\right)^{1/2}} |E_{G_j}|. \tag{6}$$

Step 5. Phase Refinement. The normalized structure-factor magnitudes $|E|$ are fixed for all reflections (including the overlapped reflections), and phase angles are the only variables in the phase refinement stage. The values of the phase angles are perturbed by a parameter-shift method to reduce the value of the minimal function $R^*(\mathbf{\Phi}) = R(\mathbf{\Phi}, |\mathbf{E}|)$. $R^*(\mathbf{\Phi})$ is initially computed on the basis of the set of phase-angle values obtained from the structure-factor calculation in Step 3. When considering a given phase angle $\phi_{\mathbf{H}}$, the values of $R^*(\mathbf{\Phi})$ are evaluated respectively with phase angles $\phi_{\mathbf{H}}$, $\phi_{\mathbf{H}} \pm 90^\circ$, and $\phi_{\mathbf{H}} + 180^\circ$ ($\phi_{\mathbf{H}}$ and $\phi_{\mathbf{H}} + 180^\circ$ for centrosymmetric space groups). Then the minimum of these R^* values is found, and the phase angle $\phi_{\mathbf{H}}$ is updated to correspond to that minimum. Consideration of $\phi_{\mathbf{H}}$ is then complete, and parameter shift proceeds to the next phase angle. Refined phase angles are used immediately in the subsequent refinement of other phase angles.

Step 6. Fourier Summation. Fourier summation is used to transform phase-angle information into an electron-density map. Normalized structure-factor amplitudes ($|E|$'s), including those newly partitioned amplitudes for the over-lapped reflections, are used at this stage. The Fourier grid size is typically chosen to be approximately one-third of the maximum resolution of the data.

Step 7. Real-Space Filtering. Image enhancement is accomplished by a discrete electron-density modification consisting of the selection of a specified number of the largest peaks on the Fourier map for use in the next structure-factor calculation. The simple choice, in each cycle, of a number of the largest peaks corresponding to the number of expected atoms has given satisfactory results.

3.3   Successful Applications of *PowSnB*

PowSnB, the computer program that implements the powder *Shake-and-Bake* procedure, has been tested successfully using the experimentally measured atomic-resolution powder diffraction data for the known organic structures of cimetidine [5] (chemical formula $C_{10}H_{16}N_6S$, space group $P2_1/n$) and fluorescein dictate [11] (chemical formula $C_{24}H_{16}O_7$, space group of $P\bar{1}$). The latter structure was one of the most difficult structures (due to lack of heavy atoms) that has been solved by *ab initio* direct-methods to powder data. We have reported [26] *PowSnB*'s efficacy, measured by the success rate (SR) or percentage of trial structures that converge to solutions, as well as solution quality, measured by (i) the mean phase error (MPE) or average absolute value of the deviations of the phase angles from their values calculated using the final refined coordinates and thermal parameters; (ii) the atom-match rate (AMR) or percentage of atoms whose coordinates are within a fixed radius of the final refined coordinates; and (iii) the average distance (DIST) between those matched atoms.

Owing to the presence of large number of heavy atoms in microporous materials, special cares need to be taken to ensure successful structure determination. We use a two-step process: first locate the heavy-atom positions (so-called heavy-atom substructure), then use them to carry out an extension and refinement. In the process of heavy-atom substructure determination, *PowSnB* uses a set of default-parameter values listed in Table 1.

Table 1. Default parameter values of *PowSnB* for heavy-atom substructure determination, where N_μ is the number of heavy atoms in the asymmetric unit.

Number of reflections	$30\,N_\mu$
Number of invariants	$300\,N_\mu$
Number of cycles	$2\,N_\mu$
Number of peaks	N_μ
Number of trials	1000
Grid size	$0.33\,\text{Å}$
Inter-peak distance	$3.0\,\text{Å}$

Table 2. Basic crystallographic information of the test structures.

Structure name	Chemical formula	Space group	Cell dimension (a, b, c, β^o)	Data resolution
EMM8 [4]	Al_4P_4	$B2/b$	22.5541, 13.7357, 14.0756, 98.6174	$1.03\,\text{Å}$
MCM-70 [7]	$Si_{12}O_{24}$	$Pmn2_1$	13.663, 4.779, 8.723, 90.0	$0.75\,\text{Å}$
EMM3 [1]	$Al_{24}P_{24}O_{96}$	$I2/m11$	10.3132, 12.6975, 21.8660, 89.656	$1.00\,\text{Å}$

We have chosen three known microporous materials as test structures. The basic chemical and crystallographic information, including structure name, chemical formula, space group, cell dimension and data resolution, is listed in Table 2. The high-resolution, synchrotron powder X-ray diffraction data were collected at the ExxonMobil beamline X-10B at the Brookhaven National Laboratory.

The integrated intensities were extracted from the index profile using Le Bail or Pawley pattern-decomposition methods implemented in the computer software *GSAS* [12]. The initial structure models were originally obtained by direct-methods software package *FOCUS* and then used for Rietveld refinements. The indexed intensities were input to *PowSnB* using default-parameter values mentioned in Table 1 to carry out structure determination process and the testing results have shown that:

1. *PowSnB* is capable of producing sets of atomic positions of heavy atoms that can be readily refined and extended to complete solutions by the Rietveld refinement method, provided that the integrated intensities have been extracted reliably to atomic resolution.
2. *PowSnB* effectively yields high quality solutions as indicated by high success rates (SR), low average mean phase errors (MPE), high atom match rates (AMR), and small average distances (DIST) between matched atoms.
3. The minimal function histogram is no longer a bimodal distribution as commonly observed in single-crystal applications. The unimodal histogram is the result of unavoidable errors in the extracted structure-factor intensities, as well as the omission of light atoms (*e.g.* O) in the structure determination process.
4. The initial heavy-atom substructure models produced by *PowSnB* were successfully used for structure extension (adding O-atoms half way between two heavy atoms) and refinements to produce finally structures. Figure 4 shows a comparison of two graphic views, one from the published paper and another from *PowSnB*, of structure EMM8 from various angles. This demonstrates that *PowSnB* is capable of solving microporous structures.

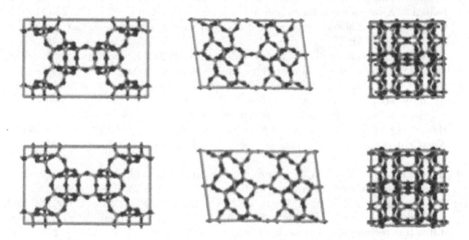

Fig. 4. Comparison of graphic views of the structure of EMM8 from three different viewing angles. Pictures in the top row are from the published paper [4] and ones in the bottom row are produced by *PowSnB* (courtesy of Douglas L. Dorset).

4   Concluding Remarks and Perspectives for Future Research

The powder *Shake-and-Bake* algorithm has been applied successfully to solve microcrystal structures from X-ray powder diffraction data at atomic resolution. We anticipate further development of the algorithm to determine microcrystal structures with low-resolution diffraction data.

Acknowledgments. This research was partially supported by the Knowledge Building Grant from ExxonMobil Research and Engineering.

References

1. Afeworki, M., Dorset, D.L., Kennedy, G.J., Strohmaier, K.G.: Synthesis and characterization of a new microporous material. 1. Structure of aluminophosphate EMM-3. Chem. Mater. **18**, 1697–1704 (2006)
2. Altomare, A., Cuocci, C., Giacovazzo, C., Moliterni, A.G.G., Rizzi, R.: The combined use of Patterson and Monte Carlo methods for the decomposition of a powder diffraction pattern. J. Appl. Crystallogr. **39**, 145–150 (2006)
3. Burla, M.C., Camalli, M., Cascarano, G., Giacovazzo, C., Polidori, G., Spagna, R., Viterbo, D.: SIR88 - a direct-methods program for the automatic solution of crystal structures. J. Appl. Crystallogr. **22**, 389–393 (1989)
4. Cao, G., Afeworki, M., Kennedy, G.J., Strohmaier, K.G., Dorset, D.L.: Structure of an aluminophosphate EMM-8: a multi-technique approach. Acta Crystallogr. B **63**, 56–62 (2007)
5. Cernik, R.J., Cheetham, A.K., Prout, C.K., Watkin, D.J., Wilkinson, A.P., Willis, B.T.M.: The structure of cimetidine ($C_{10}H_{16}N_6S$) solved from synchrotron-radiation X-ray powder diffraction data. J. Appl. Crystallogr. **24**, 222–226 (1991)
6. Cochran, W.: Relations between the phases of structure factors. Acta Crystallogr. **8**, 473–478 (1955)
7. Dorset, D.L., Kennedy, G.J.: Crystal structure of MCM-70: a microporous material with high framework density. J. Phys. Chem. B **109**, 13891–13898 (2005)
8. Frazão, C., Sieker, L., Sheldrick, G.M., Lamzin, V., LeGall, J., Carrondo, M.A.: Ab initio structure solution of a dimeric cytochrome c3 from Desulfovibrio gigas containing disulfide bridges. J. Biol. Inorg. Chem. **4**, 162–165 (1999)
9. Hauptman, H.A., Karle, J.: ACA Monograph 3. Solution of the Phase Problem. I. The Centrosymmetric Crystal. American Crystallographic Association, Michigan (1953)
10. Hendrickson, W.A.: Analysis of protein structure from diffraction measurements at multiple wavelengths. Trans. Am. Crystallogr. Assoc. **21**, 11–21 (1985)
11. Knudsen, K.D., Pattison, P., Fitch, A.N., Cernik, R.J.: Solution of thecrystal and molecular structure of complex low-symmetry organic compounds with powder diffraction techniques: fluorescein diacetate. Angew. Chem. Int. Ed. Engl. **37**, 2340–2343 (1998)
12. Larson, A.C., Von Dreele, R.B.: General structure analysis system GSAS. Los Alamos Nation Laboratory LAUR, pp. 86–748 (2004)
13. Le Bail, A., Duroy, H., Fourquet, J.L.: Ab-initio structure determination of $LiSbWO_6$ by X-ray powder diffraction. Mater. Res. Bull. **23**, 447–452 (1988)

14. Main, P.: On the application of phase relationships to complex structures. XI. A theory of magic integers. Acta Crystallogr. A **33**, 750–757 (1977)
15. Patterson, A.L.: A direct method for the determination of the components of inter-atomic distances in crystals. Z. Kristallogr. (A) **90**, 517–542 (1935)
16. Pawley, G.S.: Unit-cell refinement from powder diffraction scans. J. Appl. Crystallogr. **14**, 357–361 (1981)
17. Perutz, M.F.: Isomorphous replacement and phase determination in noncentrosymmetric space groups. Acta Crystallogr. **9**, 867–873 (1956)
18. Rappleye, J., Innus, M., Weeks, C.M., Miller, R.: SnB version 2.2: an example of crystallographic multiprocessing. J. Appl. Crystallogr. **35**, 374–376 (2002)
19. Rossmann, M.G., Blow, D.M.: The detection of sub-units within the crystallographic asymmetric unit. Acta Crystallogr. **15**, 24–31 (1962)
20. Sheldrick, G.M.: The SHELX-97 Homepage. http://shelx.uni-ac.gwdg.de/SHELX/
21. von Delft, F., Blundell, T.L.: The 160 selenium atom substructure of KPHMT. Acta Crystallogr. A **58**(Suppl.), C239 (2002)
22. Weeks, C.M., DeTitta, G.T., Hauptman, H.A., Thuman, P., Miller, R.: Structure solution by minimal-function phase refinement and Fourier filtering II. Implementation and applications. Acta Crystallogr. A **50**, 210–220 (1994)
23. Weeks, C.M., Miller, R.: The design and implementation of SnB version 2.0. J. Appl. Crystallogr. **32**, 120–124 (1999)
24. Xu, H., Hauptman, H.A.: Statistical approach to the phase problem. Acta Crystallogr. A **60**, 153–157 (2004)
25. Xu, H., Smith, A.B., Sahinidis, N.V., Weeks, C.M.: SnB version 2.3: triplet sieve phasing for centrosymmetric structure. J. Appl. Crystallogr. **41**, 644–646 (2008)
26. Xu, H., Weeks, C.M., Blessing, R.H.: Powder shake-and-bake method. Z. Kristallogr. Suppl. **30**, 221–226 (2009)
27. Xu, H., Weeks, C.M., Deacon, A., Miller, R., Hauptman, H.A.: Ill-conditioned Shake-and-Bake: the trap of the false minimum. Acta Crystallogr. A **56**, 112–118 (2000)
28. Yao, J.-X.: On the application of phase relationships to complex structures XVIII. RANTAN-random MULTAN. Acta Crystallogr. A **37**, 642–644 (1981)

Detection of Counterfeit Coins Based on Modeling and Restoration of 3D Images

Saeed Khazaee$^{(\boxtimes)}$, Maryam Sharifi Rad, and Ching Y. Suen

CENPARMI, Concordia University, Montreal, Canada
{s_khaza, sey_shar, suen}@encs.concordia.ca

Abstract. In image-based coin detection, making the image readable is an indispensable part of the feature extraction. However using a 2-D image processing approach for detecting a counterfeit coin is nearly impossible in case of destroyed coins whose textures are severely burnt, sulfated, rusted, or colored.

In this research, we used a 3-D scanner to scan and model an acceptable number of coins capturing height and depth instead of levels of color. The most important advantage of 3-D scanning is to compensate for the above-mentioned destructions of the coin surface. Despite this advantage, we had several unexpected degradations due to shiny coin images. To solve this problem, the 3-D image was decomposed column-wise to a number of separate 1-D signals, which were analyzed separately and restored by the proposed method. This approach gave remarkable results when used to extract valuable features.

Keywords: Counterfeit detection · Coin recognition · 3D-images · Restoration

1 Introduction

A counterfeit coin is an imitation of a genuine coin with a very high quality so that it can deceive many ordinary people and even experts. Nowadays, a lot of companies, museums, and government agencies have increased the demand of automatic systems to classify precious, historical, and common coins. Thanks to the increased demand on the automatic approaches for detecting counterfeit coins, coin recognition has been continuously improving in classification performance and image-based coin recognition has become a key part of coin detection.

Many automatic counterfeit coin detectors employ a primitive technology based on low-cost sensors to measure the weight, thickness, radius, conductivity, magnetic, or acoustic features of the coins [16]. However, these systems cannot distinguish fake coins from genuine ones when their physical properties are essentially the same. In order to increase the potential of fake coin detectors and image-based coin detection, several methods based on image processing techniques and classification algorithms were proposed [3, 12–14, 16] and many lectures and tutorials were devoted to them [1, 7, 14]. Some of these methods are not very sophisticated and use coins' colors and radius based features to detect counterfeit coins. Unfortunately, these approaches are incapable of distinguishing coins having the same values of the limited set of features.

In recent years, image-based counterfeit detection has expanded and many researchers have applied image processing techniques to extract effective features from

© Springer International Publishing AG 2017
R.P. Barneva et al. (Eds.): CompIMAGE 2016, LNCS 10149, pp. 178–193, 2017.
DOI: 10.1007/978-3-319-54609-4_13

the texture of the coin images [6, 15]. In particular, edge detection information has been widely used in the feature extraction process. In references [6, 11], an edge map was extracted and segmented into several parts. After that, the authors proposed a method using histogram analysis and Fourier transformation to handle and recognize rotated coin images. However, the extracted features were not very useful for noisy and degraded images such as rust, dust, or sulfated coin images. In [15], the authors extracted the letters on the coin surface and tried to recognize them. In spite of the mentioned novelties of the proposed method, it is clear that the method was not robust enough to distinguish the counterfeit from genuine coins whose images have weak or smooth edges. In reference [10] a new method using rotation-invariant region binary patterns based on gradient magnitudes was proposed. To increase the accuracy of coin recognition, it computes gradient magnitudes in a coin image and extracts rotation-and-flipping-robust features using local difference magnitude transform. Although the result is acceptable from feature extraction point of view, the time-performance of the model is not satisfactory.

Paper [16] proposes to use the sensor of an optical mouse as a counterfeit coin detector based on the image acquisition capabilities of the optical mouse sensor in short distances. The results are compared with the ones of trained and untrained users.

In this paper, a new method based on analysis and restoration of 3-D images for detecting counterfeit coins is proposed. We employed a 3-D scanner to scan a large number of coins with different qualities. The first contribution of this research is exploiting the significant advantages of the 3-D approach in coin recognition, which had not been discussed in the previous works. Considering the nature of the data captured by 3-D scanning, it was expected to extract effectively features related to height or depth instead of colors related to coin's luminance. In addition, the most important benefit of 3-D scanning is robustness of the quality of the coin surface. This capability contributed to purifying images captured from sulfated, rusted, or colored coins.

Despite the advantages of the 3-D scanning [17, 18], there is a serious challenge when shiny coins are processed [18]. Although 3-D scanning is indeed independent of the lighting condition of the environment, we had a lot of unexpected degradations and shadowing on shiny coin images. Therefore, we faced unreal values of height or depth. The next contribution of the proposed method is to eliminate this degradation. In order to restore an image, it was decomposed to many separated signals. To enrich the signals, parts of the coin images in the form of rings were converted into a long rectangle, on which all letters were placed. Then, the signal of each row was analyzed and restored by the proposed method separately. The signal restoration process was notably improved and we could use each newly restored signal to extract new effective features. The experimental results on various classifiers showed that the proposed method has an outstanding performance in terms of true positive (detection rate), false positive, precision, recall, f-measure, and feature extraction time.

The rest of this paper is organized as follows: Sect. 2 explains briefly the 3-D scanning and its advantages and challenges. Section 3 describes the proposed method to detect counterfeit coins and contains three important subsections: straightening the coin, signal restoration, and feature extraction. In Sect. 4, we present the experimental results and analyze them. Finally, we conclude in Sect. 5 and comment some future work.

2 Degradation Problem in the 3-D Images

There are many different technologies used in 3-D scanners; each technology has its own advantages, restrictions, and cost. For example, optical technologies face many difficulties when processing reflecting, transparent, or shiny objects [5, 17, 18]. Figure 1 illustrates how a 3-D scanner captures height and depth: it is robust regarding the quality of the coin surface, while the 2-D image of this coin is completely unreadable. In Fig. 2 we can see the degradation of a shiny coin image captured by the 3-D scanner which uses optical technology. As it can clearly be seen from this figure, in spite of the remarkable advantages of 3-D scanning for poor quality coins, there may be some abnormal and invalid results while scanning shiny and high-quality coins. However, the validity of the data captured by scanning is crucial for this research as any invalid value related to height and depth will adversely affect the pattern recognition process. Therefore, our proposed restoration module should enhance not only the quality of the images as a whole, but also each small part of the image must be restored as precisely as possible.

(a)   (b)

Fig. 1. (a) Twenty Kroners 1990 coin captured by a normal camera, and (b) the same coin captured by a 3-D scanner.

   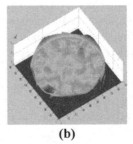

(a)   (b)

Fig. 2. (a) Degraded image of a shiny coin, (b) mesh plot of the same image.

3   Proposed Counterfeit Detection Method

In this section, we present a new method based on signal restoration using 3-D images robust enough against rust, dust, sulfate, or any other factors, which may affect the coins' quality. In addition, the proposed method eliminates the problem of the degradation related to the shiny coin images, which occurs during the 3-D scanning.

3.1   Straightening the Coin Image

In this section, we present the first innovation of our proposed method, which is very important for performing the next steps. Having in mind that our final goal is to extract features from the height and depth information, we assume that the part of the coin where the letters are placed in is more useful. Because the remaining part of the coin such as head profile or shapes placed in the center of the coin most likely would have greater height and will be scratched earlier than the letters. Therefore, the straightening coin image is proposed with the aim of reducing the complexity of processing. This algorithm can also be very useful for segmentation of the letters in 2-D approaches. For implementing this step, we proposed an algorithm mapping the ring part of the coin into a rectangle using the equation of circle in polar coordinates. The code listed below demonstrates the algorithm for straightening the coin where $C_2 - C_1$ is the radius of the part we want to process. It means that the matrix *linear_image* is created column by column. Each column is composed by the values of the pixels from the circumference toward the center as much as $C_2 - C_1$ pixels. Below, *lr* is the growth of *r* for reading the next pixel's values of the current column diagonally, and *lt* adjusts the increase of the angle for reading the next column of the *linear_image* from the original image circularly. These parameters should be adjusted for various images in different sizes on the circular images with center coordinates (a, b). Figure 3(b) shows the result of the algorithm for Fig. 3(a). As it can be seen from this figure, all letters are placed next to

(a)

(b)

Fig. 3.   (a) Original image, and (b) rectangular image after executing the straightening algorithm.

each other and can easily be processed in a simpler way than in a circular coin. This is convenient before the segmentation of the 2-D image is performed.

Straightening Algorithm:

```
Input: cropped circular coin image
Output: straightened coin image (linear image)
C₁ ← (the number of column)/2
C₂ ← (the number pixels which should be considered as the
ring part of the coin's image)
For each t between π and −π step lt=-0.001
    k← (k+1)
    i←0
    For each r in the ring part of coin image from C₁ to C₂
    step lr=0.01
        i← (i+1)
        x←round(r.*sin(t)+a)
        y←round(r.*cos(t)+b)
        if x>0 and y>0
            Linear_image(i,k)←Circular_image(x,y)
```

Pseudocode for the main part of the straightening algorithm.

In the next section, the key role of the straightening process for the proposed method will be presented. Figure 4 presents schematically the steps of the proposed method to create a record dataset, which will be used for training and testing a counterfeit coin detector.

3.2   Image and Signal Restoration

In digital image processing, restoration means recovering an image that has been degraded by using *a priori* knowledge of the degradation phenomenon. Therefore, restoration techniques are oriented toward modeling the degradation function in order to infer the inverse process and recover the original image [2].

After the straightening process, we can see an adverse effect of bad scanning on shiny coin images. For some kinds of coins, for example Danish 20 kroner 2008, all the images are totally degraded. In Fig. 5, the differences between two examples of the 1-D signal of a specific row related to a normal and a degraded image have been shown. In both images, a long rectangular image containing the letters can be seen. The signals below these images have numerous peaks showing the heights of the letters or numbers in the lines marked. Clearly, the figure shows that shininess has adversely affected the quality of the image and signal.

Our initial experiments showed that the degradation is not a random noise. We scanned three coins, four times each one of them, and we obtained the same poor results.

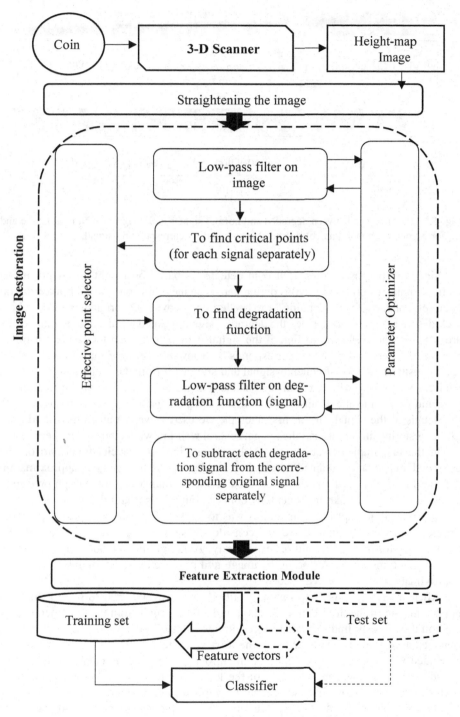

Fig. 4. Structure of the proposed method

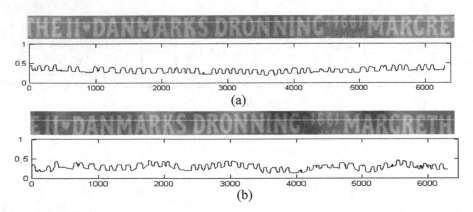

Fig. 5. The difference between degraded and normal signals and images: (a) Normal image and its corresponding signal, and (b) degraded image and its corresponding signal.

Since our proposed counterfeit coin detector is based on height and depth information, signal restoration from this digital data is an inevitable process. For given digital data set obtained from scanning, we needed to recover the original signal, which includes more precise data, since the degraded signals related to fake and genuine coins are very similar. Referring to Fig. 5, the signal demonstrates that the important information of letters and numbers corresponds to a low-frequency signal. Therefore, we have to estimate the low-frequency signal and subtract it from the original signal. Thus, we have $x = A + D$; where, x is the original and degraded signal, A is an approximation of x, and D contains the details of x. In other words, the useful information comes from the details of the signal. In our first attempt, we used wavelet transform to find A or D. For restoring all signals and the image, several well-known methods were performed.

In the first implementation, we used various kinds of wavelet transform on the degraded image and obtained interesting results by subtracting the approximation A from the original signal x. However, the obtained signals were not enough precise to allow us distinguish fake from genuine coins through height and depth information.

Our second attempt for estimating A was to apply a Gaussian low-pass filter using Fourier transform. To find A, we can consider $A_T = X * H$, where A_T, X, and H are Fourier transforms of A, x, and h, respectively. Although this method was better than the wavelet transform in restoring the image and produced a better signal, it could not approximate A as well as we needed.

After applying Gaussian low-pass filter and wavelet transform, to find an approximation of the degradation function and subtracting it from the degraded signal, we could obviously find that the restored image was still degraded. Although we had obtained acceptable image after using the low-pass filters, there had not been the best estimated values of height or depth. As mentioned before, the foundation of the proposed method is based on the height of the letters or numbers. Therefore, we need to perform a restoration method, which can be applied to each signal (row) of the image separately because the degradation function is completely abnormal and has to be considered as a 1-D signal restoration. For this purpose, we propose using a restoration module, which contains several steps as described below.

Estimating Degradation Function and Signal Restoration

For estimating more accurately the degradation function for each 1-D signal, we proposed an estimator, which uses selected points determined by the discrete derivative. Since the signals, on which the process should be performed, are not smooth enough to perform mathematic operations, we used a low-pass filter to make them smoother. For estimating the approximation of degradation function related to the original signal g, we use the formula:

$$f(x) = \frac{g_i(x_i) - g_{i+1}(x_{i+1})}{x_i - x_{i+1}}(x - x_i) + g_i(x_i) \tag{1}$$

where $g_i(x_i)$ is the closest critical selected point to x and $x_i < x < x_{i+1}$. The critical selected point is routinely an extremum point selected by the effective point selector function. This function uses a momentum to avoid local minima, which are called noises and the signal has a lot of them. In addition, if the distance between two selected points is greater than a specific threshold, the point selector will consider several points on this part of the signal instead of the critical points. Figure 6 illustrates a small part of the original signal and the approximation of degradation signal. As shown in the figure, the approximated signal is too sharp and needed to be smoothed. Consequently, we used a low-pass filter again to obtain the final signal and we have $A = inverse(F * H)$ where F and H are Fourier transforms of f, and a 1-D Gaussian filter in the order. Also, the *inverse* function returns the inverse discrete Fourier transform of the resulted vector. In addition, there is a simple parameter optimizer which regulates the value of σ in Gaussian filter, with the aim of minimizing the Euclidean distance between A and f. Figure 7 demonstrates that the original image has been restored successfully. The signal related to the marked row of the restored image shows that the heights of the letters are more precise than the original rectangular image. Figure 7 also shows that the proposed restoration produces more balanced height information for the surface of the coin.

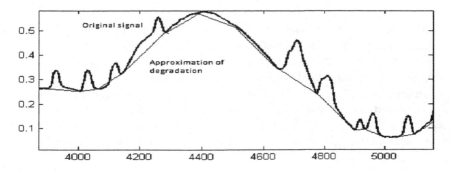

Fig. 6. Estimating the approximation of degradation function

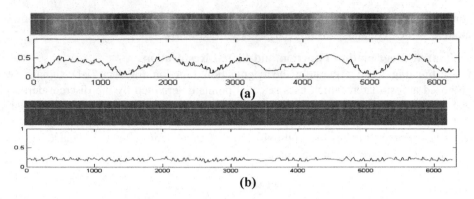

Fig. 7. The result of proposed restoration method on the images and their signals: (a) degraded image and its corresponding signal, and (b) restored image and its corresponding signal.

3.3   Feature Extraction and Creation of a Record Dataset

After performing the straightening process and restoring the images, we noticed that the signals corresponding to the genuine coin (in the same rows) are identical, while there is a significant difference between the genuine and the counterfeit coins. In Fig. 8, the 1-D signals of the same row of six genuine and fake coins are shown for comparison. Since we can easily distinguish the counterfeit coin from the genuine with a height-map signal, we concentrated on the simple features of the signals. For each rectangular image, we have 311 rows, whose signals cover all the letters and the numbers. If we consider each signal as a feature, all these features would form a substantial set of features to distinguish fake from genuine coins, whereupon we would need to consider one or more features for each signal instead of the whole signal. In this paper, we use two simple features for each signal. The first feature is the energy (L2-Norm) of each decomposed signal, which can easily be calculated using the formula [8]:

$$E_i = <x_i(n).x_i(n)> = \sqrt{\sum_{n=1}^{c}|x_i(n)|^2} \tag{2}$$

where x_i is the signal of row i and c is the number of column in the matrix related to the rectangular image of the coin. An additional feature for each signal is the percentage of the energy of wavelet details in level two, whose calculation method was mentioned before; we denote it with PE_i. In addition, each coin has two general features: *year and coin number* and a target label *class* (*Fake* or *Genuine*). Accordingly, each rectangular image is converted to a record data and thus we have a total of 624 features from 311 signals and a coin's ID are assigned for the training process. After performing the feature extraction method on all existing images, a record dataset is created.

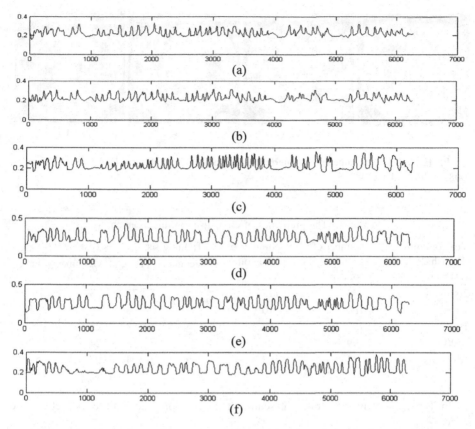

Fig. 8. Examples of 1-D signals related to the rectangular image of coins for the same row: (a, b) two genuine Danish coins from 2008 and (c) a fake Danish coin from 2008; (d, e) two genuine Danish coins from 1996 and (f) a fake Danish coin from 1996.

4 Experimental Results on Counterfeit Detection

In this section, the efficiency of the proposed system is discussed. The hardware of the test environment consisted of an i7-4500U 4.2 GHz CPU (only one core was used), DDR3 6 GB RAM, the operating system used was Windows 8.1-64 bit, and the programming environment was MATLAB 2015.

4.1 Dataset

In this research, we used a total of 322 genuine and 162 fake coins for training and evaluating the system respectively. The coins used in the research were provided by the Law Enforcement Office and we used all of them. We would note that the access to more fake coins was restricted. In Fig. 9, samples of fake and genuine coins of years 1990, and 2008 are shown. Table 1 indicates four types of Danish 20 Kroner scanned by the 3-D scanner and used in this model verification.

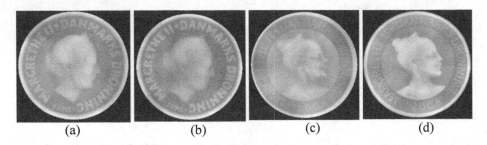

Fig. 9. Examples of genuine and counterfeit coins: (a) genuine 1990, (b) fake 1990, (c) genuine 2008, and (d) fake 2008

Table 1. The properties of coins used in this research.

Year of coin (Danish 20 Kroner)	Number of genuine coins	Number of fake coins	Training set		Test set	
			Genuine	Fake	Genuine	Fake
2008	23	113	15	75	8	38
1996	100	10	75	6	25	4
1991	100	14	75	8	25	6
1990	99	25	75	15	24	10
All coins	322	162	225	104	97	58

In Table 2 the average times of scanning, restoration, and feature extraction, are listed.

Table 2. Time spent for the steps of the proposed method.

Year of coin (Danish 20 Kroner)	Scanning time (avg/min)	Restoration time (avg/sec)	Feature extraction time (avg/sec)
2008	9.8	16.2	1.48
1996	10.2	14.4	1.46
1991	10.6	15.1	1.47
1990	10.4	13.9	1.52

As previously stated, after scanning and performing the proposed method on the coin images, each height-map image whose class (fake or genuine) was clear was converted to a record data. Therefore, the final data set contained 484 labeled records. Consequently, we used various classifiers to illustrate the fitness of the proposed method and extracted features to detect the counterfeit coins.

4.2   Evaluation Criteria

There are several standard metrics to evaluate a classification system. *True positives* or *detection rate, true negatives, false positives,* and *false negatives* are often used to evaluate a classification system [9]. For a counterfeit detector, a *true positive* or *detection rate* indicates that the system detects precisely a counterfeit coin. A *true negative* illustrates that the counterfeit detector has correctly detected a genuine coin. The *false positive* is the case when a genuine coin is falsely classified as a fake one. A *false negative* shows that the counterfeit coin detector is unable to detect the fake coin. To improve the accuracy of the evaluations, we calculated the *precision, recall,* and *F-value,* which are not very dependent on the size of the train or test dataset. They can easily be computed by Eqs. (3), (4), and (5) where *TP, FP,* and *FN* are the numbers of true positives, false positives, and false negatives, respectively. Also, β is the relative importance of precision versus recall and is usually set to one [9]. Furthermore, we considered the time of building a model for classifiers to compare their training time.

$$Precision = \frac{TP}{TP + FP} \tag{3}$$

$$Recall = \frac{TP}{TP + FN} \tag{4}$$

$$F-\text{value} = \frac{(1 + \beta^2) * \text{Recall} * \text{Precision}}{\beta^2 * (\text{Recall} + \text{Precision})} \tag{5}$$

4.3   Results and Discussion

We used several well-known classifiers available in *Weka* [4] to evaluate the performance of the proposed method. To reduce the processing complexity and improve the classification efficiency, a feature selection algorithm was employed. After using Chi-squared feature evaluation on all 624 features, 36 features were selected.

Tables 3 and 4 illustrate the results of three metrics comparing the performance of the classification process for the system using 624 extracted, and 36 selected features in the order. In terms of training and test times, *NaïveBayes,* and *Lazy.Kstar* had a shorter time than other classifiers while neural network multi-layer perceptron was the worst. *TP*

Table 3. Comparing the classification results for various classifiers using 624 features.

Classifier	Training time (sec)	Test time (sec)	True positive (%)	False positive (%)
SGD	0.67	0.23	97.8	7.8
MLP	157.9	21.3	94.9	26.7
NaiveBayes	0.03	<0.01	92	12.6
Decision tree	0.11	<0.01	90.5	38.6
Lazy.Kstar	<0.01	<0.01	83.9	83.9
Logistic	0.63	0.2	94.49	4.7

Table 4. Comparing the classification results for various classifiers using 36 features.

Classifier	Training time (sec)	Test time (sec)	True positive (%)	False positive (%)
SGD	0.05	<0.01	99.3	3.8
MLP	1.52	0.71	97.1	4.2
NaiveBayes	<0.01	<0.01	92.0	12.6
Decision tree	0.02	<0.01	91.2	31.1
Lazy.Kstar	<0.01	<0.01	95.6	15.5
Logistic	0.02	<0.01	95.6	4.5

(detection rate) and *FP* are weighted and averaged for both the genuine and fake classes. *Stochastic Gradient Descent* or SGD (with loss function: Hinge loss (SVM), learning rate: 0.01, epoch: 500, and lambda: 1.0E−4), and MLP (with learning rate: 0.2, momentum: 0.3, number of sigmoid nodes in hidden layer: 20) had better results compared to others. The *decision tree* was the weakest classifier in these metrics. These classifiers for both experiments (624 features, and 36 features) showed roughly the same superiority in the mentioned criteria, although the performance of the classification was dramatically improved after feature selection. SGD has correctly classified 99.3% of the coins by 36 features; for the other classifiers, we can see acceptable results in the metrics. Taking *precision*, *recall*, and *f-value* into consideration, SGD was still the most remarkable classifier according to Table 5. In spite of the relatively weaker results for *decision tree*, and *NaiveBayes* in the classification of the fake coins, the other classifiers could classify both fake and genuine coins satisfactory. As it can be seen clearly from Table 5, *precision*, *recall*, and *f-value* were indispensable to measure the accuracy of the classifications and the results showed that although *NaiveBayes* and *decision tree* had an appropriate outcome in the detection of the genuine coins, they were slightly poorer in detecting the counterfeit coins. Apart from the ability of classifiers in the mentioned standard metrics, all the results are satisfactory for the proposed method and demonstrate that the extracted features are very effective for any kind of classification algorithms.

Table 5. Comparing the results of classifiers in terms of precision, recall, and f-value for classifying fake and genuine coins.

Class	Metric	Classifier					
		SGD	MLP	Naïve Bayes	DT (J48)	Lazy. Kstar	Logistic
Genuine	Precision	0.991	0.991	0.973	0.933	90.966	0.991
	Recall	1	0.974	0.93	0.965	0.983	0.957
	F-value	0.996	0.982	0.951	0.949	0.974	0.973
Fake	Precision	1	0.875	0.704	0.778	0.9	0.808
	Recall	0.955	0.955	0.864	0.636	0.818	0.955
	F-value	0.997	0.913	0.776	0.7	0.857	0.875

Very few researchers have used image-based coin recognition technique to detect counterfeit or invalid coins. Therefore, the proposed method was compared with two recent counterfeit coin detection methods. Albeit the test time is not reported well in these two methods, it looks like in reference [16], researchers designed a real-time system to detect counterfeit coins, and the method has performed much better than the other methods in term of processing time. The possible metrics for more precise comparison of these methods are classification accuracy, and test data information. As Table 6 shows, the proposed method performs better than reference [16] in accuracy but its accuracy is not as precisely as reference [15]. However, with respect to the test process of these methods, there is another issue that could have a straightforward effect on the accuracy. Although the accuracy resulted in the method of reference [15] is 100%, the size of the test set for evaluation of the system was very small. Also, it is clear that our proposed method was trained and tested by four different types of coins (years 1990, 1991, 1996, and 2008) while in reference [15] only two types of coins (years 1990 and 1996) were considered.

Table 6. Comparison of different methods with the proposed counterfeit detector in accuracy.

Accuracy		Optical mouse [16]	Shape and lettering features [15]	Proposed method using SGD classifier
		97%	100%	99.3%
Testset information	Genuine coins	100	3	97
	Fake coins	96	2	58
	Type of coins	Two-Euro	20 Kroner 1990, 1996	20 Kroner 1990, 1991, 1996, and 2008

In order to indicate the impact of the restoration process proposed in Sect. 3.2 on this paper, we tested the coins with different qualities separately. As Table 7 demonstrates, the signal restoration had a significant impact on the classification of the degraded images resulted from scanning of the shiny coins. It is also worth noting that all six oxidized coins (4 fake and 2 genuine) classified correctly. It means that a 3-D scanning of an oxidized coin can produce an image as good as a normal coin.

Table 7. Impact of the restoration process on the classification of the coins with different qualities.

	Number of test samples	Correctly classified (%)	
		Before restoration	After restoration
Oxidized coins	6	100	100
Degraded images (of shiny coins)	81	59.4	98.7
Normal coins	68	100	100
Overall	155	78.9	99.3

5   Conclusion and Future Work

In this paper, a new counterfeit coin detection method based on analysis and restoration of 3D images was proposed. We used a 3-D scanner to scan a large number of coins and obtained the coin height-map images. To simplify and reduce the complexity of the proposed method, we suggested a straightening algorithm to convert each circular coin image to a linear rectangular image. Since the shiny coin images were abnormally degraded, it was impossible to use height and depth information to detect counterfeit coins. Because of that, the next step was to restore the images, processing separately signal by signal. After restoration, the 1-D signals of the image, the energy, and the percentage of the energy of the wavelet details in level two were extracted for all 311 signals. In sum, we had 624 features before the feature selection process. The comparison of the performance of several well-known classifiers in *Weka* illustrated that the proposed method gives excellent results in detecting counterfeit coins. Employing multi-processing allowed easily restoring the signals and reducing the processing time. In addition, there was no need to perform any of the well-known methods for noise reduction or and contour tracking for coin recognition. Surprisingly, the proposed method had significant results on counterfeit coin detection using four kinds of coins to train and test the method. Finally, the proposed counterfeit detection method is robust against rust, dust, and sulfation, cases of which were not considered in most of the previous works.

Our plan for future work is to focus on the drawbacks of the proposed method. In spite of the substantial precision and acceptable time for feature extraction, the time of coin scanning needs improvement. To overcome this problem, we plan using a 3-D scanner to capture only a small part of the coin and work on it. Enriching the feature extraction and using a hybrid classifier built especially for the created data set can improve the performance of the counterfeit detector.

Another direction of our research will be towards coins with non-circular shapes and ones with no text embossing.

References

1. Goldsborough, R.: Counterfeit Coin Detection (2013). http://rg.ancients.info/guide/counterfeits.html
2. Gonzalez, R.C., Woods, R.E.: Digital Image Processing, 3rd edn, p. 311. Prentice, Pearson, Upper Saddle River (2008)
3. Hassoubah, R.S., Aljebry, A.F., Elrefaei, L.A.: Saudi riyal coin detection and recognition. In: 2nd IEEE International Conference on Image Information Processing (ICIIP), Int. J. Comput. Appl., pp. 62–66, December 2013
4. http://www.cs.waikato.ac.nz/~ml/weka/
5. https://en.wikipedia.org/wiki/3D_scanner
6. Huber, R., Ramoser, H., Mayer, K., Penz, H., Rubik, M.: Classification of coins using an eigenspace approach. Pattern Recognit. Lett. **26**(1), 61–75 (2005)
7. Inksure Technologies. Coin anti-counterfeiting (2014). http://www.inksure.com/banknotesecurity/254-coin-anti-counterfeiting

8. Kaiser, J.F.: On a simple algorithm to calculate the 'energy' of a signal. In: Proceedings of the International Conference on Acoustics, Speech, and Signal Processing ICASSP 1990, Albuquerque, NM, vol. 1, pp. 381–384 (1990)
9. Khazaee, S., Faez, K.: A novel classification method using hybridization of fuzzy clustering and neural networks for intrusion detection. IJMECS 6(11), 11–24 (2014)
10. Kim, S., Lee, S.H., Ro, Y.M.: Image-based coin recognition using rotation-invariant region binary patterns based on gradient magnitudes. J. Vis. Commun. Image Res. 32, 217–223 (2015)
11. Maaten, L.J., Postma, E.O.: Towards automatic coin classification, 1st EVA-Vienna, April 2006, pp. 19–26
12. Mehta, D., Sagar, A.: An efficient way to detect and recognize the overlapped coins using Otsu's algorithm based on Hough transform technique. Int. J. Comput. Appl. 73(9), 18–21 (2013)
13. Saranya, Y.M., Pugazhenthi, R.: Harris-Hessian algorithm for coin apprehension. Int. J. Adv. Res. Comput. Eng. Technol. 2(5), 1689–1693 (2013)
14. Shen, L., Jia, S., Ji, Z., Chen, W.S.: Extracting local texture features for image-based coin recognition. IET Image Process. 5(5), 394–401 (2011)
15. Sun, K., Feng, B.Y., Atighechian, P., Levesque, S., Sinnott, B., Suen, C.Y.: Detection of counterfeit coins based on shape and letterings features. In: Proceedings of 28th ISCA International Conference on Computer Applications in Industry and Engineering, San Diego, USA, pp. 165–170, October 2015
16. Tresanchez, M., Pallejà, T., Teixidó, M., Palacín, J.: Using the optical mouse sensor as a two-euro counterfeit coin detector. Sensors (Basel, Switzerland) 9(9), 7083–7096 (2009)
17. Yoshida, T.H.: Fundamentals of Three Dimensional Digital Image Processing. Springer, New York (2009)
18. Zhang, C., Xu, J., Xi, N., Zhao, J., Shi, Q.: A robust surface coding method for optically challenging objects using structured light. IEEE Trans. Autom. Sci. Eng. 11(3), 775–788 (2014)

Automated Brain Tumor Diagnosis and Severity Analysis from Brain MRI

Sabyasachi Mukherjee[1], Oishila Bandyopadhyay[2(✉)], and Arindam Biswas[1]

[1] Department of Information Technology,
Indian Institute of Engineering Science and Technology, Howrah, India
[2] Advanced Computing and Microelectronics Unit,
Indian Statistical Institute, Kolkata, India
oishila@gmail.com

Abstract. Analysis of brain MRI is of utmost importance as it reveals the underlying details of the main controlling portion of human body. In this paper, we have proposed a fully automated approach to differentiate abnormal brain images from healthy MRI. The proposed method segments out the tumor region from the abnormal MRI by analyzing the energy profile of the image pixels. After tumor segmentation, the tumor features are analyzed to classify the degree of malignancy. This approach can be applied to segment both high grade and low grade tumors.

Keywords: Brain MRI · Tumor · Edema · Dead cell · Image energy

1 Introduction

Brain tumor segmentation using different modalities of medical imaging is the focus of the research community in the past few decades. The accurate and early detection of brain tumor helps doctors to plan for the treatment appropriately. Nowadays, magnetic resonance imaging (MRI) has become the most effective and common tool for brain tumor diagnosis. Automated diagnosis of brain tumor from MRI and analysis of the nature of tumor may have significant contribution in reducing patient's mortality rate.

In recent years, several research groups are working on different approaches for segmentation of brain tumor from MRI analysis. Symmetry based checking among left and right lobes of brain is an efficient way to detect abnormalities [6,14]. On detection of abnormality, different image segmentation approaches such as thresholding, deformable model approach, and maxima transformation are applied for segmentation of the tumor region [8,16]. Several algorithms for brain tumor segmentation apply generative and discriminative probabilistic models to perform brain tissue classification [12]. Machine learning approaches are also used for classification of brain MRI slices [15]. Color based tumor extraction with watershed segmentation and edge detection algorithm is proposed by Maiti and Chakraborty [10]. An integration of cellular automata with texture based region growing image segmentation approach is also proposed for brain MRI segmentation [5]. A survey on brain tumor diagnosis from MRI has

© Springer International Publishing AG 2017
R.P. Barneva et al. (Eds.): CompIMAGE 2016, LNCS 10149, pp. 194–207, 2017.
DOI: 10.1007/978-3-319-54609-4_14

mentioned several new approaches for segmentation of tumor and surrounding tissues, registration and modeling methods proposed in this domain [4]. Most of the semi-automated and fully-automated brain tumor segmentation approaches focus on the high grade gliomas (HGGs). Menze et al. has come up with a benchmarking system after performing detailed analysis of different brain tumor segmentation approaches [11].

MRI produces multi-spectral image sequences that can provide different and partly independent information about different tissues and tumors in the brain. Three different types of MRI sequences such as T1-weighted (T1), T2-weighted (T2), and Fluid Attenuated Inversion Recovery (FLAIR) are widely used by the doctors for brain tumor diagnosis. In this paper, we propose an efficient method for automated brain tumor diagnosis and grade analysis from T1c (T1 contrast enhanced) and FLAIR MRI sequences. In this approach, the presence of abnormality in the MRI is identified by analyzing the image energy distribution. In the abnormal image, the tumor segmentation is performed by exploring the intensity profile and connected component analysis of the region-of-interest (ROI). The method then selects the MRI sequence (T1c or FLAIR) suitable for tumor segmentation. Finally, the grade of the tumor is determined. The proposed tumor segmentation approach can be applied on both high grade and low grade gliomas.

2   Proposed Method

The proposed method consists of several phases. Figure 1 shows different phases of the proposed method.

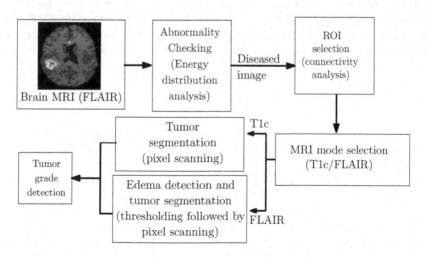

Fig. 1. Different phases of the proposed method.

In the initial phase, the energy distribution analysis is performed on the input FLAIR MR image to identify the presence of abnormality in the image. In case any abnormality is found, the region of interest (ROI) in the MR image is found by performing connectivity analysis. The next phase select the input image with appropriate contrast between T1c and FLAIR MR image of the same patient. After the selection of the proper mode, the pixel scanning phase segments the tumor affected pixels by removing the pixels belong to white-matter (WM), grey-matter (GM), and cerebrospinal fluid (CSF) from ROI of the T1c MRI. In FLAIR MRI, the method detects the presence of edema region and then apply thresholding approach to segment the tumor area. The grade detection phase performs relative intensity analysis of ROI to diagnose the grade of the tumor.

2.1 Detection of Abnormality in MRI

The proposed approach uses brain MRI FLAIR image as the input image and analyze the image energy distribution to differentiate between the healthy and diseased brain MRI. This phase plays an important role in automating the process of brain tumor diagnosis. The outcome of this phase helps the system to decide whether the image will undergo the remaining phases for tumor detection or it will be marked as a healthy brain MRI.

2.1.1 Computation of Image Energy

Image energy represents the gray level intensity distribution in the image. The higher energy value in a small region ($M \times N$) of image I implies the presence of fewer gray levels in that region [17]. The tumor region in brain MRI (FLAIR) mostly appears as a bright patch in the gray image. To detect the presence of tumor region in a brain MR image, we divide the entire image in small blocks of size $s \times s$ and compute the local energy value of each block from normalized gray value. The local energy E_k in the k-th block is represented as

$$E_k = \sum_{k=1}^{s} (P_k/G)^2 \tag{1}$$

where P_k represent the sum of gray level intensity values of all pixels of k-th block and $G = 255$ for gray image. If c_{ij} is a pixel in the k-th block, then P_k can be computed as

$$P_k = \sum_{i=1}^{s}\sum_{j=1}^{s} c_{ij} \tag{2}$$

We consider only those blocks having non-zero energy values (computed with block size 9×9). In a brain MRI with tumor, the tumor region covers small area compared to the area covered by GM, WM, and CSF. Hence the block energy values of the entire MRI have repetitions. These frequently occurring values represents the block with GM, WM, and CSF. We calculate the average of the energy value (E_{avg}) of the non-zero energy blocks. We have used block size 9×9 for energy calculations as very small block size (3×3) fails to cover the entire ROI may cause undersegmentation whereas a bigger block size (15×15) results in oversegmentation.

In Fig. 2(b), the energy values of different blocks of the input brain MRI (Fig. 2(a)) are shown in the graph. The energy values higher than E_{avg} are marked as 'red' circles and lower energy values are marked as 'green' boxes. In case of a healthy brain MRI (Fig. 2(c)), the value of E_{avg} is higher compared to a tumor affected brain (as shown in Fig. 2(d)). We compute the standard-deviation of energy values of all non-zero energy blocks of the image. As the energy block in tumor region has higher energy than its surrounding brain matter, the tumor affected MRI shows higher standard deviation (energy) than the healthy MRI (as shown in Fig. 3 for 20 images). We have utilized this feature to differentiate healthy and abnormal MRI. It may be noted that the standard deviation (energy) of healthy MRI is <3.5 whereas diseased MRI has standard deviation (energy) >10.

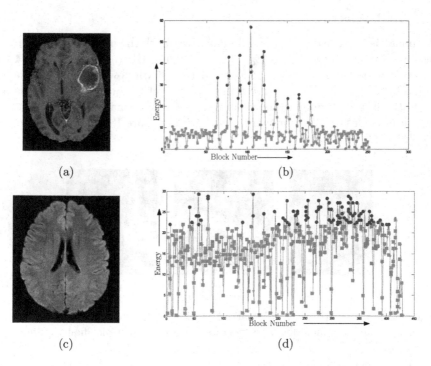

(a)

(b)

(c)

(d)

Fig. 2. (a) Input FLAIR MRI (diseased), (b) Energy distribution (energy $>E_{avg}$ are 'red', lower energy are 'green'), (c) Input FLAIR MRI (healthy), (d) Energy distribution (energy $>E_{avg}$ are 'red', lower energy are 'green'). (Color figure online)

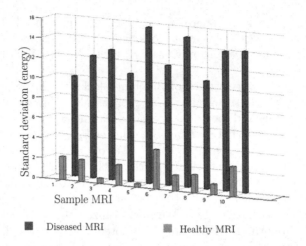

Fig. 3. Standard deviation (energy) of healthy and diseased MRI.

2.1.2 ROI Selection

If the image is identified as a diseased image, we mark the pixels belonging to the higher energy blocks (with $E > E_{avg}$) in the brain MRI (as shown in Fig. 4(b)). We also include the surrounding blocks of (6×6 window size) of the marked regions. In this marked region, we perform connected component analysis and identify the largest region that encloses the blocks with maximum energy, as the region of interest (ROI) for further investigation. Figure 4(c) shows ROI marked in 'red'.

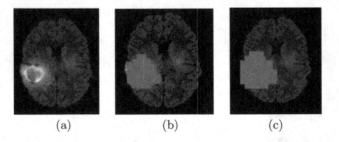

Fig. 4. (a) Input FLAIR MRI, (b) Marked blocks, (c) Marked ROI (after removal of small components and inclusion 6×6 neighbouring pixels). (Color figure online)

2.2 Tumor Detection

Different types of tumors consist of different biological tissues, a single MRI sequence cannot provide sufficient information for all kind of tumors. Hence, the collective information from different MRI sequences of the same patient helps to perform the diagnosis more accurately [18]. In the proposed approach, we have used T1c and FLAIR sequence of patient's brain MRI for segmentation of different types of tumor.

2.2.1   Selection of Proper MR Image Sequence

In order to select the suitable MRI sequence for tumor detection, the image energy analysis (as mentioned in Sect. 2.1) is performed on both FLAIR and T1c image of the same patient. The two image sequences with marked ROI are then compared to check whether the same region is marked in both the cases. If the same region is marked in both FLAIR and T1c image as ROI, then the process uses T1c image for tumor segmentation (having better contrast), otherwise, FLAIR image is used for further processing.

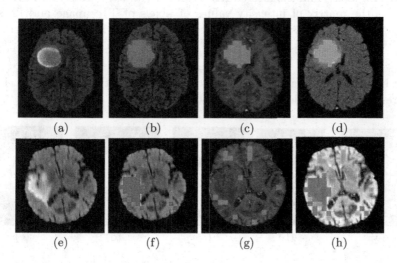

Fig. 5. (a) Input FLAIR MRI of patient1, (b) Marked blocks in (a), (c) Marked blocks in T1c of patient1, (d) Combine (b) and (c), (e) Input FLAIR MRI of patient2, (f) Marked blocks in (e), (g) Marked blocks in T1c of patient2, (h) Combine (e) and (g). (Color figure online)

Figure 5 shows the MRI sequence selection process for tumor detection. For the first scenario (Fig. 5(a)), the ROI is marked in the same region for both FLAIR (Fig. 5(b), 'red' patch) and T1c (Fig. 5(c), 'green' patch) images of patient1. The combined image (Fig. 5(d)) shows the overlapping ROI as 'yellow' region. In such cases, T1c image is used for tumor segmentation. In the second case (Fig. 5(e)), the ROI marked in FLAIR (Fig. 5(f), 'red' patch) and T1c (Fig. 5(g), 'green' patch) images of patient2 are in different portions of the input MRI. Hence, the combined image does not have any overlap (as shown in Fig. 5(h)). In such cases, the proposed method uses FLAIR image for tumor segmentation.

2.2.2   Tumor Segmentation from T1c MRI

The ROI marked in the T1c image include some part of gray-matter (GM), white-matter (WM), and cerebrospinal fluid (CSF). The proposed approach extracts the ROI from T1c MRI and then performs GM, WM, and CSF removal for segmentation of the tumor region.

2.2.2.1 GM, WM, and CSF Removal from ROI: In FLAIR MRI, GM appears with higher contrast than WM and CSF. Whereas in T1c MRI, WM appears with higher contrast than other components. We have utilized this property of FLAIR and T1c image to remove GM, WM, and CSF from the ROI. In order to compute the threshold intensity of WM, we have subtracted FLAIR image (Fig. 6(b)) from T1c image (Fig. 6(c)) of the same patient data, and then computed the threshold intensity of WM from that result (Fig. 6(c)). This threshold intensity value is applied on the ROI (extracted with some part of WM, GM, and CSF, Fig. 6(d)) and the filtered ROI will be obtained (as shown in Fig. 6(e)).

If I1 is the matrix of intensity values of pixels of T1c image and I2 is the matrix of intensity values of pixels of corresponding Flair image then subtraction can be expressed as -

$$I = I1 - I2; \tag{3}$$

The intensity values are defined in a specific range between 0 and 255. Hence in the result of subtraction, any value <0 becomes 0.

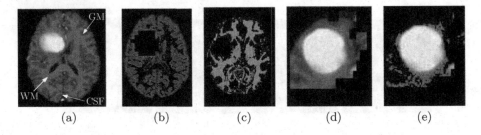

(a)   (b)   (c)   (d)   (e)

Fig. 6. (a) Different components of brain MRI, (b) FLAIR MRI (masked ROI), (c) T1c MRI (masked ROI), (d) Segmented ROI with GM, WM, and CSF, (e) ROI after removal of GM, WM, and CSF.

2.2.2.2 Pixel-Based Scanning: We have performed horizontal (left-to-right) and vertical (top-to-bottom) pixel scanning on the ROI (Fig. 6(e)) to segment the tumor pixels. The pixels of the tumor region are brighter (Fig. 7(a)) in comparison to its dark background pixels and surrounding brain matter pixels (gray). Hence, each horizontal or vertical scan line passing through the tumor region shows the maximum intensity value (I_{max}) for the tumor region pixels. These pixels also have high intensity difference from the surrounding pixels. Some tumors appear with dark center surrounded by bright outer boundary (as shown in Fig. 7(d)). In those cases, multiple intensity peaks will appear with valley region in between them. In both the scenarios, the pixel intensities preceding or following the leftmost and rightmost intensity peak, represent the pixels of the surrounding region. We compute the closeness score (C_i) of each pixel on the scan line. C_i can be represented as

$$C_i = p_i / I_{max} \tag{4}$$

where p_i is the intensity of the i-th pixel, and I_{max} is the maximum intensity of the pixels on the scan line. For the tumor having multiple pixels with higher intensity and having pixels with lower intensity in between them (two intensity peaks with a valley in between) in the scan line, I_{max} will be $I_{leftmax}$ for pixels preceding the leftmost highest intensity value pixel in the scan line and $I_{rightmax}$ for pixels following the rightmost highest intensity value pixel in the scan line. We consider the pixels with high intensity values having closeness score $C_i > 0.3$ as part of the tumor and discard all pixels with lower C_i.

Figure 7(b) and (e) show the variation of intensity and closeness values of the pixels on a scan line passing through the tumor region of the MRI (Fig. 7(a) and (d) respectively). The pixels with low intensity and closeness score are marked as white points on the curve. These points are part of the brain matter (WM, GM, and CSF) and these are removed from the segmented tumor region. The blue points on the curve (Fig. 7(b) and (e)) represent the pixels with high intensity

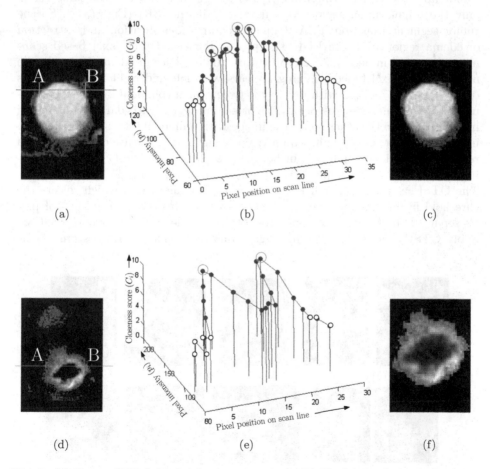

Fig. 7. (a) Tumor ROI (with bright tumor region), (b) Pixel intensity and closeness score along scan line segment AB, (c) Segmented tumor, (d) Tumor ROI (with dark center and bright boundary region), (e) Pixel intensity and closeness score along scan line AB of (d), (f) Segmented tumor. (Color figure online)

and higher closeness score. These points belong to the tumor area of the MRI. All pixels appearing in the scan line with high intensity and higher closeness score are considered as tumor region pixels.

We perform this filtering process for each scan line (in both horizontal and vertical directions) and combine the result to segment the tumor region pixels (having higher C_i values and higher intensity difference from the surrounding pixels for both horizontal and vertical scanning). The final image appears with the segmented tumor region (shown in Fig. 7(c) and (f)).

2.3   Tumor Segmentation from FLAIR MRI

In FLAIR image, peritumoral edema may appear with hyperintensity and the tumor appears with low intensity [7] (as shown in Fig. 8(a2)). In some cases, only the bright tumor region remains visible in the MRI (Fig. 8(a1)). So the tumor segmentation from FLAIR image requires identification and extraction of edema region from the ROI. The proposed method uses pixel based scanning (discussed in Sect. 2.2.2.2) to identify the presence of edema region in the image. If the pixel based scanning appears with left and right intensity peaks ($I_{leftmax}$ and $I_{rightmax}$) with a valley between them (bimodal intensity distribution), it confirms the presence of bright edema region with dark tumor region in the image. On the other hand, the image with unimodal intensity distribution implies the presence of bright tumor region (Fig. 8(a1)). In case of bright edema region with dark tumor region in between, we have used the OTSU thresholding approach [13] to compute the threshold intensity value for edema region. The OTSU method is a well known thresholding algorithm, which selects the threshold by minimizing the within class variance between two groups of pixels separated by thresholding operator. Once the edema region is extracted, we apply OTSU thresholding method again on the edema region to segment the

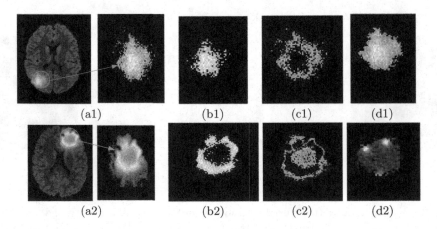

Fig. 8. (a1), (a2) FLAIR image and ROI, (b1), (b2) Extracted edema-high, (c1), (c2) Edema-low, (d1), (d2) Segmented tumor.

edema-high and edema-low region (as shown in Fig. 8(b1), (b2)). The tumor region left after edema extraction is the segmented tumor (shown in Fig. 8(d2)). If the bright edema region cannot be identified separately (unimodal intensity distribution), we apply OTSU thresholding on ROI to segment the tumor (shown in Fig. 8(d1)).

3   Tumor Grade Analysis

The most frequent primary brain tumors (gliomas) originate from the glial cells and penetrate among the surrounding tissues [11]. Gliomas are classified into four World Health Organization (WHO) grades (I, II, III, IV) based on their aggressiveness. Low grade gliomas (grade I and II) are slowest growing benign tumors whereas high grade gliomas (grade III and grade IV) are fastest growing malignant tumors [1,9]. In this work, we have identified the tumor grade by mapping the clinical observation of the MR image with the image features that appear during MRI analysis. The summary of this mapping is shown in Table 1. The hypointense tumor region appears with dark shade compared to its surrounding tissues, whereas hyperintense region appears with lighter color than its surrounding area (in T1c MRI). Necrosis implies the presence of dead cells in the tumor. During image analysis, these dead cells within tumor region appear as a black patch at the center of the tumor [9].

Low Grade Tumor: Low grade tumor shows hypodark appearance in T1C MRI. The proposed approach detects the low grade tumor by analyzing the ROI extraction from FLAIR image and T1c image. If the ROI extracted from the FLAIR image differs from the ROI of T1c image (as discussed in Sect. 2.2.1), then the tumor is classified as a low grade tumor.

High Grade Tumor: In high grade tumor, the extracted tumor region has a very different intensity profile than the surrounding tissues (GM and WM) in both T1c and FLAIR image. In such cases, the tumor is classified as high grade tumor. The proposed method classify high grade tumor by analyzing the intensity distribution (discussed in Sect. 2.2.2.2) in the segmented tumor region.

Table 1. Grading criteria.

Grade	MRI observation			Image analysis		
	Hypo intense	Hyper intense		Bright tumor region	Dark center with bright tumor border	Dark tumor region
		With Necrosis	Without Necrosis			
High	×	×	√	√	×	×
	×	√	×	×	√	×
Low	√	×	×	×	×	√

Presence of Dead Cells: Necrosis (a collection of dead cells) appears in high grade brain tumors e.g., a grade IV tumor [9]. Necrosis is manifested as a dark area surrounded by bright tumor tissues. By scanning the T1c image in horizontal and vertical directions, the proposed method determines the presence of Necrosis (discussed in Sect. 2.2.2.2). For each scan line, sudden drop of pixel intensities between two peak intensity values indicates region of dead cells.

4   Experimental Results

We have applied the proposed approach on 40 MRI images selected from the MRI available in BRATS database (https://drive.google.com/open?id=0B5M6Y0ylgFnSR1V3eXdaMTVpRkE). Table 3 shows the input FLAIR and T1c images, segmented tumor region, and the grade detected by the proposed approach for 10 such images. This method works fine for most of the cases. For few low-grade tumor MRIs, results appear with poorly segmented tumor. Image no. 4 and no. 8 of Table 3 have very poor contrast in the tumor region of the T1c image. This results in poor segmentation. The proposed method fails to identify the grade of Image no. 8 correctly due to poor contrast of the input image.

To evaluate the quality of the tumor segmentation, we have used the standard *Dice* index [2]. If TP represents the number of true positive, FP false positive, and FN false negative, then the *Dice* metric is defined as -

$$Dice = \frac{2 \times TP}{2 \times TP + FP + FN} \tag{5}$$

The value of *Dice* metric ranges from zero to one, where higher values indicate better agreement. The other segmentation quality evaluation metrics such as accuracy (Ac), sensitivity (Sn), and specificity (Sp) are defined as

$$Ac = \frac{TP + TN}{TP + TN + FP + FN}, \quad Sn = \frac{TP}{TP + FN}, \quad Sp = \frac{TN}{TN + FP} \tag{6}$$

Table 2. Tumor segmentation evaluation.

Img No.	Ac (%)	Sn (%)	Sp (%)	Dice
1	99.44	93.7	99.8	0.95
2	99	94.92	99.21	0.9
3	99.58	96.9	99.69	0.94
4	97.52	78	99.44	0.85
5	99.35	99.93	99.3	0.95
6	99.57	65.42	99.93	0.76
7	98.89	100	98.82	0.91
8	98.66	75.94	99.93	0.86
9	99.76	96.98	99.84	0.96
10	99.02	94.51	99.49	0.95
11	98.41	77.44	99.85	0.86
12	99.46	99.77	99.44	0.93
13	97.53	56.64	99.85	0.71
14	99.4	98.86	99.42	0.92
15	95.72	56.43	100	0.72
16	99.58	65.42	99.93	0.76

Img No.	Ac (%)	Sn (%)	Sp (%)	Dice
17	99.49	94.53	99.52	0.66
18	99.36	97.91	99.43	0.93
19	99.43	97.83	99.5	0.94
20	99.51	98.44	99.53	0.91
21	99.57	98.94	99.6	0.95
22	97.83	83.92	98.57	0.8
23	98.48	78.44	99.85	0.87
24	98.98	73.15	99.95	0.84
25	98.86	60.79	100	0.76
26	99.08	54.29	100	0.7
27	94.83	42.89	100	0.6
28	97.91	60.24	100	0.75
29	98.06	59.09	100	0.74
30	98.3	66.49	100	0.8
Average	98.68	80.59	99.66	0.84

Table 3. Grading results.

Sl No.	Input MRI (FLAIR)	Input MRI (T1c)	Extracted Tumor	Features Detected	Diagnosis
1.				Necrosis Present, Hyperintensive	**High Grade**
2.				No Necrosis, Hyperintensive	**High Grade**
3.				Necrosis Present, Hyperintensive	**High Grade**
4.				No Necrosis, Hyperintensive	**High Grade**
5.				No Necrosis, Hypointensive	**Low Grade**
6.				No Necrosis, Hypointensive	**Low Grade**
7.				No Necrosis, Hyperintensive	**High Grade**
8.				No Necrosis, Hypointensive (Ground truth: Hyperintensive, High Grade)	**Low Grade**
9.				No Necrosis, Hyperintensive	**High Grade**
10.				No Necrosis, Hypointensive	**Low Grade**

We have compared the segmented tumor region pixels with the ground truth tumor image available in BRATS database to compute Ac, Sn, Sp, and $Dice$ index. Table 2 shows the detail results with 30 images of BRATS database. It can be observed that a few low grade images (Sl no. 13, 15, 16 and 27 of Table 2) has low $Dice$ indices. The poor contrast difference between tumor and edema in FLAIR images results in poor segmentation of low grade images. Other tumor segmentation approaches also show low $Dice$ index values [19], [3] for low grade segmented tumor region.

5   Conclusion

This paper focuses on detection of abnormality in the brain tissues in a brain MRI and subsequent segmentation of the abnormal region. Tumor region is further investigated to identify the grade of the tumor. The proposed approach strongly ensures that healthy brain tissues must not be categorized as tumor tissues. We achieved a good rate of around 99% of specificity parameter in tumor segmentation. This work can be extended in future for detail classification of high grade and low grade tumors.

Acknowledgement. Authors would like to acknowledge Department of Science & Technology, Government of India, for financial support vide ref. no. SR/WOS-A/ET-1022/2014 under Woman Scientist Scheme to carry out this work.

References

1. Akkus, Z., Sedlar, J., Coufalova, L., Korfiatis, P., Kline, T.L., Warner, J.D., Agrawal, J., Erickson, B.J.: Semi-automated segmentation of pre-operative low grade gliomas in magnetic resonance imaging. Cancer Imaging **15**(12), 1–11 (2015)
2. Ashburner, J., Friston, K.J.: Unified segmentation. NeuroImage **26**, 839–851 (2005)
3. Bauer, S., Fejes, T., Slotboom, J., Wiest, R., Nolte, L.P., Reyes, M.: Segmentation of brain tumor images based on integrated hierarchical classification and regularization. In: Proceedings of Multimodal Brain Tumor Segmentation Challenge (MICCAI-BRATS), pp. 10–13 (2012)
4. Bauer, S., Wiest, R., Nolte, L.P., Reyes, M.: A survey of MRI-based medical image analysis for brain tumor studies. Phys. Med. Biol. **58**(13), 97–129 (2013)
5. Charutha, S., Jayashree, M.J.: An efficient brain tumor detection by integrating modified texture based region growing and cellular automata edge detection. In: Proceedings of IEEE Control, Instrumentation, Communication and Computational Technologies (ICCICCT), pp. 1193–1199 (2014)
6. Dvorak, P., Kropatsch, W., Walter, K.: Automatic detection of brain tumors in MR images. In: Proceedings of IEEE Telecommunications and Signal Processing (TSP), pp. 577–580 (2013)
7. Eis, M., Els, T., Hoehn-Berlage, M., Hossmann, K.A.: Quantitative diffusion MR imaging in cerebral tumor and edema. In: Ito, U., Baethmann, A., Hossmann, K.-A., Kuroiwa, T., Marmarou, A., Reulen, H.-J., Takakura, K. (eds.) Brain Edema IX, pp. 344–346. Springer, Vienna (1994)

8. Khotanlou, H., Colliot, O., Bloch, I.: Automatic brain tumor segmentation using symmetry analysis and deformable models. In: Proceedings of Advances in Pattern Recognition (ICAPR), pp. 198–202 (2007)
9. Louis, D.N., Ohgaki, H., Wiestler, O.D., Cavenee, W.K., Burger, P.C., Jouvet, A., Scheithauer, B.W., Kleihues, P.: The 2007 WHO classification of tumours of the central nervous system. Acta Neuropathol. **114**, 97–109 (2007)
10. Maiti, I., Chakraborty, M.: A new method for brain tumor segmentation based on watershed and edge detection algorithms in HSV colour model. In: Proceedings of Computing and Communication Systems (NCCCS), pp. 1–5 (2012)
11. Menze, B.H., et al.: The multimodal brain tumor image segmentation benchmark (BRATS). IEEE Trans. Med. Imaging **34**(10), 1993–2024 (2015)
12. Menze, B.H., et al.: A generative probabilistic model and discriminative extensions for brain lesion segmentation with application to tumor and stroke. IEEE Trans. Med. Imaging **35**(4), 933–946 (2016)
13. Otsu, N.: A threshold selection method from gray-level histograms. IEEE Trans. Syst. Man Cybern. **9**, 62–66 (1979)
14. Pedoia, V., Binaghi, E., Balbi, S., Benedictis, A.D., Monti, E., Minotto, R.: Glial brain tumor detection by using symmetry analysis. In: Proceedings of SPIE Medical Imaging, p. 831445 (2012)
15. Selvaraj, H., Selvi, S.T., Selvathi, D., Gewali, L.: Brain MRI slices classification using least squares support vector machine. Int. J. Intell. Comput. Med. Sci. Image Process. **1**(1), 21–33 (2007)
16. Somasundaram, K., Kalaiselvi, T.: Automatic detection of brain tumor from MRI scans using maxima transform. In: Proceedings of National Conference on Image Processing (NCIMP), pp. 136–141 (2010)
17. Umbaugh, S.E.: Feature Analysis and Pattern Classification. CRC Press, Boca Raton (2011)
18. Zhang, N., Ruan, S., Lebonvallet, S., Liao, Q., Zhu, Y.: Kernel feature selection to fuse multi-spectral MRI images for brain tumor segmentation. Comput. Vis. Image Underst. **115**, 256–269 (2011)
19. Zikic, D., Glocker, B., Konukoglu, E., Shotton, J., Criminisi, A., Ye, D.H., Demiralp, C., Thomas, O.M., Das, T., Jena, R., Price, S.J.: Context-sensitive classification forests for segmentation of brain tumor tissues. In: Proceedings of Multimodal Brain Tumor Segmentation Challenge (MICCAI-BRATS), pp. 1–9 (2012)

Medical Image Segmentation Using Improved Affinity Propagation

Hong Zhu[1(✉)], Jinhui Xu[2], Junfeng Hu[1], and Jing Chen[1]

[1] School of Medical Information, Xuzhou Medical College, Xuzhou, China
{zhuhong,hujunfeng,chenjing}@xzmc.edu.cn
[2] Department of Computer Science and Engineering,
State University of New York at Buffalo, Buffalo, USA
jinhui@buffalo.edu

Abstract. Affinity Propagation (AP) is an effective clustering method with a number of advantages over the commonly used k-means clustering. For example, it does not need to specify the number of clusters in advance, and can handle clusters with general topology, which makes it uniquely suitable for medical image segmentation as most of the objects in medical images are not roundly shaped. One factor hampering its applications is its relatively slow speed, especially for large-size images. To overcome this difficulty, we propose in this paper an Improved Affinity Propagation (IMAP) method with several improved features. Particularly, our IMAP method can adaptively select the key parameter p in AP according to the medical image gray histogram, and thus can greatly speed up convergence. Experimental results suggest that IMAP has a higher image entropy, lower class square error contrast, and shorter runtime than the AP algorithm.

Keywords: Medical image segmentation · Affinity propagation · Gray level histogram

1 Introduction

Medical image segmentation is a pivotal step in medical image processing/analysis and plays an increasingly significant role in many clinical practices, such as the detection and diagnosis of diseases, treatment planning, guidance of interventions, and the follow-up monitoring of patients [10, 11, 14]. It automates (or facilitates) the delineation of anatomical structures and other regions of interest making it possible for further image measurement, registration, fusion, and image comprehension. With the wide applications of various image modalities (such as Computed Tomography (CT), Magnetic Resonance (MR), and Positron Emission Tomography (PET) imaging) in

This work is supported in part by Overseas Training Program for Outstanding Young Teachers and Principals of Universities in Jiangsu Province, Natural Science Fund Project of College in Jiangsu Province (14KJB520039), the National Nature Science Foundation of China (No. 61379101), and the Basic Research Program (Natural Science Foundation) of Jiangsu Province of China (No. BK20130209).

R.P. Barneva et al. (Eds.): CompIMAGE 2016, LNCS 10149, pp. 208–215, 2017.
DOI: 10.1007/978-3-319-54609-4_15

clinical settings, there is an urgent need for more effective medical image segmentation algorithms. Due to its importance, a considerable amount of effort has been devoted to this problem and many medical image segmentation algorithms have been obtained, based on different strategies such as thresholding [13], region growing [1, 16], classifiers [2], clustering [12], artificial neural networks [7], and deformable models [3].

As one of a few commonly used approaches, clustering has demonstrated its effectiveness in segmenting various types of medical images. Quite a number of clustering approaches have been used for image segmentation such as K-means, Fuzzy c-means, Density-based method, Mean Shift method, Expectation Maximization and their improved algorithms. Unlike classifiers, clustering is an unsupervised learning method and does not require any training dataset. But often it needs to specify in advance the number of clusters (and their centers in some scenarios) to yield spheroidal segmentation. As one of the remedies, AP clustering algorithm was introduced to simultaneously consider all data points as potential exemplars. By viewing each data point as a node in a network, it recursively transmits real-valued messages along edges of the network until a good set of exemplars and corresponding clusters emerge. In this way, AP can overcome some drawbacks of K-means and Fuzzy c-means. However, at the same time, its high time complexity and the selection of a key parameter could also be problematic.

To avoid such pitfalls, we propose in this paper an improved AP algorithm called IMAP. Our algorithm adaptively selects the parameter for itself, based on the gray histogram of the input image and can significantly reduce the time complexity. Experimental results show that our algorithm can accurately obtain the segmentation results in a much shorter period of time.

2 Affinity Propagation Clustering Algorithm [6]

Affinity Propagation clustering algorithm was introduced by Frey and Dueck in 2007. It has been widely used in many areas such as searching the optimal route [9], detecting genes in microarray data [4], clustering images of faces [8] and so on. Different from other partition-based clustering methods, AP algorithm can automatically determine the proper number of clusters and their corresponding centers. In order to avoid inappropriate initialization, it first takes all data points as potential cluster centers, and then exchanges messages iteratively between any pair of points to determine whether they belong to the same cluster. This process continues until a stable clustering status is reached.

In the AP algorithm, the similarity between any two data points x_i and x_j is measured by $s(i, j) = -\|x_i - x_j\|^2$ and stored in an $N \times N$ matrix. Before clustering, AP takes as input a real number $s(k, k)$ for each data point k so that data points with larger values of $s(k, k)$ are more likely to be chosen as exemplars. These values are referred to as "preferences". If no a priori knowledge exists, all data points are equally suitable as exemplars, and the preferences should be set to a common value, called parameter p. This value can be varied to produce different numbers of clusters. During the execution of the AP algorithm, $s(k, k)$ will be recalculated. Finally, those data points with bigger $s(k, k)$ values will become clustering centers.

There are two kinds of messages exchanged between data points (see Fig. 1). The "responsibility" $r(i, k)$, sent from data point i to candidate exemplar point k, reflects the accumulated evidence for how well-suited point k is to serve as the exemplar for point i, taking into account other potential exemplars for point i. The "availability" $a(i, k)$, sent from candidate exemplar point k to point i, reflects the accumulated evidence for how appropriate it would be for point i to choose point k as its exemplar, taking into account the support from other points that point k should be an exemplar. The responsibilities and availabilities are initialized to zero at the beginning of AP algorithm, and then computed using the following rules (where s.t. is the abbreviation of "subject to"):

$$r(i,k) \leftarrow s(i,k) - \max_{k's.t.k' \neq k}\{a(i,k') + s(i,k')\} \tag{1}$$

$$a(i,k) \leftarrow \min\left\{0, r(k,k) + \sum_{i's.t.i' \notin \{i,k\}} \max\{0, r(i', k)\}\right\} \tag{2}$$

$$r(k,k) = p(k) - \max\{a(k,k') + s(k,k')\} \tag{3}$$

$$a(k,k) \leftarrow \sum_{i's.t.i' \neq k} \max(0, r(i',k)) \tag{4}$$

When updating the messages, it is important that they should be damped to avoid numerical oscillations that arise in some circumstances. Each message is set to λ times its value from the previous iteration plus $(1-\lambda)$ times its prescribed updated value A_{old}, such as: $R = (1 - \lambda)*R + \lambda*R_{old}$ and $A = (1 - \lambda)*A + \lambda*A_{old}$.

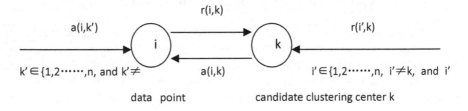

Fig. 1. Message passing procedure between data points.

3   Improved Affinity Propagation Clustering Algorithm for Medical Image Segmentation

As mentioned in the introduction section, AP suffers from several limitations. In this section, we propose an improved AP method (IMAP) in the following three aspects.

3.1   Reducing Time Complexity

As mentioned previously, AP clustering algorithm simultaneously considers all data points as potential exemplars, and transmits messages between any two data points in order to find the appropriate clustering centers and their corresponding clusters. This will incur a time complexity of $O(N*N*\log N)$, where N is the total number of pixels/voxels in the image (note that AP clustering method takes each pixel as a data point). This would be too high for some medical images, especially for those large-size images. For example, a typical medical image has 512*512 pixels. If we consider every pixel as a data point, the total number of data points is 262144. The size of similarity matrix is 262144*262144.

To reduce the time complexity, the most effective way is to reduce the number of data points. To achieve this goal, we first understand that image segmentation is the process of partitioning image into distinct regions by grouping together pixels with similar attributes such as gray value, texture, color, contrast, brightness, etc. Gray value is a feature frequently used in clustering methods for image segmentation. Different from other features, gray level has its unique property. For a typical medical image, the gray level often has a scale between 0 and 255. The number of gray levels is often independent of the size and modality of the medical images. For example, although there are CT images of sizes 256*256, 512*512, and 1024*1024, they all use roughly 256 gray levels to capture the (radiation) absorption degree of organs and tissues.

The improved Affinity Propagation clustering algorithm takes all pixels with the same gray level as a data point, instead of treating each pixel as a data point. The gray histogram is a statistical feature of the medical image that counts the number of pixels with the same gray level. If we construct the gray level histogram before executing the AP algorithm and treat each gray level as a data point, we can greatly reduce the number of data points. In this way, only 256 data points are needed to calculate the pairwise similarity. Thus the size of the similarity matrix is significantly reduced to 256*256.

It is tempting to use other features to further reduce the data size. However, they are not as friendly as gray level. Consider textural feature as an example. Although the textural feature value of a pixel is calculated according to the gray level differences between the pixel and its neighbors, it is often difficult to divide the value of textural feature into different levels and treat each level as a data point. It also does not yield good clustering result by combining gray level and textural feature as a composite feature. Thus, in this paper we use only gray level to segment medical images.

3.2   Selecting Improved Distance Metric to Replace Euclidean Distance

Although using gray levels as data points can significantly reduce the number of data points, it also causes problems for computing the similarity between two data points or between two clusters. The similarity matrix of AP is in general asymmetric, i.e., $s(i, j)$ and $s(j, i)$ could take different values, which is different from the K-means algorithm. Though, of course, expert knowledge is expected to assign different values to $s(i, j)$ and $s(j, i)$, but it is not always available. More often, like most of the clustering algorithms, AP selects the Euclidean distance to measure the similarity between any two data points. But because of the complexity of human anatomy and the irregular shapes of human issues or organs, Euclidean distance drops the shape information of interest region and is no longer sufficient to measure the similarity.

To overcome this problem, we introduce the following way to modify the similarity matrix according to the pdf (probability density function) of gray level histogram [5]. (In the pdf of gray level histogram, x-axis represents intensity and y-axis represents probability of density.)

$$s(i,j) = -\left[|d(i,j)|^2 + \left|\sum_{k=\min(i,j)}^{\max(i,j)-1} d(k, k+1)\right|^2\right]^{1/2} \tag{5}$$

where $d(i, j)$ is the Euclidean distance between point i and j along the x-axis, $d(k, k + 1)$ is the Euclidean distances between point k and $k + 1$. So. $\sum_{k=\min(i,j)}^{\max(i,j)-1} d(k, k + 1)$ is the sum of local Euclidean distances using all points between i and j, and it is the geodesic distance between point i and j. In this way, similarity between any two points is described in the non-Euclidean space by taking both intensity- and probability-based measures into account.

3.3   Calculating Parameter *p* According to Gray Level Histogram

In the AP algorithm, two parameters are critical to the performance. One is the preference p, and the other is λ (which is related to the convergence speed). In this paper, we will focus only on the parameter p. At the beginning of AP, we need to select an initial value for the parameter p. If p is set to be the mean value of similarity between any two points, the number of clusters will be medium. The larger the value is, the more the clusters are, and vice versa. Thus, it is important for us to select an initial value for preference p. It is obvious that we could not obtain a correct image segmentation result without an appropriate p.

IMAP selects p according to the medical image gray level histogram. Unlike the k-means method in which the number of clusters is always a fixed number, our IMAP algorithm can adjust the number of clusters according to the initial value of p during its execution.

It is worthy pointing out that some other clustering algorithms also initialize their parameters using gray level histogram. Ding proposed an improved Fuzzy C-means method to segment gray images [17]. It can automatically obtain the cluster number by analyzing the peaks and dips of the gray level histogram and taking the mean value of adjacent dips in the gray level histogram as the initial cluster centers for the segmentation. This method yields good results only for those images whose gray level histogram is smooth and continuous and has obvious peaks and dips. But in reality the curve of gray level histogram is not always smooth and has no obvious peaks and dips. Thus, it cannot be used in many cases.

Unlike K-means and other clustering algorithms which need the exact parameter values such as cluster number and centers before their execution, AP algorithm can adjust cluster number and centers automatically in each iteration of the algorithm. It can obtain relatively stable clustering results even if p varies in a certain range.

In 0–255 image grayscale display system, human beings can recognize at most 50 gray levels and the correct recognition rate of human eye is only 45.31 percent when the gray level reaches 32 [15]. So, IMAP selects p using the following formula:

$$\frac{p}{n} = \frac{p_{max}}{32} \tag{6}$$

In formula 6, p_{max} is the maximum value in the similarity matrix and n is the number of peak values in the gray histogram.

In this way, IMAP can obtain a rough value of p. Then it takes all data points as potential cluster center first, and exchanges messages iteratively between any two points until we get final cluster result.

4 Experimental Results

In this work, experiments are conducted on five 256-level images. The source images are from the website of the Atlas project (http://www.med.harvard.edu/AANLIB).

If we take each pixel as a data point, AP algorithm will breakdown for medical image segmentation (for the reason introduced in Sect. 3.1). So, in the experimental AP algorithm, we take each gray level as a data point under the condition that every medical image has 256 gray levels.

The original image, segmentation results of AP method and IMAP method are listed in Fig. 2.

We compare IMAP with AP and K-means in three aspects:image entropy (EN), classes square error contrast (SEC), and running time (RT). The result is shown in Table 1.

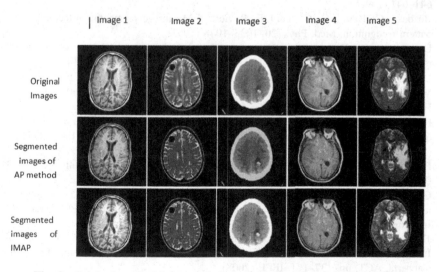

Fig. 2. Original medical image and segmentation results of AP and IMAP.

Table 1. Performance of different methods in image 1 to 5.

	AP			IMAP		
	RT (s)	EN	SEC	RT (s)	EN	SEC
Image 1	0.5586	4.6690	2516	0.3256	4.6732	2381
Image 2	1.0355	4.0563	1535	0.6687	4.0926	1036
Image 3	0.8349	4.0143	1966	0.6735	4.0436	1694
Image 4	0.5802	4.4940	2772	0.5735	4.50154	2554
Image 5	0.5746	4.0025	2439	0.5515	4.2518	2130

The experiment results show that IMAP has better performance than AP algorithm. Its runtime is shorter, EN is larger, and SEC is lower than those of AP.

5 Conclusion

In this paper, we propose an improvement (called IMAP) for the AP algorithm. It significantly reduces the time complexity by taking the same gray level as a data point. A corresponding distance metric and a new way of selecting the value of parameter p are introduced to accommodate such a change. Experiments show that IMAP outperformances AP in running time, image entropy and classes square error contrast. It remains open to incorporate other features into IMAP for better segmentation result.

References

1. Adams, R., Bischof, L.: Seeded region growing. IEEE Trans. Pattern Anal. Mach. Intell. **16**, 641–647 (1994)
2. Bezdek, J.C., Hall, L.O., Clarke, L.P.: Review of MR image segmentation techniques using pattern recognition. Med. Phys. **20**, 1033–1048 (1993)
3. Davatzikos, C., Bryan, R.N.: Using a deformable surface model to obtain a shape representation of the cortex. IEEE Trans. Med. Imaging **15**, 785–795 (1996)
4. Dueck, D., et al.: Constructing treatment portfolios using affinity propagation. In: Vingron, M., Wong, L. (eds.) RECOMB 2008. LNCS, vol. 4955, pp. 360–371. Springer, Heidelberg (2008). doi:10.1007/978-3-540-78839-3_31
5. Foster, B., Bagci, U., Ziyue, X.: Segmentation of PET images for computer-aided functional quantification of tuberculosis in small animal models. IEEE Trans. Biomed. Eng. **61**, 711–724 (2014)
6. Frey, B.J., Dueck, D.: Clustering by passing messages between data points. Science **315**, 972–976 (2007)
7. Gelenbe, E., Feng, Y., Krishnan, K.R.R.: Neural network methods for multispectral magnetic resonance images of brain using artificial neural networks. IEEE Trans. Med. Imaging **16**, 911–918 (1997)
8. Jia, S., Qian, Y., Ji, Z.: Band selection for hyperspectral imagery using affinity propagation. In: Proceedings of the 2008 Digital Image Computing: Techniques and Applications, Canberra, ACT, pp. 137–141. IEEE (2008)

9. Kelly, K.: Affinity program slashes computing times. Accessed 15 Dec 2007. http://www. news.utoronto.ca/bin6/070215-2952
10. Lee, L.K., Liew, S.C., Thong, W.J.: A review of image segmentation methodologies in medical image. In: Sulaiman, H.A., Othman, M.A., Othman, M.F.I., Rahim, Y.A., Pee, N.C. (eds.) Advanced Computer and Communication Engineering Technology. LNEE, vol. 315, pp. 1069–1080. Springer, Heidelberg (2015). doi:10.1007/978-3-319-07674-4_99
11. Masood, S., Sharif, M., Masood, A., Yasmin, M., Raza, M.: A survey on medical image segmentation. Curr. Med. Imaging Rev. **11**, 3–14 (2015)
12. Ravindraiah, R., Tejaswini, K.: A survey of image segmentation algorithms based on fuzzy clustering. IJCSMC **2**, 200–206 (2013)
13. Sahoo, P.K., Soltani, S., Wong, A.K.C.: A survey of thresholding techniques. Comput. Vis. Graph. Image. Process. **41**, 233–260 (1988)
14. Sharma, N.J., Ray, A.K., Shukla, K.K.: Automated medical image segmentation techniques. J. Med. Phys. **35**, 3–14 (2010)
15. Xie, W., Qin, A.: The gray level resolution and intrinsic noise of human vision. Space Med. Med. Eng. **4**, 51–55 (1991)
16. Zhang, X., Li, Y.: A medical image segmentation algorithm based on bi-directional region growing. Optik **126**, 2398–2404 (2015)
17. Zhen, D., Zhongshan, H., Jingyu, Y.: A image segmentation method base on Fuzzy c-means. Comput. Res. Dev. **7**, 536–541 (1997)

Simple Signed-Distance Function Depth Calculation Applied to Measurement of the fMRI BOLD Hemodynamic Response Function in Human Visual Cortex

Jung Hwan Kim$^{(\boxtimes)}$, Amanda Taylor, and David Ress

Department of Neuroscience, Baylor College of Medicine,
1 Baylor Plaza, S104, Houston, TX 77030, USA
{junghwan.kim, atl4, ress}@bcm.edu

Abstract. Functional magnetic resonance imaging (fMRI) often relies on a hemodynamic response function (HRF) elicited by a brief stimulus. At conventional spatial resolutions (≥ 3 mm), signals in a voxel include contributions from various tissue types and pia vasculature. To better understand these contributions, full characterization of the depth dependence of the HRF is required in gray matter as well as and its apposed white-matter and pial vasculature. We introduce new methods to calculate 3D depth that combines a signed-distance function with an algebraic morphing definition of distance. The new scheme is much simpler than methods that rely upon deformable surface propagation. The method is demonstrated by combining the distance map with high-resolution fMRI (0.9-mm voxels) measurements of the depth-dependent HRF. The depth dependence of the HRF is reliable throughout a broad depth range in gray matter as well as in white-matter and extra-pial compartments apposed to active gray matter. The proposed scheme with high-resolution fMRI can be useful to separate HRFs in the gray matter from undesirable and confounding signals.

Keywords: Cerebral hemodynamic response function · Signed-distance function · Cortical thickness · Brief stimulus-evoked neural activity · fMRI

1 Introduction

The gray matter of cerebral cortex is anatomically divided into a series of laminae (layers) with varying thicknesses. The overall thickness of gray matter varies across the surface of the brain, being typically thicker on the gyri and thinner in the sulci [10, 27]. One approach to dealing with these thickness variations is to assume that the laminae of the gray matter scale in proportion to its thickness, as suggested by functional measurements [27]. Thus, we want to define a depth coordinate that is normalized to the local thickness of the gray matter. More precisely, we need a smooth function, $D(x,y,z)$, defined in the gray-matter volume between an outer surface S_1 (i.e., the pia) and an inner surface (i.e., the gray-white interface) such that $D(S_1) = 0$ and $D(S_2) = 1$ and the isosurfaces are approximately parallel to the physical laminar structure. Since the isosurfaces of D smoothly interpolate their geometry from S_1 to S_2 in a similar fashion

© Springer International Publishing AG 2017
R.P. Barneva et al. (Eds.): CompIMAGE 2016, LNCS 10149, pp. 216–228, 2017.
DOI: 10.1007/978-3-319-54609-4_16

to the cortical laminar structure, this seems likely to more accurately represent the variations in thickness within the gyri and sulci of the brain's convolutions.

We propose a novel method to compute $D(x, y, z)$ starting from a function w based on the interpolation of a signed distance function (SDF) calculated between the surfaces [1]. The approach is similar to our previously published work [17], but offers several advantages. This method also provides estimates of tissue thickness values useful in the analysis of functional activity observed in laminated brain structures.

We demonstrate our method by application to high-resolution blood oxygen-level dependent (BOLD) functional magnetic resonance imaging (fMRI). In particular, the BOLD signal generated by brief neural activity is called the hemodynamic response function (HRF), and typically consists of a sequence of three temporal periods: initial latency with little signal or possible signal decrease, then a peak that likely corresponds to a transient period of intravascular hyper-oxygenation, followed by an undershoot [22, 24].

Conventional human BOLD fMRI studies with relatively large voxel sampling size ($\sim$3–6 mm) can blur data over a range of white matter (WM), gray matter (GM) and pial vasculature [16, 28]. This is of particular concern in the convoluted human cerebral cortex, which has a thickness in the range of 1.5–4.5 mm, with the thinnest GM typically found in the depths of the sulci. A partial sampling between adjacent sides of a sulcus can confuse matters by mixing contributions from disparate portions of the cortical surface. Moreover, the pial vessels that supply and drain cortical blood are expected to have their own dynamics of coupling with the HRF in the parenchyma [31]. A conventional fMRI voxel may therefore contain significant fractions of undesirable signals, which have the potential to alter the HRF considerably from that found in a localized region of GM where neuronal activity occurs.

It has often been assumed that signals outside of the GM are either weak or unrelated to stimulus-evoked activity in the GM, and therefore, most previous studies have attempted to examine HRF only in the GM [4, 18, 19, 25, 29, 30]. However, because of the limited spatial resolution of fMRI studies, partial volume effects will inevitably mix potential signals from WM and the pial vasculature with gray-matter signals. To understand such partial volume effects, it is necessary to characterize the HRFs with respect to depth in the GM as well as in WM and the pial vasculature.

Here, we utilize our 3D depth-mapping approach for analysis of high-resolution (0.9-mm voxels) HRF measurements in GM, and spatially apposed WM and pial vasculature. The depth coordinate allows us to separate the GM signal from signals in WM and the pial vasculature, and to obtain a detailed assessment of the depth-dependence of the HRF in human early visual cortex (areas V1–3) at 3T field strength. We demonstrate significant HRFs that are spatially distinct from the GM, carefully avoiding partial-volume effects. The measurements show reliable HRFs and their parameters varying with respect to depth in the GM, and demonstrate disparate characteristics of HRFs outside of the GM. The new depth-resolved results provide further insight into the character and mechanism of the fMRI BOLD response. In particular, the work demonstrates the confounding effects of the signal from the pial vasculature, which increase noise and reduces the spatial reliability of the fMRI signal.

2 Methods

Imaging experiments were performed on a 3T GE Signa HD12 scanner. Six subjects (nine sessions) participated in the experiments; all gave informed consent under procedures reviewed and authorized by the UT Austin Institutional Review Board.

2.1 3D Depth-Mapping with Signed-Distance Function

We obtain 0.7-mm isovoxel volume anatomies for all subjects using an inversion-prepared fast RF-spoiled gradient-recalled echo (fSPGR) sequence with an 8-channel GE-product head-coil. The FreeSurfer software suite is used to initially segment the WM and GM at 1-mm sampling, which is then upsampled to 0.7-mm. The WM segmentation is then edited manually across early visual cortex to improve accuracy and reduce errors.

Our surface-based analysis utilizes an inner (GM/WM) and outer (pial) surface of the cortex. We produce a GM/WM interface mesh on the white-matter volume by isosurface contour extraction followed by mesh refinement using a volume-preserving deformable-mesh algorithm [1, 17]. For the pial surface, we use the mesh produced from the FreeSurfer segmentation, edited manually by visual inspection to correct occasional errors in early visual cortex.

There are several ways to construct the signed-distance functions. As a "gold standard," we utilize a method that formally solves the Eikonal equation [1]. This method has the drawback that it fails for regions where one or the other of the two bounding surfaces suffer from topological defects, typically manifest as inappropriate surface normals. Such defects seem to occur most frequently in the FreeSurfer generated pial meshes that define the outer surface of cerebral cortex. However, in two subjects we were able to find surfaces that were free of defects, and we used these as a point of comparison to two other methods that avoid the sensitivity to topological defects. In both methods, signs are assigned based on the segmentations (e.g., for S_{gw}, negative in the WM, positive elsewhere), while distances were calculated in two ways. First, distance magnitudes are obtained by calculating point-to-triangle distances [7] for each point in the volume. Second, we use a similar method where the distances were calculated using a nearest-neighbor Euclidean metric obtained quickly by binary search. We compare these two approaches to the "gold standard" to justify use of the point-to-triangle method.

We construct a self-reciprocal coordinate system in the space between the GM/WM interface and pial surfaces using a signed-distance interpolated between them. Let S_{gw} be the signed distance above the GM/WM interface (positive outside the surface, negative within) and S_p be the signed distance above the pial surface. We then define a normalized distance parameter w based on:

$$(1 - w)S_{gw} + wS_p = 0 \tag{1}$$

In most regions, the above equation is solved explicitly except where the denominator is close to zero ($<10^{-4}$). In these regions, Eq. (1) is solved directly using a root-solving algorithm [3].

This approach is similar to our previously published work [17], but offers two advantages. First, it avoids the computational intensity of the deformable surface evolution by substituting the much simpler algebraic solution of Eq. (1). Second, it utilizes a direct computation of signed-distance function that is based upon the segmentation, rather than the topology of the surface representations. This avoids the pitfalls of occasional topological defects that often occur in the pial surfaces. The new approach also permits sensible calculations of depth that extend beyond the two defining surfaces.

The normalized distance w is zero on the GM/WM interface, goes to unity at the pial surface, and forms a self-reciprocal depth coordinate in the vicinity of the GM that is normalized by the local GM thickness. We use the term "extra-pial" (EP) for the space between the pial surface and skull, which has values of $w > 1$ and includes the highly vascular sub-arachnoid space [6].

The normalized coordinate w is also used to obtain the physical thickness of the gray matter. First, we form $\vec{\nabla}w$ accurately by convolution with 5-point kernels in each cardinal direction. We then do ray tracing along $\vec{\nabla}w$, gradually stepping from the gray-white to the pial surface in small increments to form a series of trajectories (Fig. 1A). The tracing process is initialized along the inward surface normal from each

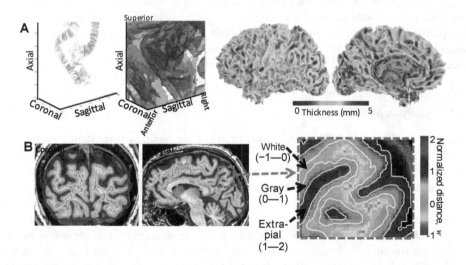

Fig. 1. (**A**) Thickness measurement. Left panel shows trajectories that measure thickness in calcarine sulcus; center shows same trajectories with gray-white and pial surfaces; right shows thickness map on gray-white surface. (**B**) Normalized depth map on coronal and sagittal slices for one subject. Coronal slice location is in posterior occipital lobe close to the functional slice prescription for HRF experiments. Sagittal slices are near the mid-sagittal plane to show the calcarine sulcus. Right: enlargement of sagittal slice showing normalized depth coordinates for white matter, gray matter, and extra-pial tissue.

node, and distances are obtained along each trajectory. Occasional topological incon-
sistencies are detected if the trajectory leaves the domain between the two surfaces (the
gray matter) too quickly (<0.25 mm). The terminal distance values obtained at the
endpoint of the trajectory are regridded onto the gray-white surface to create a physical
thickness map.

2.2   High Resolution fMRI HRF Measurements

A 1.7-s duration stimulus of flickering (4 Hz) randomly positioned dots is presented to
the subject to induce a hemodynamic response. This is followed by a 26.3-s blank
period to record the subsequent HRF. This 28-s duration trial is repeated 18 times in
each run. To improve temporal resolution, the stimulus onset is randomly jittered
(delayed) by 0, 0.5, or 1 s for every trial, and the analysis accounts for this jittering (see
below for details). Subjects are required to perform a demanding fixation-point
color-detection task throughout each run to control attention and encourage stable
fixation. Each scan typically consists of 4–5 such runs. Functional data is acquired
using a custom 7-channel surface coil array (ScanMed, Omaha NE) fabricated on a
flexible former so that it can be worn closely against the head. To obtain high-
resolution images (0.9-mm sampling), we use a three-shot (TR = 500 ms) spiral
acquisition [11, 27] acquiring eight 0.9-mm-thick slices to cover portions of early
visual areas (V1–3) and their adjacent WM and EP compartments every 1.5 s. A long
(9-ms) sinc excitation enables us to acquire such thin slices.

Visual-area boundaries are estimated using population receptive-field mapping
methods [5, 12] obtained for each subject in separate scans. The data are analyzed in
the portion of the prescription that sampled both in area V1, and in the combination of
areas V2 and V3 (V23) to sample extra-striate tissue. We aim to obtain 18 HRF
measurements (2 ROIs × 9 sessions) for six subjects.

2.3   Analysis of HRFs

The first 28-s-duration HRF is discarded to avoid various onset-transient effects. We
perform a slice-acquisition timing correction for the rest of functional data, and then
correct for subject motion using a robust intensity-based registration approach [23]
applied to the temporally smoothed (5-frame boxcar) version of the data. We next
transform the data into the volume anatomy, using the same intensity-based registration
method, and then correct for stimulus-onset jittering by cubic temporal interpolation.

We obtain four parameters from the HRFs. The peak amplitude is identified as the
global maximum of signal amplitude; the time-to-peak is the time at this peak.
Full-width-at-half maximum (FWHM) is the time span over which the signal exceeds
half of the peak amplitude. The time-to-offset is defined as the time at half maximum
after the hyperoxic peak.

We need to identify voxels in the WM and EP compartments that are in the vicinity
of the gray-matter ROI. However, to minimize partial volume effects, we choose voxels
within the WM and EP compartments that are spatially distinct from the GM. In the

WM, we use a depth range of −1.8 mm to −0.9 mm from GM/WM interface so the selected voxels are separated from the GM with at least one-voxel distance (0.9 mm). The WM voxels in this depth range are identified by computationally tracking ∇w from voxels of each gray-matter ROI. For the extra-pial compartment, we use the same approach with depth range 0.9–1.8 mm from pial surface.

We want to select strong responses without regard to the depth of their maximum response in each compartment. Therefore, we rank the responses by a depth-averaged contrast-to-noise ratio metric. First, the time-series data is averaged throughout the selected voxels in each compartment [17, 27]. We then calculate a contrast-to-noise ratio (CNR) by treating each HRF event time series as a vector. The mean HRF vector across events is computed, then the dot product is taken of each event with the mean to generate a univariate "contrast" value for each event [26]. The CNR is next estimated as the ratio of the mean contrast to its standard deviation. Based on the CNR values, we then choose the top 50% of the responses in each WM and EP compartment corresponding to each ROI in the GM compartment. The HRFs are obtained by spatially averaging the responses in this set of voxels in each compartment and temporally averaging over the many stimulus events.

The GM HRFs are calculated in an analogous fashion. First, responses are depth averaged over a normalized depth range of 0.2–0.8. Then, we again select the top 50% voxels based on CNR.

To permit comparison of responses in the WM and EP compartments to responses in the GM compartment in a fashion that removes session-to-session variations in mean gray-matter response amplitude, we normalize the WM and EP HRFs with the peak amplitude of their corresponding GM HRFs.

To test how WM and EP HRFs vary with GM HRFs, we obtain linear correlation coefficients, R, of peak amplitude and time-to-offset between the GM compartment and the WM and EP compartments. Positive correlations provide evidence that similar hemodynamic mechanisms underlie HRFs in WM and EP compartments.

We examine HRF variations as a continuous function of normalized depth. First, we select voxels that encompass the top 50% response in the GM ($0.2 \leq w \leq 0.8$) to obtain a depth-averaged ROI. This ROI is then extended into a depth range from the WM (−0.5–0) through the GM (0–1) and into the EP compartment (1–1.5) by tracking from each voxel of the ROI. Normalized depth bins of width 0.25 (one quarter of local gray-matter depth, mean thickness 0.58 mm) were defined on the range of −0.5 to 1.5 at increments of 0.1 to generate depth profiles of HRF parameters.

The noise in fMRI data is known to have a non-Gaussian distribution [14, 20]. We therefore use bootstrapping to estimate the distribution within each scan [8, 9]. Time-series from the many individual trials within a scan (typically 85) are resampled with replacement, and then averaged together. This procedure is repeated 2,000 times to estimate the statistical distribution of amplitude values at each time point of the HRF. A similar procedure is used to estimate the distributions for each of the HRF parameters. We then measure 68% confidence intervals for each parameter from their corresponding distributions. Note that this scheme implicitly accounts for multiple comparisons because it obtains the distribution based upon the entire sample set. We define the mean difference between the confidence intervals and the signal as the "variability," equivalent to the standard-error-of-the-mean for normally distributed data.

Absolute values of the HRFs and their parameters varied substantially across sessions. To permit averaging across sessions, we normalize the HRF parameters from each subject by dividing by the parameter's mean through the gray-matter normalized depth range of 0.2–0.8. After averaging across sessions, we "denormalize" by multiplying the result by the mean depth-averaged parameter in GM across sessions, thus returning the data to physically meaningful units.

3   Results

3.1   Normalized Depth Map

The 3-D signed-distance depth-mapping scheme provides unique local trajectories from one point on the GM/WM interface to another point on the pial surface (Fig. 1A). Thickness maps generated from the trajectories show the expected range of 1.5–4.5 mm, with thinner gray matter in the sulci and thicker gray matter on the gyri. The normalized distance map shows a smooth progression between the bounding surfaces, and nicely delineates the WM, GM and EP compartments (Fig. 1B).

The errors produced by our choice of distance metric by comparison to more accurate solutions are shown in Fig. 2. We separate our comparisons to inside and outside the gray matter. Inside the gray matter, there was excellent agreement between

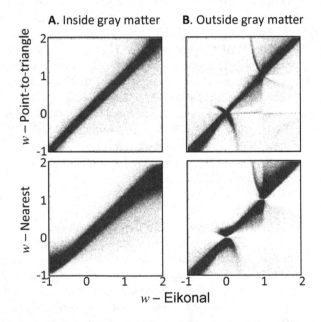

Fig. 2. Comparisons of signed-distance calculations obtained by different methods. The Eikonal-equation method was taken as the "gold standard", and differences provide an estimate of errors. (**A**) Comparisons inside the gray matter show small errors for both the point-to-triangle (upper) and nearest neighbor (lower) methods. (**B**) Comparisons outside the gray matter show larger errors for both methods, but they are still acceptably small for the point-to-triangle method.

the point-to-triangle method (root-mean-square 0.024; median 0.008 mm). Error is only somewhat worse for the nearest-neighbor approach (RMS 0.052; median 0.026 mm), with the largest errors near the bounding surfaces. So, even this very fast and simple approach can be satisfactory. However, outside the gray matter, the errors are larger, rising to RMS 0.064, median 0.011 mm for the point-to-triangle method, and RMS 0.092, median 0.034 mm for nearest neighbor. Generally, the point-to-triangle approach seems a good compromise and was used for the HRF depth calculations shown below.

3.2 HRFs in White Matter, Gray Matter and Extra-Pial Compartments

Results for all subjects confirm the existence of reliable HRFs in the WM (blue lines), and EP (red) compartments (Fig. 3). These HRFs exhibit reliable hyperoxic peak amplitudes for all 18 measurements.

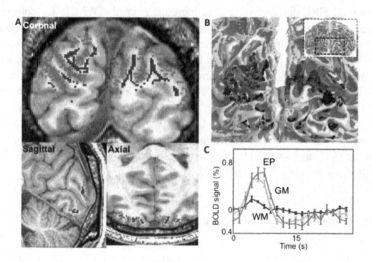

Fig. 3. (A) Example of the region for top 50% of strong response voxels in each compartment on coronal, sagittal and axial slices: WM (blue), GM (green), EP compartments (red). To avoid partial volume effects, the voxels in the WM and EP compartments are separated by at least 0.9 mm from GM/WM interface and pial surface respectively. (B) Overlaid selected voxels in each compartment onto 3D GM/WM interface. (C) Mean time series of the selected region in each compartment. Error bars are 68% confidence intervals obtained using a bootstrapping scheme. (Color figure online)

The HRF in each compartment shows the stereotypical initial delay and hyperoxic peak, but their amplitude and temporal characteristics are noticeably different among the compartments; 68% confidence intervals are marked to show time-series variability. HRF measurements for each compartment and each ROI show their considerable variability across measurements.

We normalize these HRFs with the peak amplitude of the corresponding GM HRF across 18 measurements in each compartment. The means of the parameters in the GM across measurements (green shaded bars) are $0.9 \pm 0.3\%$ (peak amplitude), 6.1 ± 1.1 s (time-to-peak), 5.4 ± 0.6 s (FWHM), and 8.9 ± 1 s (time-to-offset). The normalized WM HRFs show similar relative peak amplitude, but following the hyperoxic peak, the HRFs are comparatively weak and noisy. The normalized EP HRFs are stronger but much more variable than the WM HRFs. Relative peak amplitudes are often larger than those observed in the GM, but show large measurement-to-measurement variation. The GM HRFs, after amplitude normalization, show much greater temporal stability across measurements. However, the HRFs in all compartments show much greater variability at later times following the hyperoxic peak.

Peak amplitude increases monotonically from WM through GM to EP compartments. Normalized hyperoxic peak amplitudes in the WM are significantly weaker than in the GM ($p < 0.01$) but still very reliable ($0.35 \pm 0.1\%$), while peak amplitude ($1.08 \pm 0.35\%$) is stronger (20%) in the EP compartment than the GM ($p = 0.05$).

All temporal HRF parameters tend to increase from the WM to EP compartments. Time-to-peak goes from 5.5 ± 1 s in the WM, to 6.1 s in the gray matter, and 6.4 ± 0.9 s in the EP compartment. Time-to-peak in the WM differs significantly from that observed in both GM ($p = 0.01$) and EP ($p < 0.01$) compartments. Time-to-peak in the EP compartment shows only a trend to be slower than in the GM ($p = 0.25$). FWHM is 4.1 ± 1.1 s in the WM, 5.4 s in the GM, and 6.4 ± 0.9 s in the EP compartment. FWHM increases significantly both from WM to GM ($p < 0.01$) and from GM to EP ($p = 0.03$). The time-to-offset is the most reliable and distinctive temporal HRF parameter between the compartments. It increases from 7.9 ± 1.4 s in the WM, to 8.9 s in the GM, and 9.8 ± 0.9 s in the EP compartment. Differences are significant between all three compartments ($p \leq 0.01$).

3.3   Depth-Dependent Analysis of the HRFs

We present a continuous depth analysis from the WM to EP compartments ($-0.5 \leq w \leq 1.5$) for the peak amplitude. Individual measurements of the HRF are highly reliable throughout this depth range. Example depth profiles of peak amplitude (Fig. 4A) increase strongly from the WM through the GM, and reach a peak within the GM (cyan & magenta) or in the EP compartment (green & gray). Generally, peak amplitudes have small and consistent confidence intervals across most of the depth range. Depth profiles of peak amplitude (Fig. 4B) increase strongly from the WM through the GM, and reach a peak within the GM or in the EP compartment. The measurements begin to diverge in the superficial GM, and show strong variation in the extra-pial compartment. This behavior is quantified by the standard deviation of the peak-amplitude reliability across measurements (red-dashed lines in Fig. 4B). Generally, peak amplitudes have small and consistent confidence intervals across most of the depth range. We have termed the width of these confidence intervals as variability, which is plotted in Fig. 4C. Variability for the examples (colored lines) is very low across the depth range. In fact, variability for most measurements (gray lines) is low

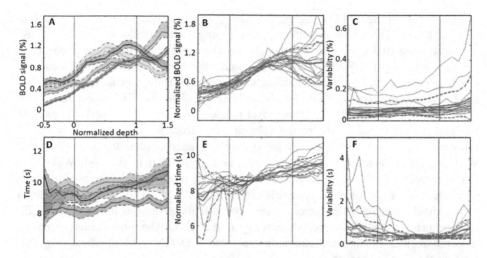

Fig. 4. Depth analysis for HRF measurements. (A) Examples (colored lines) of variation of depth profiles for peak amplitude. Shaded regions show 68% confidence interval within each example. Depth profiles of (B) normalized peak amplitudes, and (C) their variabilities. (D) Examples of variation of depth profiles for time-to-offset. (E) normalized time-to-offset, and (F) their variabilities. The measurement throughout the depth range are normalized by mean of corresponding parameter in the GM (gray lines): mean (red) and standard deviation (dashed red) across measurements. (Color figure online)

both within and adjacent to the GM. Variability is somewhat larger in the EP compartment.

Depth profiles of time-to-offset, the most reliable temporal HRF parameter, are shown in Fig. 4D–F. Depth profiles exhibit a monotonic increase from deep to superficial GM (Fig. 4D–E). The slopes of the depth trends are consistent only in the GM compartment, and become particularly variable in the WM. Variability of the examples (Fig. 4F, colored lines) is lowest within the GM, particular in the superficial GM; most of measurements also show similar depth trends (gray lines). Variability is somewhat larger in the deeper WM, and EP compartments.

4   Discussion

High-resolution fMRI, aligned with a carefully segmented high-resolution structural reference volume, is used to quantify depth-dependent variations in the HRF at 3T. Our 3D depth-mapping scheme enable to delineate white matter and extra-pial compartments from gray matter resulting in a detailed characterization of the HRF in the vicinity of human early visual cortex.

The HRFs in the WM and EP compartments have not been fully characterized. In fact, signals from these regions are usually assumed to be caused by either partial volume effects, or as noise created by vascular functions that are unrelated to a given task or stimulus. Therefore, most depth-dependent HRF studies have not examined the signals outside of the GM [18, 25, 29, 30]. Here, we show the existence of the HRF in the WM and EP compartments by combining two methods. First, we use high-resolution (0.9-mm) fMRI. Although we average signals from many voxels to obtain a meaningful HRF in each compartment, the high-resolution fMRI data still provides good depth resolution, enabling the separation of the signals in the WM and EP compartments from the GM. To obtain HRFs in WM and EP compartments, we only use voxels at least 0.9-mm (our voxel resolution) away from GM/WM interface and pial surface to reduce the partial volume effect from the GM. Second, we use a self-reciprocal, single-valued coordinate system that enables identification of apposed WM and EP compartments that correspond to a given ROI in the GM. Within the folded complexity of the GM, spatial relationships can be obscure, so the establishment of a self-reciprocal coordinate system is critical [15, 17, 32]. The use of such a depth metric should provide better correspondence of vascular responses in the WM and EP compartments to stimulus-evoked activity in the GM. The projections of strong response regions in all three compartments onto 3-D GM/WM interface (Fig. 2B) confirm their spatial registration.

Generally, the HRF is weaker in the WM: peak amplitude $\sim 39\%$ of that observed in the GM, but normalized peak amplitudes are still reliable ($0.35 \pm 0.1\%$). The weaker peak amplitudes may reflect the much lower vascular density of the WM [6, 21], or that blood flow to the WM is less sensitive to neural activity in the apposed GM. HRF responses exhibit slightly faster temporal delay parameters (time-to-peak and time-to-offset) in the WM than in the GM. The WM HRF shows early temporal behavior that is similar to that seen in the GM HRF, which indicates that the WM may be driven by the same stimulus-evoked blood flow dynamics as the GM. This finding is consistent with previous WM HRF measurements performed at conventional resolution [13].

The HRF peak amplitude is stronger and more variable in the EP compartment than the GM. Our results are thus consistent with analyses that emphasize the confounding effects of BOLD responses from these superficial vascular structures [18, 31]. The substantial spatial variation of the pial vascular density along the cortical surface [6] may contribute to the large variation of the peak amplitude in the EP compartment. HRF temporal delay parameters (time-to-peak and time-to-offset) in the EP compartment are relatively reliable and the timings are slightly slower than the GM, which is consistent with previous measurements outside the GM [13]. The combination of their strong variability, spatial separation from the neural activity in the GM, yet good temporal synchronization with the GM signals indicates that these EP-compartment signals will both degrade and confound fMRI experiments carried out at conventional spatial resolutions.

We performed continuous depth analysis throughout WM, GM and EP compartments ($-0.5 \leq w \leq 1.5$). For the peak amplitude, all measurements show significant monotonic depth trends from deep to superficial GM, which is consistent with previous studies [18, 29, 30]. The depth profiles of the deeper activity, from superficial WM to central GM, are relatively consistent among the measurements. However, there are substantial measurement-to-measurement variations for the peak amplitude in the superficial GM and EP compartment. Thus, the results indicate that measurement-to-measurement variation is much larger than the variability within each measurement, producing large depth-trend variations in the superficial GM and EP compartments. This can be substantial spatial variation of the superficial pial vascular density across [6]. Session-to-session variations of the slice prescription will sample the pial surface differently, and thereby contribute to these measurement-to-measurement variations. This could directly cause the EP-compartment variations and also impact measurement of the superficial GM through partial-volume effects.

Although there are some variations of absolute timing for time-to-offset, normalized depth profiles confirm significant linear trends across measurements (1.12 ± 0.63 s/cortical thickness). The variability within each measurement is lowest in the central to superficial GM resulting in the largest temporal reliability. This depth range usually has high vascular density [2, 6, 21], which may suggest particularly tight, regional neurometabolic and neurovascular coupling in the central-superficial GM.

Altogether, we use signed-distance depth-mapping scheme with high-resolution fMRI to characterize depth-dependent HRFs and demonstrate distinct differences between HRF parameters in the WM, GM and EP compartments. HRF peak amplitudes in the WM and EP compartments are statistically significant, and both amplitude and timing parameters are tightly coupled with corresponding GM HRF parameters, suggesting that HRFs in all compartments are stimulus evoked. Our data demonstrate and quantify BOLD fMRI HRFs not only in the GM, but also in apposed WM and EP compartments. Accordingly, high-resolution fMRI coupled with these depth-analysis methods can be useful to separate the desirable GM signals from nearby confounding signals.

Acknowledgements. We thank Evan Luther, Andrew Floren, and Clint Greene for assistance with experiments and analysis procedures. This work was supported by NIH R21HL108143.

References

1. Bajaj, C.L., Xu, G.-L., Zhang, Q.: Higher-order level-set method and its application in biomolecular surfaces construction. J. Comput. Sci. Technol. **23**(6), 1026–1036 (2008)
2. Cassot, F., et al.: A novel three-dimensional computer-assisted method for a quantitative study of microvascular networks of the human cerebral cortex. Microcirculation **13**(1), 1–18 (2006)
3. Chapra, S., Canale, R.: Numerical Methods for Engineers, vol. 9, 6th edn. McGraw-Hill, New York (2009)
4. De Martino, F., et al.: Cortical depth dependent functional responses in humans at 7T: improved specificity with 3D GRASE. PLoS ONE **8**(3), e60514 (2013)
5. Dumoulin, S.O., Wandell, B.A.: Population receptive field estimates in human visual cortex. NeuroImage **39**(2), 647–660 (2008)
6. Duvernoy, H.M., Delon, S., Vannson, J.: Cortical blood vessels of the human brain. Brain Res. Bull. **7**(5), 519–579 (1981)
7. Eberly, D.: Distance between point and triangle in 3D. Magic Software (1999). http://www.magic-software.com/Documentation/pt3tri3.pdf
8. Efron, B., Tibshirani, R.J.: Bootstrap methods for standard errors, confidence intervals, and other measures of statistical accuracy. Stat. Sci. **1**(1), 54–75 (1986)

9. Efron, B., Tibshirani, R.J.: An Introduction to the Bootstrap. CRC Press, Boca Raton (1994)
10. Fischl, B., Dale, A.M.: Measuring the thickness of the human cerebral cortex from magnetic resonance images. Proc. Natl. Acad. Sci. U.S.A. **97**(20), 11050–11055 (2000)
11. Glover, G.H.: Simple analytic spiral K-space algorithm. Magn. Reson. Med. **42**(2), 412–415 (1999). Official Journal of the Society of Magnetic Resonance in Medicine/Society of Magnetic Resonance in Medicine
12. Greene, C.A., et al.: Measurement of population receptive fields in human early visual cortex using back-projection tomography. J. Vis. **14**(1), 17 (2014)
13. Hall, D.A., et al.: A method for determining venous contribution to BOLD contrast sensory activation. Magn. Reson. Imaging **20**(10), 695–706 (2002)
14. Holmes, A., et al.: Statistical modelling of low-frequency confounds in fMRI. NeuroImage **5**, S480 (1997)
15. Jones, S.E., Buchbinder, B.R., Aharon, I.: Three-dimensional mapping of cortical thickness using Laplace's equation. Hum. Brain Mapp. **11**(1), 12–32 (2000)
16. Kamitani, Y., Tong, F.: Decoding the visual and subjective contents of the human brain. Nat. Neurosci. **8**(5), 679–685 (2005)
17. Khan, R., et al.: Surface-based analysis methods for high-resolution functional magnetic resonance imaging. Graph. Models **73**(6), 313–322 (2011)
18. Koopmans, P.J., Barth, M., Norris, D.G.: Layer specific BOLD activation in human V1. Hum. Brain Mapp. **31**(9), 1297–1304 (2010)
19. Koopmans, P.J., et al.: Multi-echo fMRI of the cortical laminae in humans at 7T. NeuroImage **56**(3), 1276–1285 (2011)
20. Kruger, G., Glover, G.H.: Physiological noise in oxygenation-sensitive magnetic resonance imaging. Magn. Reson. Med. **46**(4), 631–637 (2001)
21. Lauwers, F., et al.: Morphometry of the human cerebral cortex microcirculation: general characteristics and space-related profiles. NeuroImage **39**(3), 936–948 (2008)
22. Menon, R.S., Kim, S.-G.: Spatial and temporal limits in cognitive neuroimaging with fMRI. Trends Cogn. Sci. **3**(6), 207–216 (1999)
23. Nestares, O., Heeger, D.J.: Robust multiresolution alignment of MRI brain volumes. Magn. Reson. Med. **43**(5), 705–715 (2000). Official Journal of the Society of Magnetic Resonance in Medicine/Society of Magnetic Resonance in Medicine
24. Ogawa, S., et al.: Intrinsic signal changes accompanying sensory stimulation: functional brain mapping with magnetic resonance imaging. Proc. Natl. Acad. Sci. U.S.A. **89**(13), 5951–5955 (1992)
25. Polimeni, J.R., et al.: Laminar analysis of 7 T BOLD using an imposed spatial activation pattern in human V1. NeuroImage **52**(4), 1334–1346 (2010)
26. Ress, D., Backus, B.T., Heeger, D.J.: Activity in primary visual cortex predicts performance in a visual detection task. Nat. Neurosci. **3**(9), 940–945 (2000)
27. Ress, D., et al.: Laminar profiles of functional activity in the human brain. Neuroimage **34**(1), 74–84 (2007)
28. Seiyama, A., et al.: Circulatory basis of fMRI signals: relationship between changes in the hemodynamic parameters and BOLD signal intensity. Neuroimage **21**(4), 1204–1214 (2004)
29. Siero, J.C.W., et al.: Cortical depth-dependent temporal dynamics of the BOLD response in the human brain. J. Cereb. Blood Flow Metab. **31**(10), 1999–2008 (2011)
30. Siero, J.C.W., et al.: BOLD specificity and dynamics evaluated in humans at 7 T: comparing gradient-echo and spin-echo hemodynamic responses. PLoS ONE **8**(1), e54560 (2013)
31. Turner, R.: How much cortex can a vein drain? Downstream dilution of activation-related cerebral blood oxygenation changes. NeuroImage **16**(4), 1062–1067 (2002)
32. Waehnert, M., et al.: Anatomically motivated modeling of cortical laminae. Neuroimage **93**, 210–220 (2014)

A Study of Children Facial Recognition for Privacy in Smart TV

Patrick C.K. Hung[1,2(✉)], Kamen Kanev[1,3,4], Farkhund Iqbal[5],
David Mettrick[1], Laura Rafferty[1], Guan-Pu Pan[2], Shih-Chia Huang[1,2],
and Benjamin C.M. Fung[6]

[1] Faculty of Business and Information Technology,
University of Ontario Institute of Technology, Oshawa, Canada
{patrick.hung,david.mettrick,laura.rafferty}@uoit.ca,
kanev@rie.shizuoka.ac.jp
[2] Department of Electronic Engineering,
National Taipei University of Technology, Taipei, Taiwan
{t105369012,schuang}@ntut.edu.tw
[3] Graduate School of Science and Technology,
Shizuoka University, Hamamatsu, Japan
[4] Lassonde School of Engineering, York University, Toronto, Canada
[5] College of Technological Innovation, Zayed University, Dubai, UAE
farkhund.iqbal@zu.ac.ae
[6] School of Information Studies, McGill University, Montreal, Canada
ben.fung@mcgill.ca

Abstract. Nowadays Smart TV is becoming very popular in many families. Smart TV provides computing and connectivity capabilities with access to online services, such as video on demand, online games, and even sports and healthcare activities. For example, Google Smart TV, which is based on Google Android, integrates into the users' daily physical activities through its ability to extract and access context information dependent on the surrounding environment and to react accordingly via built-in camera and sensors. Without a viable privacy protection system in place, however, the expanding use of Smart TV can lead to privacy violations through tracking and user profiling by broadcasters and others. This becomes of particular concern when underage users such as children who may not fully understand the concept of privacy are involved in using the Smart TV services. In this study, we consider digital imaging and ways to identify and properly tag pictures of children in order to prevent unwanted disclosure of personal information. We have conducted a preliminary experiment on the effectiveness of facial recognition technology in Smart TV where experimental recognition of child face presence in feedback image streams is conducted through the Microsoft's Face Application Programming Interface.

Keywords: Smart TV · Digital imaging · Face recognition · Picture-based age estimation · Privacy

© Springer International Publishing AG 2017
R.P. Barneva et al. (Eds.): CompIMAGE 2016, LNCS 10149, pp. 229–240, 2017.
DOI: 10.1007/978-3-319-54609-4_17

1 Introduction

Nowadays Smart TV is becoming very popular in many families. Smart TV provides computing capabilities and connectivity to online services, such as on demand streaming media, interactive media, and even mobile healthcare applications [32]. For example, Google Smart TV, which is based on Google Android, can be automated, in particular, to perform complex computing applications [8, 32]. Smart TV establishes itself into the users' daily physical activities with ability to perceive context information on the surrounding environment in order to react accordingly [24]. Perception is carried out through sensors on the device such as a microphone, camera, or accelerometer. With respect to such context information, perception is fundamental to the device's ability to make timely and context-sensitive decisions. In this regard issues of the Smart TV's social context have been discussed in [17] touching upon: (i) how people would perceive the area where Smart TV is being used, (ii) what expectations for privacy in a room with Smart TV would users have, and (iii) how other people within viewing range of the Smart TV would feel.

While Smart TV is becoming increasingly integrated with various mobile devices, it is certainly bringing new privacy and security risks to its users. Gigglier et al. [11, 12], for example, demonstrate how Smart TV can be used by broadcasters and neighbors to track users if no privacy protection system is instated. Obviously data that is collected to personalize the experience with Smart TV often contains sensitive information that needs to be kept private from unwanted third parties. So, information privacy is indeed of great concern to many users who are becoming increasingly worried about how their personal data is being collected and managed by Smart TV. Personal behavioral data is particularly sensitive, as it can be used to infer a significant amount of private information about a user, such as movement and lifestyle patterns, workplace behavior, and others. Thus, it is important to account for the specific privacy concerns, laws, and regulations in respect to the Smart TV use. This becomes of particular concern when underage users such as children who may not fully understand the concept of privacy are involved in using the Smart TV services. In this study, we consider digital imaging and ways to identify and properly tag children pictures in order to prevent unwanted disclosure of personal information. We conducted a preliminary experiment on the effectiveness of facial recognition technology in Smart TV where experimental recognition of child face presence in feedback image streams is conducted through the Microsoft's Face Application Programming Interface (API). The rest of this paper is organized as follows. Section 2 provides an introduction and discussion of related works. Section 3 presents the concepts of privacy policies and context awareness. Section 4 outlines the face recognition experimental results. And Sect. 5 concludes the paper and outlines future work.

2 Related Works

Face recognition is commonly based on pattern recognition approaches applied to typical scenarios as follows. Face images of different people are collected and registered in a database of a face recognition system. When a test image is presented for recognition, the system compares and decides whether a good match to any of the database images could be found. Anggraini [1], for example, has proposed a method that uses Principal Component Analysis (PCA) and Self-Organizing Maps (SOM) for both feature extraction and clustering. The feature extraction can obtain the characteristic of a human, and clustering can gather a set of objects in the same class. Next, Horiuchi and Hada [14] have discussed a face recognition technology that accounts for face changes over the years, for various angles of photo taking, for changes of facial expressions, and for wearing accessories which is suitable for recognition accuracy evaluations with respect to criminal investigations. Huang and Chen [15] have used the Local Vector Pattern (LVP) feature to calculate the distances between an input face image and all enrolled face images, and thus determine the best possible face candidates. Then, a feature-point Bilateral Recognition (BR) program generates the final face recognition result. The face recognition becomes even more challenging problem in the case of streaming video with partial occlusions. Ragashe et al. [23], for example, have applied Canny edge, Viola-Jones Face Detection, and AdaBoost learning algorithm to boost the face recognition accuracy for different patterns of face expressions. Further on, Soldera et al. [28] have proposed a method that can handle sparse and dense face image representations. The sparse representation helps compensating for landmark location uncertainties during face image feature extraction using interpolated landmarks, and thus provides for improved accuracies with high-resolution color face images. Xi et al. [31] have proposed a method for face recognition involving deep learning, named Local Binary Pattern Network (LBPNet), which achieves high recognition accuracy without employment of costly model learning approaches on massive data. With respect to the employment of face recognition to interact with Smart TV systems some works need special mention. Lee et al. [18], in particular, have devised a face recognition system for viewer authentication, and Nguyen et al. [21] have proposed and experimented with a gaze detection method based on head pose estimation in a smart TV.

While most users appreciate the value of targeted services in Smart TV, they still express concern over how their data is collected and managed without their knowledge. For example, Cherubini et al. [5] has identified privacy as a barrier to the adoption of mobile phone context services. About 70% of the consumers have stated that it is important to know exactly what personal information is being collected and shared [19], while 92% of the users have expressed concern about applications collecting personal information without their consent [10]. Mobile applications have adapted a countless number of services to better analyze context data and to provide custom services that will be of higher value to users, based on what they are most likely to need. While allowing context data to be collected for services can prove to be of great benefit to users, there is always a tradeoff between utility and privacy [4] so many users do not even know that sensitive information might be stored in some cases even after they explicitly declined to give a permission. This clearly goes against the privacy principle of obtaining user consent before collecting such information. Other discrepancies and possible privacy violations occur when unnecessary large volumes of detailed and highly accurate data is collected and stored for longer than needed periods of time. Such discrepancies often take place because most of the permission details with respect to information collection are buried in lengthy, default-enabled policies that users can hardly understand.

With respect to the practical employment of face recognition in Smart TV applications, the Samsung platform provides for various good references. Faces, for example, can be saved to "Samsung Accounts" which can be used to sign into personalized apps such as Skype and Facebook. The Samsung's privacy policy regarding the facial recognition feature is that it can be used as a supplementary security measure in addition to passwords. We have also noted the Smart TV "Kids Service" which is supposed to filter and tune streamed content making it suitable for children, although for now it appears not to be directly linked to the facial recognition feature (http://www.samsung.com/uk/info/privacy-SmartTV.html). To complement the review we have summarized in a table format some representative mobile face recognition apps on the Android (Table 1) and iOS (Table 2) platforms.

Table 1. Google Play Apps.

App name	Developer name	Description
Face recognition	Lakshmanan Anbalagan	Experimental app using OpenCV's face recognition functions. Reviews claim it works poorly
Face recognition	SeakLeng	Facial recognition software capable of being trained to better recognize a person
AppLock face/voice recognition	Sensory TrulySecure	Facial and voice recognition technologies applied to make unauthorized access to an Android device more difficult
Face recognition-FastAccess	Sensible Vision, Inc.	Facial recognition technology applied to authorize access to saved passwords on Android devices

Table 2. Apple App Store.

App name	Developer name	Description
Face recognition FastAccess	Sensible Vision Inc.	Facial and voice recognition technologies applied to make unauthorized access to an iPhone device more difficult
BioID facial recognition authentication	BioID GmbH	Facial recognition used as a biometric security feature for accessing iPhone devices

Our review of related works has touched upon the current state of the art in face recognition with respect to Smart TV and its implications on user privacy. Face recognition is clearly an important topic that attracts a lot of attention and active research. None of the related works, however, have considered the specific issues of face-based age discrimination for enhancement of child face recognition and addressing related privacy concerns in the context of Smart TV which are the focus of our current work.

3 Privacy Policies and Context Awareness

Privacy protection is often addressed by adopting privacy policies as ways to communicate to end users how their data is collected, managed, shared, and retained, although there are still issues with the effective enforcement of such privacy policies. A privacy policy should include a standard description of what information is collected from users, for how long the information will be retained, what the information will be used for, whether and how the information will be shared with third parties and so on. However, to the best of our knowledge, there is no current standardization effort for a privacy protection policy with respect to Smart TV. Google Android based Smart TV users constitute a large segment of the consumer population and are of particular interest to market researchers that collect their personal data and usage patterns for targeted advertising [25, 27]. Third party advertisers can further infer additional person-related knowledge based on context information and thus build detailed behavioral profiles that may be used for unknown or unwanted purposes. Personal data can come in many forms including browsing history, friends list, location information, etc. Other examples of relevant context information [26] may include verbal context, roles of communication partners, goals of the communication and involved individuals in respect to the social environment, as well as spatial, chemical and other characteristics of the physical surroundings. Gathered information may seem trivial and often may not be perceived as particularly sensitive by the user, but in practice, when properly processed it may actually reveal a lot of important personal details.

The World Economic Forum [30] defines three types of context data, as categorized by the way it is collected, namely *volunteered*, *observed*, and *inferred*. Volunteered data is explicitly provided by the user and can include personal profile information or preference settings. Observed data on the other hand is not directly given by the user- it is rather gathered through device sensors such as GPS, camera, clock, etc. And finally inferred data is programmatically generated by analyzing volunteered and/or observed data, e.g., where a user is likely to be going based on typical behavior, etc. A lot of user sensitive information can be created through data analytics and inferences and, since it is neither volunteered nor observed (and thus not directly obtained), it may be difficult to enforce its full disclosure to the user. On the one hand, inferences on volunteered and observed data can make predictions with very high levels of certainty that may raise safety concerns, e.g., predictions of trip destinations and timing based on driving habits [7]. In Smart TV, in particular, personalized services are provided to the user based on context data collected and inferred through embedded sensors and other environment data both volunteered and observed. In addition, Smart TV often involves a networked environment, which introduces further user privacy and security concerns, particularly related to the context information the Smart TV is processing.

Considering the amounts of collected information, tradeoffs between disclosing sensitive data and receiving context-aware services in Smart TV are often required.

Obviously, in order to provide highly relevant services to the user, more and more personal and context information has to be collected, which however raises concerns of privacy. For example, a service in Smart TV can send special promotions and coupons to users depending on what is most relevant to them. In order to identify the most relevant promotions, the service will need to collect certain context data such as user locations, and also profile information such as age and gender to help determine what their interests might be. The application may even collect and retain historical data on the users such as previous movement patterns to determine where they are likely to be at certain times, or earlier interactions with the application to confirm interest in certain previously served promotions. In this example, it is clear that the more information is collected on the users, the more relevant services can be provided to them but nevertheless users may not be comfortable with the level of data collected and inferred on them. An application knowing where you are and what you are likely to be doing at any given time raises not only privacy but also security concern. That is why context data is in the core of privacy concerns with respect to Smart TV's applications.

Privacy goals must, therefore, be defined to ensure that sensitive data is indeed managed in a responsible and secure way. Detailed analysis should be carried out to identify and establish a security policy to ensure that the sensitive details of the user behavior cannot be derived from such retained data. While there are different approaches which aim to preserve the privacy of sensitive context data in general, in the following sections we will discuss the specific types of data that must be considered when evaluating the scope of privacy with respect to Smart TV. Generally speaking Smart TV's applications must operate in a controlled environment and must protect data and resources from other untrusted applications that may be running on the device and could, intentionally or unintentionally, violate the safety policy. When it comes to any information technology, privacy and security are at the core of ensuring that goals are achieved effectively and without compromise of personal data. The three concerns of security are *confidentiality*, *integrity*, and *availability*. Confidentiality means that access to information is restricted only to intended parties. Integrity means that data is accurate and consistent and has not been tampered with, while availability means that resources and data remain available when needed by the legitimate parties. A foundation of a security mechanism is required for privacy enforcement.

Information privacy is defined by Hung and Cheng [16] as "an individual's right to determine how, when, and to what extent information about the self will be released to another person or to an organization." In particular, personally identifiable information is any type of information that can be linked to an individual, including their activities, preferences, history, conversations, etc. In a Smart TV environment, personally identifiable information is also likely to be gathered from context data, as described in the previous section. Information privacy goals can be achieved through privacy preserving mechanisms such as access control, privacy policies, and privacy preferences. User sensitive data leakage occurs when user related information, including such derived or inferred from the Smart TV context, is exposed to individuals who are not supposed to have access to it. For the purpose of this research, we assume that although the information disclosure rules might have been stated in the privacy policy, and the users might have provided their consent, they might still not be aware of all the subtle details and thus might lack full understanding. Indeed, while consent might be provided by the user either implicitly or explicitly, in many cases such consent is just implied by using the services. As for explicit consent, it can be given if, for example, the user is required to click "I agree" in regards to the privacy policy terms and conditions in order to receive services. Another related issue is content unawareness which occurs when users are not aware of the context related information that is collected on them, for example their location information. In fact, the IETF's RFC6973 on Privacy Considerations [6] clearly identifies information disclosure and content unawareness as specific secondary threats of high importance.

Different countries and legislations have different laws for privacy protection, and there are also many international guidelines and industry regulations which outline privacy best practices. These laws and regulations can also differ depending on what type of information is being collected (e.g. health information), or who the users are (e.g. children under the age of 13). In Taiwan, the Personal Information Protection Act (PIPA) regulates the collection, processing and use of personal information and personal data from citizens or non-citizens by government and non-government entities. The PIPA requires notification before personal information is collected, processed or used. On the other side, Canada's privacy laws are outlined in the Personal Information Protection and Electronic Documentation Act (PIPEDA) [2], which governs how personal information can be collected, used, and disclosed in commercial business. PIPEDA is based on the 10 principles of privacy outlined in the Canadian Standards Association's (CSA) Model Code for the Protection of Personal Information [13], which has been recognized as a national standard as of 1996 [3]. This model code is representative of principles behind privacy legislation in many countries, including the United States and the European Union. It also bears similarities to the Organization for Economic Cooperation and Development (OECD) Guidelines for the Protection of Privacy and Trans-border Flows of Personal Data [22] which have been adopted by member countries of the European Union [9]. In respect to privacy, this research focuses on a preliminary study with a technical framework for children facial recognition in Smart TV.

Referring to Fig. 1, the children (users) may interact with Smart TV services possibly with Smart Phones in a physical and social environment. Usually Smart TV is equipped with different components, such as a microphone, camera, and sensors. As Smart TV is able to collect a variety of data such as text, picture, video, sound, location, and sensing data, this makes the context of Smart TV far more complex than traditional televisions when employed by users of different ages in varying physical and social environments. The context data collected by Smart TV may include information sufficient to identify and track individual users based on their physical and social environment (e.g., location, human faces, sound, environmental conditions and infrastructure, etc.). Since identifying and tracking children is of particular concern, in this paper, we propose a novel type of parental control that should be integrated in Smart TV services to allow parents manage and restrict the information about their children collected by the Smart TV. This is being implemented through a rich data visualization model supporting text, picture, video, sound, location, and sensing data, etc. in the form of a dashboard for parental control through the common Application Programming Interface (API) for Smart TV services. The dashboard serves as an integration, validation, and visualization tool to investigate and identify privacy policies, preferences, and rules which apply to Smart TV.

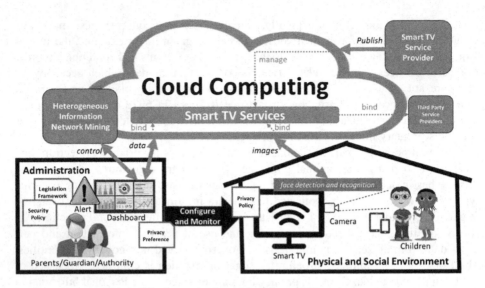

Fig. 1. The conceptual model of Smart TV environments.

The involvement of a trusted third party (legal authority), who controls the enforcement of the privacy rules in accordance with the United States Federal Trade Commission Children's Online Privacy Protection Act (COPPA) is also being considered [29]. COPPA protects the online privacy of children under the age of 13 and stipulates that child's personal information cannot be collected without parental consent. In 2010, an amendment to COPPA further elaborated that personal information includes geo-location information, photographs, and videos. The proposed rich data visualization model will therefore provide parents and guardian with an efficient and easy to understand dashboard representation of such information and related activities of their children. Such a dashboard should support the following major functionalities:

- Privacy Preferences: An authorized user should be able to define preferences for how and which of their data will be collected, shared, retained, and so on. Such preferences could then be incorporated in the privacy policy of Smart TV by generating a set of privacy rules in a knowledge database. The rules relate active subjects and operations that they perform to objects acted upon and clarify the purpose of the operations, the specific recipients, related obligations, and applicable retention policies;
- Alert Mechanisms: Referring to the privacy preferences established by an authorized user, the system should inform the user of any privacy violations or suspicious events in the corresponding physical and social environments.

4 Face Recognition Experimental Results

In our proposed experimental setup, we assume that children do not understand the concept of privacy and that third-party services running on Smart TV could, intentionally or unintentionally, collect childrens' information, e.g., their pictures. Breaches of privacy can result in physical safety concerns of children, for example, due to the possible information leakage to potential child predators. While the parents and legal guardians of a child strive to ensure their child's physical and online safety and privacy, there is no common approach for them to control the related information flow. In this context, we aim to investigate the threats of information disclosure and content unawareness in collecting child pictures by current face recognition technologies. Information leakage occurs when user's personal information is exposed to individuals who are not supposed to have access to it. We have thus conducted a set of experiments to explore the effectiveness of facial recognition technologies and their possible employment to detect underage faces and respectively limit the information collected via Smart TV. We employed Microsoft's Face Application Programming Interface (API) [20] and the experiments were organized following the three stages outlined below:

1. *Image Collection Stage*: Facial recognition engines require advance training for targeted image analysis, classification, and recognition. At this collection stage we build a target library of images that are used at the consequent stages of the experiment for training and verification purposes.
2. *Image Processing Stage*: At this stage a number of images from the constructed library are presented for processing to the Microsoft's Face API along with their characteristic parameters to train the corresponding recognition engines. The training of the Microsoft's Face API recognizer required five steps as shown in Fig. 2.
3. *Image Classification Stage*: Test images not present in the previously used training sets were presented to the Microsoft's Face API recognizer for analysis and classification. Obtained results were interpreted in terms of true/false positives and negatives as discussed in the following subsections.

Fig. 2. Process of training Microsoft's Face API.

4.1 Face-Based Age and Gender Recognition

This experiment was based on the image recognition library demos of the employed product so no specific target images were used to train the recognizers. Nevertheless at the image classification stage we were able to obtain fairly accurate gender and age estimates of a number of famous persons based on their pictures. The Microsoft's Face API properly classified 20 of the 25 presented images thus reaching a success rate of 80%.

4.2 Face Recognition in a Multiplicity of Facial Expression

Changing facial expressions are a common obstacle on the road to achieving reliable face recognition and identification. To investigate this issue we have collected 200 images of Felix Kjellberg, a popular comedy YouTuber, better known by his screen name "PewDiePie." Images of Mr. Kjellberg are abundant on the internet and come with a large variety of facial expressions from his comedy shows. Such a multiplicity of facial expressions is particularly suitable for training and accuracy testing of image recognition engines. In our experiment we used 180 of the collected images for training while retaining a diversified subset of 20 images for testing. The latter subset was constructed from individual and group images including faces at different distances and viewing angles. The Microsoft's Face API was able to properly recognize 17 of the 20 analyzed images which constitutes an 85% success rate. Additional verification was conducted with 8 faces of different persons that looked quite similar to Mr. Kjellberg, all of them being rejected by the trained recognizer (false acceptance rate of 0%).

4.3 Child Growth and Face Recognition

Since children faces change quickly with the age we wanted to investigate how the proper classification rates of a trained face recognition system might change as a child grows. We thus collected 141 pictures of Macaulay Culkin as a child (the boy from "Home Alone" movie series) and used them to train the facial recognition engines. Additional 20 pictures of the grown up Macaulay were then presented for analysis and 12 (60%) of them were properly recognized by the Microsoft's Face API. The high number of false negatives, 8 (40%) in this case, is attributed to the significant age difference between the faces in training set and in the employed test images. In a practical system this problem could be tackled, however, by retraining the Smart TV recognition engine either as planned maintenance or incrementally while a child grows.

5 Conclusion and Future Work

This paper presented a study of digital imaging in the context of identifying and properly tagging pictures of children in order to prevent unwanted disclosure of personal information. It also described our preliminary investigations on the effectiveness of the facial recognition technologies in Smart TV where experimental recognition of children faces in feedback image streams were conducted through the Microsoft's Face API. Based on the obtained results we conclude that more work must be done for the more reliable timely identification of children faces in images extracted from continuous video streams. Our continuing work will therefore aim to formalize the process and develop extensions of the current approach that are specifically designed for facial recognition and tracking of children. We will also continue our research and implementation of a theoretical model and a technical framework for enforcing and managing user level privacy policies in Smart TV environments.

Acknowledgements. This work was supported by the Ministry of Science and Technology (MOST), Taiwan under MOST Grants: 105-2923-E-002-014-MY3, 105-2923-E-027-001-MY3, 105-2221-E-027-113, and 105-2811-E-027-001; the Research Office- Zayed University, Abu Dhabi, United Arab Emirates under Research Projects: R15048 and R16083; the Natural Sciences and Engineering Research Council of Canada (NSERC) under Discovery Grants Program: RGPIN-2016-05023; and the 2016 Cooperative Research Project at Research Center of Biomedical Engineering with RIE Shizuoka University.

References

1. Anggraini, D.R.: Face recognition using principal component analysis and self organizing maps. In: Proceedings of 2014 Third ICT International Student Project Conference (ICT-ISPC), pp. 91–94 (2014)
2. Government of Canada: Personal Information Protection and Electronic Documents Act. http://laws-lois.justice.gc.ca/eng/acts/P-8.6/
3. Canadian Standards Association: Archived - Appendix 3: Model Code for the Protection of Personal Information (1996). http://cmcweb.ca/epic/internet/incmc-cmc.nsf/en/fe00076e.html
4. Chakraborty, S., Raghavan, K.R., Johnson, M.P., Srivastava, M.B.: A framework for context-aware privacy of sensor data on mobile systems. In: Proceedings of Fourteenth Workshop on Mobile Computing Systems and Applications (ACM HotMobile 2013), New York (2013)
5. Cherubini, M., de Oliveira, R., Hiltunen A., Oliver, N.: Barriers and bridges in the adoption of today's mobile phone contextual services. In: Proceedings of 13th International Conference on Human Computer Interaction with Mobile Devices and Services (MobileHCI 2011), pp. 167–176, Stockholm (2011)
6. Cooper, A., Tschofenig, H., Aboba, B., Peterson, J., Morris, J., Hansen, M., Smith, R.: RFC 6973: Privacy Considerations for Internet Protocols. IETF (2013)
7. Dewri, R., Annadata, P., Eltarjaman, W., Thurimella, R.: Inferring trip destinations from driving habits data. In: Workshop on Privacy in the Electronic Society, Berlin (2013)
8. Enck, W., Ongtang, M., McDaniel, P.: Understanding android security. IEEE Secur. Priv. **7** (1), 50–57 (2009)
9. WIPO: Directive 95/46/EC of the European Parliament and of the Council of 24 October 1995 on the protection of individuals with regard to the processing of personal data and on the free movement of such data. http://www.wipo.int/wipolex/en/details.jsp?id=13580
10. GSMA: User Perspectives on Mobile Privacy - Summary of Research Findings (2011). http://www.gsma.com/publicpolicy/wp-content/uploads/2012/03/futuresightuserperspectivesonuserprivacy.pdf
11. Ghiglieri, M.: I know what you watched last Sunday: a new survey of privacy in HbbTV. In: Workshop of Web 2.0 Security and Privacy 2014 in Conjunction with the IEEE Symposium on Security and Privacy (2014)
12. Ghiglieri, M., Tews, E.: A privacy protection system for HbbTV in Smart TVs. In: IEEE 11th Consumer Communications and Networking Conference (CCNC), pp. 648–653 (2014)
13. Government of Canada: Schedule 1 (Section 5) Principles Set out in the National Standard of Canada Entitled Model Code for the Protection of Personal Information, Personal Information Protection and Electronic Act (PIPEDA) (2000)

14. Horiuchi, T., Hada, T.: A complementary study for the evaluation of face recognition technology. In: Proceedings of 47th International Carnahan Conference on Security Technology (ICCST) (2013)
15. Huang, Y.S., Chen, S.Y.: A geometrical-model-based face recognition. In: Proceedings of IEEE International Conference on Image Processing (ICIP), pp. 3106–3110 (2015)
16. Hung, P.C.K., Cheng, V.S.Y.: Privacy and trust. In: Liu, L., Tamer Özsu, M. (eds.) Encyclopedia of Database Systems, pp. 2136–2137. Springer, New York (2009)
17. Landau, S.: What was Samsung thinking? IEEE Secur. Priv. **13**(3), 3–4 (2015)
18. Lee, S.H., Sohn, M.K., Kim, D.J., Kim, B., Kim, H.: Smart TV interaction system using face and hand gesture recognition. In: Proceedings of 2013 IEEE International Conference on Consumer Electronics (ICCE), pp. 173–174 (2013)
19. MEF Global Privacy Report 2013, MEF (2013)
20. Microsoft Cognitive Services: Face API. https://www.microsoft.com/cognitive-services/en-us/face-api
21. Nguyen, D.T., Shin, K.Y., Lee, W.O., Oh, C., Lee, H., Jeong, Y.: Gaze detection based on head pose estimation in Smart TV. In: Proceedings of 2013 International Conference on Information and Communication Technology Convergence (ICTC), pp. 283–288 (2013)
22. OCED: The OECD Privacy Framework. http://www.oecd.org/sti/ieconomy/oecd_privacy_framework.pdf
23. Ragashe, M.U., Goswami, M.M., Raghuwanshi, M.M.: Approach towards real time face recognition in streaming video under partial occlusion. In: Proceedings of 2015 IEEE 9th International Conference on Intelligent Systems and Control (ISCO) (2015)
24. Saha, D.: Pervasive computing: a paradigm for the 21st century. IEEE Comput. **36**(3), 25–31 (2003)
25. Salomon, D.: Privacy and trust. In: Salomon, D. (ed.) Elements of Computer Security. Undergraduate Topics in Computer Science, pp. 273–290. Springer, London (2010)
26. Schmidt, A.: Interactive context-aware systems interacting with ambient intelligence. In: Riva, G., Vatalaro, F., Davide, F., Alcaniz, M. (eds.) Ambient Intelligence, pp. 159–178. IOS Press, Amsterdam (2005)
27. Shabtai, A., Fledel, Y., Kanonov, U., Glezer, C.: Google Android: a comprehensive security assessment. IEEE Secur. Priv. **8**(2), 35–44 (2010)
28. Soldera, J., Behaine, C.A.R., Scharcanski, J.: Customized orthogonal locality preserving projections with soft-margin maximization for face recognition. IEEE Trans. Instrum. Meas. **64**(9), 2417–2426 (2015)
29. United States Federal Trade Commission: Children's Online Privacy Protection Act of 1998. http://www.coppa.org/coppa.htm
30. World Economic Forum: Personal Data: The Emergence of a New Asset Class (2011). http://www3.weforum.org/docs/WEF_ITTC_PersonalDataNewAsset_Report_2011.pdf
31. Xi, M., Chen, L., Polajnar, D., Tong, W.: Local binary pattern network: a deep learning approach for face recognition. In: Proceedings of 2016 IEEE International Conference on Image Processing (ICIP), pp. 3224–3228 (2016)
32. Yusufov, M., Paramonov, I., Timofeev, I.: Medicine tracker for Smart TV. In: Proceedings of 14th Conference of Open Innovations Association (FRUCT), pp. 164–170 (2013)

Scrambling Cryptography Using Programmable SLM-Based Filter for Video Streaming Over a WDM Network

Yao-Tang Chang[1(✉)], Yih-Chuan Lin[2], Yu-Chang Chen[3],
and Yan-Tai Liou[1]

[1] Department of Information Technology,
Kao Yuan University, Kaohsiung 82151, Taiwan
t10066@cc.kyu.edu.tw
[2] Department of Computer Science and Information Engineering,
National Formosa University, Yunlin 63201, Taiwan
[3] Department of Electrical Engineering, National Cheng Kung University,
Tainan 701, Taiwan

Abstract. The traditional chaotic-based enciphering technique uses pseudo-random codes to implement encryption directly (e.g., based on XOR – exclusive-or operation), which results in the violation of media compression format compliance for video streaming transmissions. In contrast to the conventional cryptography in application layers, the spatial light modulator-based (SLM-based) method uses a programmable optics wavelength filter and has been studied in order to implement a scrambling cryptography on WDM networks. In the suggested encryption scheme, we propose to use chaotic-based time series of secret keys because of their pseudo randomness and maximal period properties. The proposed scrambling SLM-based decryption is configured with an approximate symmetric scheme to perform the decryption when the initial conditions and control parameters of the chaotic sequence are communicated a priori on a private channel as individual secure keys. The experimental results showed that the proposed scrambling cryptography efficiency for video streaming is high enough to achieve secure transmission in the physical layer in terms of a peak signal-to-noise ratio (PSNR) and visual perceptual quality from the perspectives of authorized and unauthorized users. The scrambling performance of wavelength hopping is evaluated with entropy value analysis from the eavesdroppers' perspective.

Keywords: Scrambling cryptography · Chaotic based secret sequence · Spatial light modulator based (SLM based) filter · Wavelength division multiplexing (WDM) network

1 Introduction

Shake has suggested a fundamental methodology including three possible approaches to secure any network against attacks from eavesdroppers. These approaches can change the signature codes assigned to users [2–4, 9, 10, 15]. The traditional chaotic-based

© Springer International Publishing AG 2017
R.P. Barneva et al. (Eds.): CompIMAGE 2016, LNCS 10149, pp. 241–250, 2017.
DOI: 10.1007/978-3-319-54609-4_18

enciphering technique uses pseudo-random codes to implement encryption directly (e.g., based on XOR – exclusive-or operation), which results in the violation of media compression format compliance for video streaming transmissions. In contrast to the conventional cryptography in application layers, the spatial light modulator-based (SLM-based) method uses a programmable optics wavelength filter and has been studied in order to implement a scrambling cryptography on WDM networks. (See [1, 5, 6, 14] for related works.)

Parker et al. introduced a ferroelectric liquid crystal (FLC) SLM that can act as a wavelength filter [7]. However, his work gives only a conceptual idea that the SLM-based filter can be applied to a WDM network. Moreover, the configuration of the programmable SLM-based filter has never been considered applicable to scrambling cryptography.

In the present paper, we propose to select control parameters and initial values for the chaotic sequence that will be used as a secret key triggering a programmable SLM-based filter. Many random wavelength values have been obtained through the programmable wavelength filter and then the proposed cryptography scheme has been implemented. We supposed that through a dynamic and flexible wavelength would allow easily combating eavesdropping.

The rest of this paper is organized as follows. Section 2 introduces the configuration of the proposed scrambling-based cryptography. The proposed method uses a programmable SLM-based filter and an initial value and control parameter of a chaotic tent map as a secret key to implement a WDM network. Section 3 shows the efficiency of the proposed scrambling-based cryptography evaluated through the peak signal-to-noise ratio (PSNR) and a visual perspective of the encrypted videos. Further, Sect. 4 shows the scrambling wavelength hopping performance of the proposed cryptography approach defined as entropy value and investigates the uniform distribution resulting in unidentifiable effect for unauthorized users. Finally, Sect. 5 provides our concluding remarks.

2   The Proposed Scrambling Cryptography Scheme

Figure 1 shows that by applying the spatial light modulator-based (SLM-based) programmable wavelength filter to construct a tunable wavelength filter array, the proposed scrambling SLM-based cryptography is implemented through a wavelength division multiplexing (WDM) network to achieve an undistorted high resolution and privacy preservation for video streaming [12, 13]. First, the encrypting central control module is configured and initialized with N hologram patterns and is stored in the data table, namely a computer generated hologram (CGH). By triggering electrically each pixel of SLM to create the variety of spatial periods of grating that result from the various refractive indexes by the CGH pattern, the variable wavelength is filtered by a programmable SLM-based filter.

Figure 2 shows that the hologram pattern is defined as an $H_i, g(\lambda) = H_{i+1}, g(\lambda) = \{H_1, H_2, \ldots, H_g\}$ set and is stored in the database. It generates the corresponding $W_i, g(\lambda) = W_{i+1}, g(\lambda) = \{W_1, W_2, \ldots, W_g\}$ set by trigging each SLM-based filter in different i and $i + 1$ phase. Here, the subscripts i and $i + 1$ of H (or W) set denote the unique hologram pattern (or the corresponding unique wavelength) in different i and

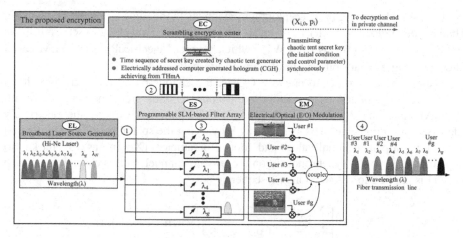

Fig. 1. The proposed scrambling encryption by programmable SLM-based filter.

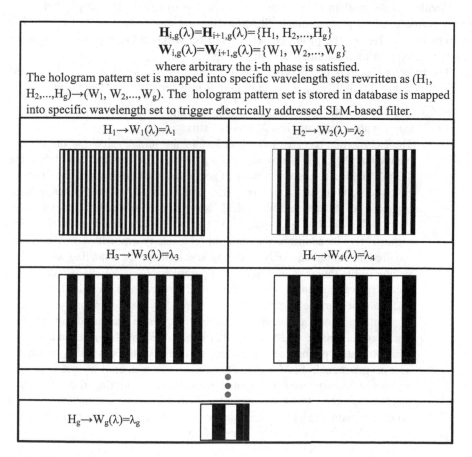

Fig. 2. The specific hologram pattern with binary bit stream embedded in database is used as to trigger each SLM-based filter for programming scrambling cryptography.

$i + 1$ phases while scrambling cryptography happens in the same hologram H database. That is, the hologram pattern set maps into a specific wavelength set as follows $(H_1, H_2, \ldots, H_g) \rightarrow (W_1, W_2, \ldots, W_g)$, which triggers electrically each SLM-based filter by using a binary bit stream of a hologram pattern.

Subsequently, in order to create g CGHs, the chaotic-based time sequence for a secret key is generated by the proposed chaotic tent/hologram mapping algorithm (THmA). Here, the algorithm selects random CGH patterns to trigger the electrically addressed programmable SLM-based array by varying the spatial period of grating and transmitting a pair of initial values and control parameters $(X_{i,0}, p_i)$ as a secret key to implement decryption synchronously on the private channel. In this case, the tent map Eq. (1) is applied to create a chaotic sequence set,

$$X_{i,e} = \begin{cases} X_{i,e}/p & \text{if } 0 \leq X_{i,e} \leq p \\ \frac{1-X_{i,e}}{1-p} & \text{if } p \leq X_{i,e} \leq 1 \end{cases} \text{ where } X_{i,e} \in [0, 1] \qquad (1)$$

Finally, a new random visiting rule is developed to select randomly g hologram patterns to trigger the programmable SLM-based wavelength filters of $(\lambda_1, \lambda_2, \ldots, \lambda_g)$ simultaneously. Hence, the proposed scrambling SLM-based cryptography can work out the encryption of incoming video streaming.

The proposed scrambling encryption is comprised of four steps and is summarized in the following description by using the circular number markers ①–④ (Fig. 1).

① The EL module provides the broadband light source generator (e.g., Hi-Ne laser) whose spectrum has N wavelengths.

② The EC module receives the secret key of the initial value and the control parameter $(X_{i,0}, p_i)$, following the supervisor on demand. Applying the THmA yields a series of chaotic tent sequences that are mapped into a series of corresponding hologram patterns, and each SLM-based filter is then triggered to filter specific optical wavelengths of authorized users as $[U_1(\lambda_2), U_2(\lambda_3), U_3(\lambda_1), U_4(\lambda_4) \ldots U_g(\lambda_g)]$ in video channels CH1-CHg, respectively. In addition, the scrambling strategy embedded in the EC is required for simultaneously updating the transmitted changing key information of the initial value and control parameter $(X_{i,0}, p_i)$ in advance over a secure (private) channel. The proposed scheme can use the initial condition and control parameter $(X_{i,0}, p_i)$ as secret keys synchronously, without transmitting numerous secure protocols in a private channel continuously.

③ Each programmable SLM-based filter (i.e., in ES module) depends on the various spatial periods of the grating to filter out the desired wavelength. The programmable SLM-based filter arrays are triggered by the computer and result in programmable spatial periods of the grating (i.e., hologram pattern), namely a CGH. Based on the THmA, which generates the various hologram patterns and filters the desired wavelengths, the authorized user is converted from the current state of $[U_1(\lambda_2), U_2(\lambda_3), U_3(\lambda_1), U_4(\lambda_4) \ldots U_g(\lambda_g)]$ to the new state of $[U_1(\lambda_1), U_2(\lambda_4), U_3(\lambda_3), U_4(\lambda_2) \ldots U_g(\lambda_g)]$ while the variation occurs from $(X_{i,0}, p_i)$ to $(X_{i+1,0}, p_{i+1})$.

④ In order to perform E–O modulation, the scrambling functions involved in the encryption are implemented using the WDM scheme, and the H.264/AVC videos are then modulated and combined as $[U_1(\lambda_2), U_2(\lambda_3), U_3(\lambda_1), U_4(\lambda_4)\ldots U_g(\lambda_g)]$ in CH1-CHg, respectively.

The decryption process (Fig. 3) uses a symmetric decryption scheme with a photodetector and a reconfigurable mechanism to synchronize with the encryption through a public or private channel. On the decryption end, the programmable SLM-based filter is triggered using the same control parameter and initial value of a chaotic sequence used on the encryption end. The decryption procedure is comprised of four steps, which are summarized in the following description scheme by using the circular number markers ①–④ (Fig. 3).

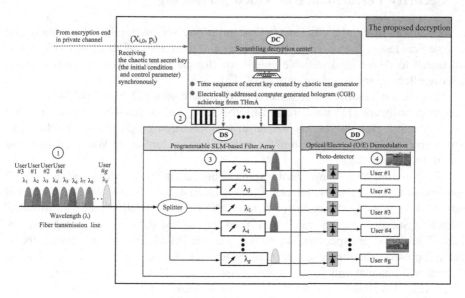

Fig. 3. The proposed scrambling decryption by programmable SLM-based filter.

① The summed optical signal $[U_1(\lambda_2), U_2(\lambda_3), U_3(\lambda_1), U_4(\lambda_4)\ldots U_g(\lambda_g)]$ in CH1-CHg mixes the wavelength-coded signals of all simultaneously active users transmitting in a public network, and then arrives at the proposed decryption.

② When the secret key of the initial value and control parameter $(X_{i,0}, p_i)$ are transmitted from the EC module of the proposed encryption scheme, the DC module obtains the specific hologram pattern through the THmA. The specific CGH pattern is composed of a binary bit sequence, thus resulting in a different spatial periods of the grating of the SLM-based filter, which are achieved by varying the diffraction angle of the SLM. The corresponding wavelength of an authorized user is then triggered.

③ The corresponding wavelength of an authorized user in the DS module is the same as the wavelength hopping pattern produced in scrambling the encryption of $[U_1(\lambda_2), U_2(\lambda_3), U_3(\lambda_1), U_4(\lambda_4)...U_g(\lambda_g)]$ to the new state of $[U_1(\lambda_1), U_2(\lambda_4), U_3(\lambda_3), U_4(\lambda_2)...U_g(\lambda_g)]$ while the variation occurs from $(X_{i,0}, p_i)$ to $(X_{i+1,0}, p_{i+1})$. Each SLM-based filter of the DS module performs a complete correlation to synchronously demultiplex the matched wavelength by using the THmA and the secret key $(X_{i,0}, p_i)$.

④ Through optical/electrical demodulation processing, the desired data bit of each authorized user is recovered using a photo-detector and a decision unit in the proposed scrambling decryption.

3 Security Performance of Video Streaming

In the current study, the efficiency of the proposed scrambling-based cryptographic scheme was tested from a visual perspective, and the PSNR of encrypted videos was calculated to detect unidentifiable videos for unauthorized users that, consequently, achieve effective image cryptographs [12, 13].

This section describes a series of experiments conducted using H.264/AVC JM Reference software [11] implemented by the Joint Video Team to generate the test video streams. This program is written in C application programming interface [8], which gives access to a large part of the underlying structure of the Windows operating system to programmers.

The proposed scrambling-based cryptographic scheme in the aforementioned $g = 8$ configuration (i.e., {User#1, User#2, User#3, User#4, ..., User#8}) was applied, and eight videos, namely "Container," "Crew," "Foreman," "Hall Monitor," "Mobile," "News," "Soccer," and "Stefan," were used as the test data set in all experiments. Table 1 presents the experimental parameters of the JM 16.2 reference software codec. Table 2 presents the format specification of the test videos. For instance, Fig. 4 is the "Container" channel and Fig. 5 is the "Crew" channel that shows the difference in visual video streaming for authorized and unauthorized users, respectively.

Table 3 shows the effectiveness of the proposed secure transmission scheme verified according to the PSNR scale. The experimental results showed that the average

Table 1. Experimental parameters for H.264/AVC codec.

Parameters	Information
Profile	Baseline
IDR period	15
Reference frames	5
Frames encoded	300
Search range	32
Quantization parameter	28
FMO map type	Dispersed
Slice groups	8

Table 2. Test video format parameters.

Parameter	Information
Video format	CIF
YUV format	4:2:0
Frame size	352 × 288
Frame rate	30 fps

difference in the peak signal-to-noise ratio (PSNR) between authorized and unauthorized users exceeded 25 dB or more at quantization parameter of 28 (QP = 28). Thus, unauthorized users may receive an incomprehensive video stream. In other words, the security of the proposed system will increase by scrambling encryption. For instance, the perceptual quality of H.264/AVC streaming of "Container" and "Crew" standard test videos are shown in Figs. 4 and 5, respectively.

Fig. 4. Container: left panel shows unscrambled image (detected by an authorized user); right panel show image scrambled by the THmA (detected by an unauthorized user).

Fig. 5. Crew: left panel shows unscrambled image (detected by an authorized user); right panel show image scrambled by the THmA (detected by an unauthorized user).

Table 3. Average scrambled PSNRs depending on various H.264/AVC video stream statistical processing methods.

	Original (authorized user)	Proposed THmA algorithm (unauthorized user)
Container	36.50	10.78
Crew	38.04	12.48
Foreman	37.07	11.11
Hall monitor	38.03	11.52
Mobile	35.23	10.22
News	38.89	9.40
Soccer	36.85	11.73
Stefan	36.85	12.29
Average PSNR	37.18	11.19

4   Scrambling Performance of Wavelength Hopping

Figure 6 shows that the eavesdropper has tracked on the different eight channels (proposed THmA) and the application patterns (wavelengths) were counted for 600 frames. In addition, the results in these channels lead to the regularity of pattern distribution that can be found by long-term observing.

In the above-mentioned case, eavesdroppers obtain a lower PSNR because the original video stream cannot be detected correctly due to the proposed scrambling-based cryptography. Note that, it is assumed that eavesdroppers track and monitor the same wavelength continually with each authorized user facing increased exposure risk of the video stream because there is an insufficient variety of wavelength hopping patterns, which are easily predicted by an eavesdropper. Hence, a higher entropic value of wavelength hopping implies a higher scrambling effect to achieve better security for the proposed scheme.

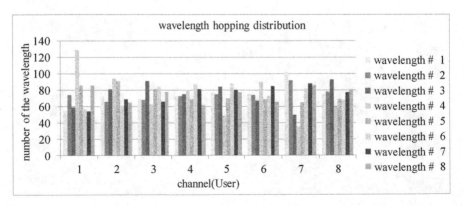

Fig. 6. Each pattern detected by an eavesdropper in 600 frames for proposed THmA in each channel.

In information theory, entropy is a measure of unpredictability or information content. In this paper, it is regarded as a criterion of different algorithms about distribution probability of each pattern and defined by a specific function as below:

$$H(x) = -\sum_{i=1}^{n} p(x_i) \log_2 p(x_i) \tag{2}$$

where $p(x_i)$ is the probability density function of each pattern in a particular channel.

Table 4 shows that after calculating the entropy function of each channel, respectively, the average entropy value is 2.974 in THmA, approaching the entropy value 3 of uniform wavelength hopping distribution. In wavelength #4 of channel 1 and 7, exists the outliers to be detected by an eavesdropper. In the original definition, the probability of each item is equal and the entropy value is much larger. The maximum entropy value is 3 when the probability of each pattern is equal (i.e., $p(x_i) = 0.125$). Thus, the proposed THmA is more unpredictable for an eavesdropper.

Table 4. Entropy analysis in scrambling wavelength hopping.

	Entropy value
Container	2.932
Crew	2.983
Foreman	2.989
Hall monitor	2.993
Mobile	2.982
News	2.991
Soccer	2.934
Stefan	2.989
Average entropy value	2.974
Uniform distribution	3

5 Conclusion

This study proposed scrambling-based cryptographic scheme applying a programmable SLM-based filter array in WDM networks. The experiment results showed that the average difference in the PSNR between authorized and unauthorized users exceeded 25 dB or more at quantization parameter of 28 (QP = 28). The average entropy value is 2.974 in THmA approaching the entropy value 3 of uniform wavelength hopping distribution. Both PSNR and entropy analysis achieve distortion and diffusion performance to enhance the security of the proposed scrambling wavelength hopping. Moreover, because of its high scrambling efficiency, the proposed scheme can produce unidentifiable videos for unauthorized users. Hence, this scheme is suitable for secure H.264/AVC streaming in a cross-layer manner (application and physical). In future work, we will search for any kind of the mapping algorithm which generates more different patterns (i.e., code family) to reduce PSNR unauthorized use and make the entropy values approach uniform wavelength hopping distribution.

Acknowledgment. This study was supported under grant No. MOST 105-2221-E-244-004 from the Ministry of Science and Technology, Taiwan.

References

1. Baugher, M., McGrew, D., Naslund, M., Carrara, E., Norman, K.: The Secure Real-Time Transport Protocol (SRTP), IETF RFC 3711, March 2004
2. Chang, Y.-T., Lin, Y.-C.: Dynamic reconfigurable encryption and decryption with chaos/M-sequence mapping algorithm for secure H.264/AVC video streaming over OCDMA passive optical network. Multimedia Tools Appl. **74**(15), 1931–1948 (2016). [and **75**(16), 9837–9859]
3. Chang, Y.-T., Sue, C.-C., Huang, J.-F.: Robust design for reconfigurable coder/decoders to protect against eavesdropping in spectral amplitude coding optical CDMA networks. IEEE J. Lightwave Technol. **25**(8), 1931–1948 (2007)
4. Chang, Y.-T., Tsailin, C.W.: Dynamic scrambling scheme of arrayed-waveguide grating-based encryptors and decryptors for protection against eavesdropping. Comput. Electr. Eng. **49**(1), 184–197 (2016)
5. Chiaraluce, F., Ciccarelli, L., Gambi, E., Pierleoni, P., Reginelli, M.: A new chaotic algorithm for video encryption. IEEE Trans. Consum. Electron. **48**(4), 838–844 (2002)
6. Mao, Y., Wu, M.: A joint signal processing and cryptographic approach to multimedia encryption. IEEE Trans. Image Process. **15**(7), 2061–2075 (2006)
7. Parker, M.C., Cohen, A.D., Mears, R.J.: Dynamic digital holographic wavelength filtering. J. Lightwave Technol. **16**(7), 1259–1270 (1998)
8. Petzold, C.: Programming Windows®, 5th edn. Microsoft Press, Redmond (1998)
9. Shake, T.H.: Security performance of optical CDMA against eavesdropping. IEEE J. Lightwave Technol. **23**(2), 655–670 (2005)
10. Shake, T.H.: Confidentiality performance of spectral-phase-encoded optical CDMA. IEEE J. Lightwave Technol. **23**(4), 1652–1663 (2005)
11. Sühring, K.: H.264/AVC Reference Software Group, January 2009. http://iphome.hhi.de/suehring/tml/, Joint Model 12.2 (JM12.2)
12. Wang, Y., Ostermann, J., Zhang, Y.Q.: Video Processing and Communications. Prentice Hall, Upper Saddle River (2001)
13. Wang, Y., O'Neill, M., Kurugollu, F., O'Sullivan, E.: Privacy region protection for H.264/AVC with enhanced scrambling effect and a low bitrate overhead. Signal Process. Image Commun. **35**, 71–84 (2015)
14. Wiegand, T., Sullivan, G., Bjontegaard, G., Luthra, A.: Overview of the H.264/AVC video coding standard. IEEE Trans. Circuits Syst. Video Technol. **13**, 560–576 (2003)
15. Wu, B.B., Narimanov, E.E.: A method for secure communications over a public fiber-optical network. Opt. Express **14**(9), 3738–3751 (2006)

An Accelerated H.264/AVC Encoder on Graphic Processing Unit for UAV Videos

Yih-Chuan Lin[✉] and Shang-Che Wu

Department of Computer Science and Information Engineering,
National Formosa University, Yunlin 63201, Taiwan
lyc@nfu.edu.tw

Abstract. With regards to the nature of high intensive computation for motion estimation with an H.264/AVC encoder, this paper presents a parallel block-matching algorithm implemented on a general purpose graphics processing units (GPU) to speed up the execution of UAV video coding. Traditional parallel block-matching algorithms are primarily used to leverage the huge number of computational cores in graphic processing units, which can be used to compute the block-matching operation at each candidate position in a search range by an independent thread of kernel computation. In realistic scenarios, the time used to transfer pixel values among the various memory modules to fulfill the operation in a GPU system is much higher than the computation time used for computing each block-matching operation by the kernel threads. This leads to a performance improvement bottleneck for GPU algorithm design. The proposed algorithm exploits the characteristics of distinct memory modules on the data transfer speed for the block-matching algorithm and proposes a feasible mechanism to reduce the bandwidth of data transmission required for the parallel block-matching algorithms implemented on GPU system. With experiments on GPU systems, the proposed parallel block-matching algorithm gains up to 99% execution reduction of motion estimation compared to the host processor only motion estimation process.

Keywords: Unmanned aerial vehicle · Aerial video coding · GPU · Motion estimation · H.264/AVC

1 Introduction

This paper addresses the requirement of real-time communications for aerial surveillance videos transmitted from an unmanned aerial vehicle (UAV) to a ground control station [4]. With the demand for high video resolution under a constrained-bandwidth data link, standard video encoders were designed with more complex coding tools to fully exploit the data redundancy of videos to lower the transmission data rate. However, this leads to a drastically high computation time by the video encoder on an

This work is supported in part by the Ministry of Science and Technology, Taiwan, under the contract MOST 104-2221-E-150-029.

R.P. Barneva et al. (Eds.): CompIMAGE 2016, LNCS 10149, pp. 251–258, 2017.
DOI: 10.1007/978-3-319-54609-4_19

UAV to compress aerial videos. Motion-compensated video encoders, such as H.264/AVC [12] or HEVC [10], which require a motion estimation module to capture the temporal correlation between neighboring video frames, are considered for compressing UAV video. In the implementation of motion estimation, a block-matching algorithm requires time-consuming processes to find the best motion vector, which expend about 65,536 arithmetic operations if the block size is 8×8 and the search range is 32×32. These calculations have to be repeated 32,400 times in the 1080p resolution (1920×1080). Due to high computational complexity, some studies have focused on fast motion estimation by using fast mode decision [1, 13, 14] to reduce about 50% of the encoding time, but it is still challenging in real-time motion estimation even though the computing ability of the central processing unit (CPU) is much better than before. Therefore, some research has focused on parallel motion estimation algorithms in the GPU [2, 3, 8]. With the growth of the semiconductor industry, the computation capability of GPUs is much higher than in CPUs. Consequently, more and more non-graphical computations can be offloaded to GPUs. The programming model for GPUs, Compute Unified Device Architecture (CUDA), is a parallel computing platform created by NVIDIA, which is designed for GPGPU usage [9].

In this paper, general data-parallelism motion estimation [2, 7] is used for UAV video compression by an H.264/AVC encoder. In the implementation of data-parallelism motion estimations, three kernel functions are designed to be launched in sequence during execution along with massive intermediate data communication in-between to accomplish the motion estimation. This leads to a limited speedup of execution time for the parallel motion estimation approach. To handle this issue, a modification of the general parallel motion estimation is proposed to take advantage of special characteristics in most UAV videos for reducing the amount of memory access requirements between kernel functions.

2   The Proposed Method

Motion estimation belongs to the prediction processing of motion-compensated video encoders for inter-frame coding, which requires dividing the incoming frame into a series of variable-sized blocks and computing the best-matched motion vectors of each block within a search area of the preceding frames. If full search algorithms are used to find the best motion vectors within a 32×32 search range, there are 1024 candidate sum-of-absolute-differences (SAD) per block needed. Single thread computation encoders need the calculations of candidate SADs and the comparison among these candidates one after another to find the minimum SAD cost. We propose a parallel block-matching algorithm to perform the computation of SADs for each block and to calculate the best-matched motion vector (MV). Figure 1 depicts the flow of parallel computation threads on the GPU and the interactions between host CPUs and GPUs.

The device denoted in Fig. 1 represents one or more GPUs that are equipped with a host machine to compute motion estimation process. Each incoming video frame is divided into square blocks and all of them are sent to the GPU device's texture memory. The host machine launches a SAD kernel function, which is going to be performed in parallel threads on the GPU device to associate each square block with a

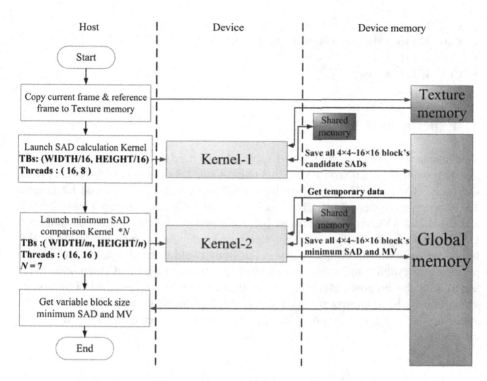

Fig. 1. Sequential kernel stages from the proposed parallel algorithm.

set of threads. Each thread in the set targets for computing the SAD at a specified position (or MV) within the search range. That is, all possible search candidate positions are evaluated in parallel and a set of SADs are output for finding the best MV. These computed SADs are stored in the device's global memory for later use. To determine the best MV of a square block, the second kernel function is invoked to concurrently compare these stored SADs for finding the minimum SAD and the associated searched position. If there are N types of variable block size considered, N kernel functions are needed for determining the best MVs of blocks with the considered block sizes, each of which is responsible one type of block size. The N is 7 because we consider the block sizes, $4 \times 4, 4 \times 8, 8 \times 4, 8 \times 8, 8 \times 16, 16 \times 8$, and 16×16. In the final stage, the computed SADs and MVs for the considered block types are transferred to the host machine to be used by the subsequent video encoding process. To reduce the memory consumption and its access time, the types of block partitions can be further omitted in the algorithm. Therefore, the algorithm can further be modified to be specific for UAV video. In a UAV video, the motion of most objects is slightly changed so that each block in the frame tends to be a larger partition (e.g., $16 \times 16, 16 \times 8, 8 \times 16$). Therefore, four cases of the partition sets are defined in the proposed algorithm for the performance on UAV videos coded in the reduced set of partition sizes:

- Case 1: $\{16 \times 16, 16 \times 8, 8 \times 16, 8 \times 8, 8 \times 4, 4 \times 8, 4 \times 4\}$;
- Case 2: $\{16 \times 16, 16 \times 8, 8 \times 16, 8 \times 8\}$;
- Case 3: $\{16 \times 16, 16 \times 8, 8 \times 16\}$;
- Case 4: $\{16 \times 16\}$;

3   Experimental Results

To evaluate the speed of our proposed approach, we used the following computational platform with: (1) Intel Xeon E3-1230v3 Quad-Core 3.3 GHz host CPU; (2) NVIDIA GeForce GTX Titan Black GPU device with 2,880 CUDA cores; (3) 64-bit Ubuntu 14.04 operating system; (4) CUDA Toolkit 6.5 with CUDA compute capability 3.5; and (5) H.264/AVC reference software [6]. Although in the actual micro- or mini-level UAV, the configuration for the above-mentioned computation platform might be different, the experimental results shown in the study could be used as reference data for performance evaluation. Three aerial videos [5, 11], and one general video (Fig. 2) are tested with the proposed algorithm on motion estimation performance. The general video is used for comparison with a UAV video in the four cases. We evaluate the speed-up with Eq. (1), where *ME* denotes the total execution time of motion estimation.

$$Speedup = \frac{ME_{CPU} - ME_{GPU}}{ME_{CPU}} \times 100\% \tag{1}$$

(a)   (b)

(c)   (d)

Fig. 2. Test videos; (a) Video-I: 1080p, 1300 frames; types of block sizes: Case1; (b) Video-II: 480p, 20000 frames; types of block sizes: Case1; (c) Video-III: 1080p, 500 frames; types of block sizes: Case1–Case4; (d) Video- IV: 1080p, 500 frames; types of block sizes: Case1–Case4.

We integrate the proposed GPU motion estimation scheme with the H.264/AVC's reference software (JM) to evaluate performance. Table 1 lists the total execution time of motion estimation performed by the proposed GPU scheme and by the CPU only, respectively, on the same test videos. Based on the results, the proposed GPU motion estimation has effectively gained a speed-up ratio up to 99%.

Table 1. Performance of the proposed parallel motion estimation process

	Video-I	Video-II
CPU only	8.12 h	26.55 h
Proposed GPU	5.01 min	15.42 min
Speed-up ratio	98.9%	99.0%

Tables 2 and 3 show the coding efficiency. ΔPSNR, and ΔBitrate are calculated by Eqs. (2–3) below, where the *CaseN* denotes the Case 2, Case 3, or Case 4.

$$\Delta PSNR = PSNR_{CaseN} - PSNR_{Case1} \tag{2}$$

$$\Delta Bitrate = \frac{Bitrate_{CaseN} - Bitrate_{Case1}}{Bitrate_{Case1}} \times 100\% \tag{3}$$

Although the motion of object is slow in the UAV video, it does not show that an UAV video that can use the single largest block partition. With the experiment, it was found that the coding efficiency of the video degrades more for Case 4 because the UAV video has multiple small objects, which can cause the smaller partition in the encoder.

Table 2. The coding efficiency of UAV video: video-III

	QP	PSNR (dB)	Bitrate (kbps)	ΔPSNR (dB)	ΔBitrate (%)
Case 1	22	42.207	16700.62	N/A	N/A
	27	38.365	6575.50		
	32	34.862	2829.33		
	37	31.890	1453.92		
Case 2	22	42.203	16828.75	−0.004	0.76
	27	38.363	6642.05	−0.002	1.01
	32	34.859	2843.85	−0.003	0.51
	37	31.892	1456.46	0.002	0.17
Case 3	22	42.198	17041.89	−0.009	2.04
	27	38.364	6707.24	−0.001	2.00
	32	34.858	2853.69	−0.004	0.86
	37	31.891	1457.06	0.001	0.21
Case 4	22	42.143	18368.16	−0.064	9.98
	27	38.319	7316.41	−0.046	11.26
	32	34.794	3057.35	−0.068	8.05
	37	31.840	1519.72	−0.050	4.52

Table 3. The coding efficiency of non-UAV video: video-IV

	QP	PSNR (dB)	Bitrate (kbps)	ΔPSNR (dB)	ΔBitrate (%)
Case 1	22	40.265	33621.88	N/A	N/A
	27	37.877	11169.24		
	32	35.788	4903.36		
	37	33.568	2576.15		
Case 2	22	40.251	34059.46	−0.014	1.30
	27	37.880	11330.47	0.003	1.44
	32	35.785	4948.05	−0.003	0.91
	37	33.564	2592.64	−0.004	0.64
Case 3	22	40.244	34251.68	−0.021	1.87
	27	37.880	11422.25	0.003	2.26
	32	35.787	4975.26	−0.001	1.46
	37	33.564	2603.92	−0.004	1.07
Case 4	22	40.166	35313.86	−0.099	5.03
	27	37.852	12045.99	−0.025	7.84
	32	35.741	5262.47	−0.047	7.32
	37	33.520	2746.15	−0.048	6.59

Figures 3 and 4 show the comparisons of degradation of H.264/AVC coding efficiency between UAV and non-UAV videos. As seen from the figures, the UAV videos coded with Cases 2 and 3 feature less degradation on coding efficiency than the non-UAV video, especially for the higher quantization parameter (QP) settings. That means that if lower bit-rate encoding is used; the proposed encoder can be speeded up with less degradation for the UAV videos.

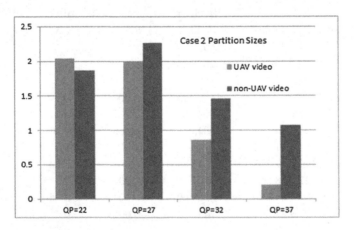

Fig. 3. Degradation of coded bit-rates with the UAV and non-UAV videos encoded by H.264/AVC encoder using Case 2 reduced set of block partition sizes.

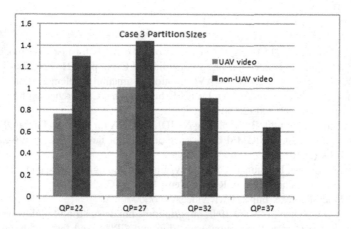

Fig. 4. Degradation of coded bit-rates with the UAV and non-UAV videos encoded by H.264/AVC encoder using Case 3 reduced set of block partition sizes.

Table 4 shows the execution time of proposed algorithm encodes a 1080p video frame with a single reference frame in different cases. The column of total time represents the execution time of a kernel function and data transmission.

Table 4. Performance of individual kernels under the four cases of block-size set.

	Kernel-1 (ms)	Kernel-2 (ms)	Total time (ms)
Case 1	10.78	17.45	41.30
Case 2	8.59	4.51	26.17
Case 3	6.91	2.58	22.62
Case 4	2.46	0.52	16.10

4   Conclusion

A block-matching algorithm is a computationally intensive process for finding the best possible matched block of an incoming anchor image block among a set of candidate image blocks. There are many image/video processing and computer vision applications requiring this fundamental process. We presented a modified parallel block-matching algorithm targeting the speedup of motion estimation module in a UAV video encoder on GPUs, such that small and light-weight computer hardware for video capture can be easily equipped with the small-sized UAVs. The proposed parallel motion estimation algorithm has also been integrated into an H.264/AVC standard software encoder to validate its degradation of coding efficiency for encoding UAV and non-UAV videos. With the experimental results, the proposed algorithm can be dramatically accelerated while maintaining acceptable increases of encoded bit-rate and degradation of video quality by using Case 2 and Case 3 in the motion estimation for UAV videos.

References

1. Chen, B.-Y., Yang, S.-H.: Using H.264 coded block patterns for fast inter-mode selection, In: IEEE International Conference on Multimedia and Expo, pp. 721–724. Hannover (2008)
2. Chen, W.-N., Hang, H.-M.: H.264/AVC motion estimation implementation on compute unified device architecture (CUDA). In: IEEE International Conference on Multimedia and Expo (ICME), pp. 697–700 (2008)
3. Colic, A., Kalva, H., Furht, B.: Exploring NVIDIA-CUDA for video coding. In: Proceedings of the First Annual ACM SIGMM Conference on Multimedia systems, pp. 13–22. Phoenix, Arizona, USA (2010)
4. Colomina, I., Molina, P.: Unmanned aerial systems for photogrammetry and remote sensing: a review. ISPRS J. Photogramm. Remote Sens. **92**, 79–97 (2014)
5. Free HD Stock Footage. https://www.videezy.work/
6. H.264/AVC Reference Software 19.0. http://iphome.hhi.de/suehring/tml/download/
7. Lin, Y.C, Wu, S.C.: Parallel motion estimation and GPU-based fast coding unit splitting mechanism for HEVC. In: IEEE High Performance and Extreme Computing Conference (HPEC 2016), Boston, USA, 13–15 September 2016
8. Moriyoshi, T., Takano, F., Nakamura, Y.: GPU acceleration of H.264/MPEG-4 AVC software video encoder. In: Asia-Pacific Signal and Information Processing Association Annual Summit and Conference (APSIPA), Xi'an, China (2011)
9. NVIDIA CUDA C programming guide. http://docs.nvidia.com/cuda/cuda-c-programming-guide/
10. Sullivan, G.J., Ohm, J.-R., Han, W.-J., Wiegand, T.: Overview of the high efficiency video coding (HEVC) standard. IEEE Trans. Circuits Syst. Video Technol. **22**(12), 1649–1668 (2013)
11. UCF Aerial Action Data Set. http://crcv.ucf.edu/data/UCF_Aerial_Action.php
12. Wiegand, T., Sullivan, G.J., Bjøntegaard, G., Luthra, A.: Overview of the H.264/AVC video coding standard. IEEE Trans. Circuits Syst. Video Technol. **13**(7), 560–576 (2003)
13. Yang, F., Ma, H.: Aerial video encoding optimization based on x264. Open J. Appl. Sci. **3**(1B), 36–40 (2013)
14. Yang, J., Chen, Y.: A novel fast inter-mode decision algorithm for H.264/AVC based on motion estimation residual. In: WASE International Conference on Information Engineering (ICIE), vol. 1, pp. 153–156, Taiyuan, Shanxi (2009)

Author Index

Printed in the United States
By Bookmasters